T0250279

Advances in
Ergonomics in
Manufacturing

Advances in Human Factors and Ergonomics Series

Series Editors

Gavriel Salvendy
Professor Emeritus
School of Industrial Engineering
Purdue University

Chair Professor & Head
Dept. of Industrial Engineering
Tsinghua Univ., P.R. China

Waldemar Karwowski
Professor & Chair
Industrial Engineering and
Management Systems
University of Central Florida
Orlando, Florida, U.S.A.

3rd International Conference on Applied Human Factors and Ergonomics (AHFE) 2010

Advances in Applied Digital Human Modeling
Vincent G. Duffy

Advances in Cognitive Ergonomics
David Kaber and Guy Boy

Advances in Cross-Cultural Decision Making
Dylan D. Schmorrow and Denise M. Nicholson

Advances in Ergonomics Modeling and Usability Evaluation
Halimahtun Khalid, Alan Hedge, and Tareq Z. Ahram

Advances in Human Factors and Ergonomics in Healthcare
Vincent G. Duffy

Advances in Human Factors, Ergonomics, and Safety in Manufacturing and Service Industries
Waldemar Karwowski and Gavriel Salvendy

Advances in Occupational, Social, and Organizational Ergonomics
Peter Vink and Jussi Kantola

Advances in Understanding Human Performance: Neuroergonomics, Human Factors Design, and Special Populations
Tadeusz Marek, Waldemar Karwowski, and Valerie Rice

4th International Conference on Applied Human Factors and Ergonomics (AHFE) 2012

Advances in Affective and Pleasurable Design
Yong Gu Ji

Advances in Applied Human Modeling and Simulation
Vincent G. Duffy

Advances in Cognitive Engineering and Neuroergonomics
Kay M. Stanney and Kelly S. Hale

Advances in Design for Cross-Cultural Activities Part I
Dylan D. Schmorrow and Denise M. Nicholson

Advances in Design for Cross-Cultural Activities Part II
Denise M. Nicholson and Dylan D. Schmorrow

Advances in Ergonomics in Manufacturing
Stefan Trzcielinski and Waldemar Karwowski

Advances in Human Aspects of Aviation
Steven J. Landry

Advances in Human Aspects of Healthcare
Vincent G. Duffy

Advances in Human Aspects of Road and Rail Transportation
Neville A. Stanton

Advances in Human Factors and Ergonomics, 2012-14 Volume Set:
Proceedings of the 4th AHFE Conference 21-25 July 2012
Gavriel Salvendy and Waldemar Karwowski

Advances in the Human Side of Service Engineering
James C. Spohrer and Louis E. Freund

Advances in Physical Ergonomics and Safety
Tareq Z. Ahram and Waldemar Karwowski

Advances in Social and Organizational Factors
Peter Vink

Advances in Usability Evaluation Part I
Marcelo M. Soares and Francisco Rebelo

Advances in Usability Evaluation Part II
Francisco Rebelo and Marcelo M. Soares

Advances in
Ergonomics in
Manufacturing

Edited By
Stefan Trzcieliński
and
Waldemar Karwowski

CRC Press
Taylor & Francis Group
Boca Raton London New York

CRC Press is an imprint of the
Taylor & Francis Group, an **informa** business

CRC Press
Taylor & Francis Group
6000 Broken Sound Parkway NW, Suite 300
Boca Raton, FL 33487-2742

Version Date: 20120529

International Standard Book Number: 978-1-4398-7039-6 (Hardback)

Visit the Taylor & Francis Web site at
http://www.taylorandfrancis.com

and the CRC Press Web site at
http://www.crcpress.com

Table of Contents

viii

Section III: Human Factors in Work Systems

Preface

People like friendly environment and products. Manufacturing is this area of human being activities that delivers a great number of products and in this way influences strongly on the environment and satisfaction of the products users. Ergonomics factors belong to these which are crucial for achieving an improvement in this matter.

A lot of investigation is led by researchers in different countries to improve the ergonomics of products and the natural and work environment. Some results of their findings are presented in this book. We decided to collect them and published because we believe that they can inspire or support other researchers to faster their own investigation or even can be implemented in the practice. This means that the book is addressed to the both researchers and practitioners. With this presumption this book has been arranged in four sections.

The first section covers variety of topics that refers to human oriented organization. It starts form a general view point on socio-technical systems including organizational innovativeness and agility of enterprise. It is followed by issues about designing ergonomic production systems with taking into consideration: workforce diversity, high-wage countries, work related risk, work environment factors, ICT, and demographic features. The last thematic part of this section is focused on assembly planning and production inventories management.

The second section presents effects of work study concerning improvement of skills, quality and effectiveness of workers. Its beginning part depicts the influence of workers experience and the technology they use on the work effectiveness. Next, the comparison of non-expert and expert work is studied to find patterns that can be used to improve the technique of performing different tasks by the less skilled employees. Also some experimental methods using the virtual agent and augmented reality-based assistance for organizational learning as well as a method for forecasting the learning time are presented.

The third section deals with some outcomes on human factors in work systems. It consists with four thematic subsections. The first one discusses the human-robot interactions, designing and implementation of assembly systems as well as the role of the operator and his expectations in robotic systems. The second subsection presents the influence of work environment factors on the efficiency, health and well-being of the operators. The third one depicts such problems like fatigue of the operator during short-cycle work, the cognitive workload and a method of biomechanical evaluation of dynamic and asymmetric lifting. The fourth part is sacrificed for methods supporting eye-gaze location and writing and visual recognition of object while walking. The fifth one discusses the use of participatory ergonomics to improve the quality of work.

The fourth section is focused on shaping an ergonomic product. It includes four thematic parts. The first part involves problems like ergonomics cost in product development and the product features reducing the risk for the user. The second part concerns the method of identification of the components of the products which are crucial for value and satisfaction of the user. The third part presents some examples of factors influencing the product life cycle connected with the customer satisfaction and preferences. And the last part depicts an example of application of anthropometrical methods of measurement in the sportswear and shoe manufacturers industry.

The contents of this book required the dedicated effort of many people. Firstly, we must thank the authors, whose research and development efforts are recorded here. Secondly, we wish to thank the following Editorial Board members for their diligence and expertise in the stage of selecting and reviewing the papers:

S. Bagnara, Italy
T. Bikson, USA
A. Chan, Hong Kong
Y. S. Chang, Korea
F. Daniellou, France
P. Dawson, UK/Australia
C. Drury, USA
E. Fallon, Ireland
E. Gorska, Poland
A. Gramopadhye, USA
W. Grudzewski, Poland
I. Hejduk, Poland
M. Helander, Singapore
A. Herman, Poland

S. Hsiang, USA
R. Lifshitz, Israel
A. Madni, USA
N. Marmaras, Greece
A. Matias, Philippines
P. Ordonez de Pablos, Spain
A. Polak-Sopinska, Poland
A. Sage, USA
C. Schlick, Germany
H. Schulze, Switzerland
M. Soares, Brazil
J. Stahre, Sweden
J. Wilson, UK
K. Zink, Germany

March 2012

Waldemar Karwowski
University of Central Florida
Orlando, Florida, USA

Stefan Trzcieliński
Poznan University of Technology
Poznań, Poland

Editors

Section I

Human Oriented Organization

CHAPTER 1

Organizational Innovations and Knowledge Based Enterprises. Theoretical Postulates and Empirical Issues

Edmund Pawłowski

Poznan University of Technology
Poznan, Poland
Edmund.Pawlowski@put.poznan.pl

ABSTRACT

The nature and the set of futures describing a knowledge based enterprise does not form a unified concept. The conception of knowledge based enterprise evolved from a learning organization to a competitive learning organization, and finally to an intelligence organization, and agile organization. Theoretical postulates of the organizational innovations in knowledge based enterprises can be treated as a normative model, aimed toward maximum freedom. It is an extremely flat organization, free of official hierarchy, based on horizontal coordination relationships and variable hierarchy of goals, blended into external economic networks, completely decentralized, based on wide specialization of employees, and very low level of standardization and formalization; it uses most of the modern techniques of management. The empirical research of the organizational innovations in knowledge based enterprises (case studies, two special studies in Poland, and some issues from European Community Innovation Survey) indicate, that in practice more diversified organizational solutions are being used, and it is rather hard to find the enterprise on that extremely theoretical level. The gap between theoretical and practical models led to the following conclusions: 1. the research models were too simplified – in some aspects, 2. One of the most important aspect was: the research model should be adjusted to the particular areas of business processes of an enterprise (marketing, R&D, production etc.)

Keywords: Organizational innovation, knowledge based enterprise, organizational structure

1   INTRODUCTION

This paper is a part of a larger research project called "Adjustment of enterprises' management systems to knowledge-based economy". The project, undertaken at the Faculty of Engineering Management at Poznan University of Technology, started in 2009 with the aim to define:

1. Model solutions (best practices) in regard to changes in enterprises regarding: strategy and organizational structure, human capital, innovations, ICT systems, relationships with institutional-legal environment, which enable them to realize the knowledge-based organization model.

2. Mechanisms of enterprises' behavior, ignoring or blocking the influence of changes occurring in the environment, which result in maintaining the organization inadequate to occurring opportunities and efficient competing.

3. Barriers existing outside and inside the enterprises, which neutralize or make negative the relationship between changes in the environment, reflecting the knowledge-based economy and changes in enterprises describing the knowledge-based organization.

This paper focuses on organizational innovation aspects in knowledge based enterprises with the aim of summarizing the actual research issue both: theoretical postulates and their empirical verification.

2   THE CHARACTERISTIC AND MODELS OF KNOWLEDGE BASED ENTERPRISES

The nature and features constituting the knowledge-based enterprises has evolved for years and still does not form a unified concept. The characteristic of this development has been described in literature many times (Senge, 2003, Grudzewski, Hejduk, 2004, Perechuda, 2005, Mikuła, 2006, Jashapara, 2006, Mikuła et al, 2007, Davenport, 2007, Stabryła, ed., 2009).

The logic of development of knowledge based organization derives from the concept of "organizational learning" As a result of this learning process, an organization acquires new characteristics which constitute the notion of "learning organization". Ashok Jashapara (2006, pp. 303-324) indicates three parallel trends in the development of a learning organization:

- American approach, represented by the concept of fifth discipline and learning organization by P. Senge, and the concept of organizational learning by d. Garvin;

- British approach, focused around 3 level model of a learning organization by

B.Garratt, and a model of a learning enterprise by M Pedler, J. Bugoyne, T. Boydell;

- Japanese approach, identified with a knowledge generating enterprise (I.Nonaka, 1991).

These three approaches indicate a variety of possible ways of learning and creating a learning organization, through: individual mastery, team learning, knowledge sharing, experimenting, systemic thinking, using heuristic techniques and quality management methods.

Introducing the concept of a knowledge generating enterprise, I.Nonaka (1991) emphasized the role of organizational learning focused on innovations and gaining competitive advantage. It is not enough for the organization just to learn. This learning should be purposeful and oriented at gaining competitive advantage. A. Jashapara (2006, p.317-321) believes, that a competitively learning organization, unlike a learning organization, reaches a higher level of: learning pace, orientation of learning, level of communication, information flow and organizational efficiency.

Another concept of a knowledge based organization is an intelligent organization. It is treated as another, higher level of a knowledge based organization, sometimes presented as a perfect enterprise. While describing an intelligent enterprise, the following features are indicated (Romanowska 2001, Grudzewski, Hejduk 2004, Krupski 2005, Mikuła i inni 2007):

- it is a self-learning and improving organization
- it has adaptation abilities
- it is flexible
- it should be an intelligent innovator
- using intelligence and professional knowledge are crucial
- it has skills in knowledge management and gathering intellectual capital
- it is built on foundations of competencies supported with curiosity, trust and joint actions
- doesn't need to have a legal personality; it can operate within external economic networks, also as a virtual enterprise.

The last model related to knowledge management is the concept of agile organization. Agility of an organization depends on the knowledge, experience and innovativeness of its members and their access to information. An agile organization, through its structure and management process, quickly and smoothly activates its social capital to generate the value for a customer when a market opportunity occurs (Goldman, Nagel, Preiss, 1995, p. 42-43). A current summary of a development of a concept of an agile enterprise is the model proposed by S. Trzcielinski (2005, p.12-16), which includes four dimensions:

- acuteness of the enterprise, which is a function assigning to the turbulent environment a string of potential market opportunities,
- resource flexibility of the enterprise, which transforms the string of potential opportunities into a string of resource available opportunities
- enterprise's intelligence, comprehended as an ability to understand a situation and find deliberate reactions to them, that is to activate proper resource to weaken the threats or use the opportunities,

- smartness of the enterprise, as an ability to quickly use the opportunities in a benefit brining manner.

Models of intelligent and agile organizations are not just a simple expansion of a learning organization; they are a result of an integration with new concepts of organizational structures, development of IT and lean management philosophy. Therefore an attempt to define a list of features of knowledge based organizations is not an easy task, and interpretations are not always unambiguous.

The most common features of the knowledge –based enterprises are (Mikula et al, 2007, p.33-38):

1. Structure of resources and investments in intangibles, which constitute a majority of organization's market value. It refers particularly to intellectual capital, which consists of:

- human capital (people and their knowledge, skill, values, norms, attitudes, views, emotional intelligence, etc.),

- structural capital - understood mainly as the organizational capital,

- customer capital - created by customers, which reflects their potential value of purchase of products and services offered by the organization;

- intellectual property, including: patents, licenses, copyrights, trademarks, secrets, projects, etc.

2. Knowledge management – understood as conscious and deliberate management of knowledge, including the aspects of strategy, structure, culture, technology and people.

3. Shaping the relationship with environment in order to, by using one's knowledge, gain an advantageous location in economic network.

4. Organizational structure is characterized by: high flexibility, openness to environment within network and virtual structures, wide use of temporary task teams and creation of positions or teams responsible for knowledge management.

5. Organizational culture adjusted to new conditions and favoring knowledge management.

6. Specific roles and responsibilities of people. Gradual diminishing of difference between workers and management. Extending the range of people's activities and encouraging initiative and searching for possible system improvements.

3   THEORETICAL POSTULATES OF THE ORGANIZATIONAL INNOVATIONS IN KNOWLEDGE BASED ENTERPRISES AND THEIR EMPIRICAL VERIFICATION

Organizational innovation means an implementation of a new organizational method in company's principles of operation, in workplace organization or in relationships with company's environment (Oslo, 2008, p. 53). Organizational innovation can be differentiated in two types: structural, and procedural. They can be further differentiated in an intra – organizational and inter – organizational dimension. Structural organizational innovations change and improve an organizational structure. Procedural innovations affect routines, processes and operations of a company. While intra – organizational innovations occur within an organization, inter – organizational innovations include new organizational structures and procedures beyond a company's border (Armbruster at al., 2006, pp. 20-21)

Stuctural innovations will be analyzed in context of five dimensions of organizational structure: configuration, centralization, specialization, standardization and formalization.

1. A configuration of a structure in a knowledge based enterprise should lead toward maximal flexibility. There are many ways of increasing this flexibility. First of them is flattening the organizational structure by maximally decreasing the number of hierarchical levels, decreasing the administrative role of a manager and replacing the traditional forms of imperative and functional coordination with a horizontal-process coordination. (Stabryła 2009, Zgrzywa Ziemak 2009, Probst at al. 2002). It leads toward transformation from traditional mechanistic structure to organizational structures with an increased role of informal organizational relationships, face to face communications and two loops of communications downwards and upwards (Ahmed, 1998). I. Nonaka and H. Takeuchi (2000, pp. 195-231) in their critique of traditional bureaucratic structures, indicate task based structures as dynamic and flexible solutions. However, they also notice that teams are temporary in character, and the knowledge generated within a team is not easily transferable to other organization's members when the task is terminated. For knowledge creating enterprises, they propose a model of a hypertext organization, which combines the traditional bureaucracy with task teams. A hypertext organization consists of three layers: a business system layer, a project teams layer and a knowledge resources layer. Replacing the rigid organizational structures with teams and projects, replacing individual and group work with teamwork, replacing fixed functions with project work, replacing hierarchy resulting from positions with relationships resulting from the hierarchy of goals - these are the postulates of many authors (Mikuła et al, 2007, .Zgrzywa – Ziemak, 2009, Stabryła et al., 2009). The most far going postulate is introducing a network organization, which elements are connected with weak coordination relationships and then fuzzing the borders of an organization, increasing the strength of relationships with the environment (customers and cooperators) and increasing the importance of inter-organizational relationships (Stabryła et al., 2009, pp. 179-181). An important postulate for activation of knowledge based organization at the highest level is expanding the types of organizational relationships. C. L. Wang i P.K. Ahmed (2003, pp. 57-60) notice that simple increasing of informal relationships importance is insufficient.. It is necessary to introduce three new dimensions of organizational relationships: trust - based relationship, externally-oriented interactive relationship, and emotionally-inclusive relationship. In Poland, empirical studies related to context variables in knowledge based enterprises, have been conducted twice. At the beginning of

2000, A. Zgrzywa Ziemak conducted a study on the factors which influence the enterprise's learning ability. The results of this study confirmed that learning is enhanced by flat structures, replacing hierarchical relationships with cooperation relationships and increasing importance of goal hierarchy over the hierarchy of positions (Zgrzywa-Ziemak, Kamiński, 2009, pp. 155-156). A second study, concerning the organizational structures in knowledge based economies, was conducted by the team of A.Stabryła in 2008-2009. The results showed that enterprises with the highest index of knowledge potential showed tendencies similar to theoretical postulates, however at a much lower level than theoretical expectations): structures are rather flattered (usually 3 leveled); a basic form of coordination is still a hierarchical organization (although a half of surveyed enterprises also uses a coordination through plans and goals and through regulations and procedures), the scope of multiple subordination increases (Stabryła et al, 2009, pp.296-312).

2. Specialization as a dimension of an organizational structure of a knowledge based organization is rarely discussed in literature. A general tendency (not only in knowledge based organizations) to expand the level of specialization and versatile education necessary in teamwork is indicated. A hypothesis: the lower level of specialization, the more innovative organizational structure, has not been fully confirmed by the studies of A. Zgrzywak-Ziemak (for the detailed distribution of tasks the hypothesis was insignificant, however the flexibility of task distribution was significant) The study of A. Stabryła confirms that in 68% of surveyed enterprises which belonged to a group with a highest knowledge potential, a high specialization of work occurred.

3. Centralization is a dimension which describes the level of concentration or dispersion of decision making authority in organizations. Theoretical postulates are usually consistent: increasing decentralization increases utilization of employees intelligence, which also increases their participation in creation and usage of knowledge. The study of A. Stabryłą confirm that in the group with the highest knowledge potential, 78% of enterprise declares high level of decentralization. In the study of A. Zgrzywak-Ziemak, in turn, this factor is considered irrelevant to the learning ability of an organization. Certain doubts regarding the significance of centralization are also expressed by G. Probst, S. Raub and K. Romhardt (2002, p.284). They point out that decentralization leading toward the freedom of action may have positive influence on the development of internal knowledge, however the empowerment of company's units may impede the usage of knowledge in the company as a whole.

4. Standardization and formalization unifies and consolidates the processes and human behaviors in organizations. It enables gaining high repeatability of processes' flow and results of these processes. On the other hand it limits the use of workers' initiative and intelligence. High standardization and formalization are associated with a model of a fixed, bureaucratic organization, very distant from knowledge based organizations. The study of A. Zgrzywak-Ziemak confirm a thesis that the lower the level of standardization and

formalization, the more likely we are to encounter an innovative organizational structure. Organizational learning is also favored by acceptance of different ways of activity and behavior. The studies of A. Stabryła prove however that the majority of enterprises with the highest index of knowledge potential have full organizational documentation (statute, organizational regulations, organizational structure chart, scopes of activity for workstations, employment plan and documents flow regulations), and the convergence of tasks of organizational units actually performed with the tasks described in an organizational regulations is very high (95% of companies).

Summarizing the theoretical postulates regarding the structure of a knowledge based organization and the concepts of intelligent organization in particular, we get the picture of an organization with a maximal internal freedom. It is an extremely flat organization, almost devoid of the chain of command, based on relationships of horizontal coordination and changing hierarchy of goals, blended into external economic networks, totally decentralized, based on wide specialization of workers, characterized by a very low level of standardization and formalization (Pawłowski et al, 2011). However, the empirical studies reveal the reality quite distant from a theoretically-normative model. A general relation between organizational structure innovativeness and enterprise's learning ability is confirmed. Structures are more flat, the importance of horizontal coordination processes increases, the role of task teams and project teams supplementing traditional organizational structure also increases. A level of centralization, standardization and formalization has no unequivocal interpretation in empirical studies.

Procedural organizational innovations change or implement new procedures and processes within the company, such as simultaneous engineering or zero buffer-rules. They may influence the speed and flexibility of production (e.g. just in time concepts) or quality of production (e.g. continuous improvement process, quality circles) (Armbruster at al. 2006, p. 20). Procedural innovations are usually a result of implementation of new management concepts, methods and techniques. In theories of knowledge based organizations there are no explicit postulates regarding the list of management methods and techniques which should be implemented. Some methods, such as learning methods, knowledge management methods, heuristic methods and techniques, methods connected with processes improvement and quality management, are inevitably associated with knowledge based organizations. This list may be expanded however, having in mind specific features of intelligent and agile organization. On the other hand, such list should not be identical for industrial, trade and service companies. Empirical studies conducted by the team of A. Stabryłą, were focused on twelve methods. In a group of enterprises with the highest knowledge potential, the following management methods and techniques were used (Stabryła et al, 2009, p. 300): Controlling (in 75% of enterprises from this group), Budgeting (50%), Scenario planning (30%), TQM (30%), Outsourcing (30%), Reengineering (22%), Lean Management (22%), Outsourcing the organizational units (22%), JIT (18%), BSC (18%), Benchmarking (18%), Outplacement (8%) In the enterprises with lower knowledge potential, these methods were used in much smaller degree. Nevertheless, the level of procedural innovation for the group of enterprises with the highest knowledge potential is surprisingly low.

4   ANALYSIS OF A GAP BETWEEN THEORY AND PRACTICE OF KNOWLEDGE BASED ENTERPRISES

Theoretical models of knowledge based enterprises create postulated view of an enterprise with the highest level of organizational innovativeness. Empirical studies show the organizational reality rather distant from such extreme normative model. Basic empirical trends of enterprises development in the process of adjustment of knowledge based economy confirm major theoretical assumptions, however not in their radical version.

While searching for the sources of these discrepancies, we should mention well known publication from a trend based on case studies (Peters and Waterman, 2000, Hammer and Champy, 1995, Senge et al, 2002, Probst et al, 2002, Jashapara, 2006, Nonaka, Takeuchi, 2000, Grudzewski, Hejduk 2004). They present various organizational innovation, however none of tchem confirms a comprehensive implementation of theoretical normative models. Successful American enterprises from the 80s were characterized by: using flexible and fixed organizational forms at the same time, autonomy and initiative of some departments and rigid discipline in other, centralization in some areas and decentralization in other (Peters and Waterman, 2000, PP. 46-50). Similar examples are given by Hammer and Champy (1995, pp. 51-64): hybrid centralization and decentralization of operations, multi-variant standards of process, control limited to economically justified level, limiting excessive distribution of tasks and specialization.

Another contribution to explanation of the analyzed gap between theory and practice is the result of empirical studies over the scope of implementation and scope of usage of organizational innovations. The analysis of the German Manufacturing Survey 2003 shows that only a small proportion of the companies that make use of a certain organizational innovation have fully implemented this organizational innovation in all business areas. For example: more than 60% of all firms claim to have implemented team work, however, only 10% say that they have fully exploited the potential of this organizational innovation, task integration has been realized by more than 60%, but only 7% have implemented this innovation throughout the whole corporation (Armbruster at al, pp. 34-35).

We may therefore form a hypothesis that organizational innovations in an enterprise are not evenly distributed between all areas of enterprise's activity, and so, drawing conclusions based on the average results of entire enterprise does not reflect the actual innovativeness of key areas of the enterprise. To verify this hypothesis I conducted a preliminary study in the second half of 2011 on a group of 30 Polish enterprises. These enterprises were selected according to four criteria, which enabled their differentiation based on: economic sector (production, trade, services), size of an enterprise, dependence of the international corporations (independent enterprise, subsidiary or division), enterprise's age (year of creation).

A research tool was a structuralized interview with the companies' Management, based on a questionnaire and verification of source documentation. Questions were constructed in a manner, which allowed gaining information confirming (or not) the existence of symptoms of a given characteristic (such as organizational structure or management method characteristic) and not only the declaration of managers. Major conclusions drawn from this study are the following:

1. There is a positive relationship between the size of an enterprise and the level of organizational innovativeness. A similar relationship exists between internationalization of the enterprise and organizational innovativeness. Large and medium sized enterprises are dominant, particularly the divisions of international corporations. Small companies use a number of innovative organizational solutions, however they do it rather intuitively and not are not always able to name them.

2. Small companies more often and more easily enter the external, network organizational structures.

3. For large and medium sized organizations, a hypothesis of diversified level of organizational innovativeness in various areas of enterprise's activity was confirmed. The most flexible solutions in regard to the dimensions of organizational structure were present in the following areas: marketing and sales, products development and logistics. The most rigid structural solutions were found in accountancy, human resources and production.

5 SUMMARY AND FINAL CONCLUSIONS

Knowledge based enterprises should be innovative by definition. Along with this assumption, theoretical normative models of a learning enterprise, competitively learning enterprise, intelligent and agile enterprise are constructed. Organizational innovation in such enterprise is identified with the highest possible level. Empirical studies do not confirm these expectations, indicating intermediate solutions between traditional and intelligent enterprise. Case studies and empirical studies over the scope of implementation of organizational innovations in enterprises have shown that the scope of innovations' implementation is different in different areas of the enterprise. I also confirmed this thesis in my own pilot research.

A conclusion, that the model of knowledge based enterprise presented at the level of general concept related to the characteristic of the entire enterprise does not yet explain the internal logic of functioning of such enterprise, seems reasonable. Internal organizational innovations, both structural and procedural, should be distinguishable in particular areas of enterprise's functioning.

12

REFERENCES

Ahmed P.K. 1998. Culture and Climate for Innovation. European Journal of Innovation Management. Vol. 1 No 1, pp. 30-43

Armbruster H., Kirner E., Lay G., 2006. Patterns of Organizational Change in European Industry. Fraunhofer. Institute Systems and Innovation Research, Karlsruhe

Isaacs D. 2002. Połączyć najlepsze z obu światów. Podstawowe procesy w organizacjach jako wspólnotach; [in]: Senge P.M. et al, Piąta dyscyplina. Materiały dla praktyka, Oficyna Ekonomiczna, Kraków

Davenport T.H. 2007. Zarządzanie pracownikami wiedzy, Oficyna Wolters Kluwer, Kraków

Goldman S., Nagel R., Preiss K.. 1995. Agile Competitors and Virtual Organization. Strategies for Enriching the Customer, Van Nostrad Reinhold, New York

Grudzewski W.M., Hejduk I.K.. 2004. Zarządzanie wiedzą w przedsiębiorstwach, Warszawa

Hammer M., Champy J.. 1995. Reengineering the Corporation. A Manifesto for Business Revolution, London, Nicolas Brealey Publishing Limited

Jashapara A. 2006. Zarządzanie wiedzą, PWE, Warszawa

Krupski R. 2005. Elastyczność organizacji. [In:] Krupski R. (Ed.). Zarzadzanie przedsiębiorstwem w turbulentnym otoczeniu. PWE Warszawa

Mikuła B. 2006. Organizacje oparte na wiedzy, Wydawnictwo Akademii Ekonomicznej w Krakowie, Seria specjalna Monografie, nr 173, Kraków

Mikuła B., Pietruszka – Ortyl A., Potocki A., (ed.). 2007. Podstawy zarządzania przedsiębiorstwami w gospodarce opartej na wiedzy

Nonaka I. 1991. The Knowledge - creating Company, Harvard Business Review, 1991, No 69,

Pawłowski E., Trzcielinski S., Włodarkiewicz Klimek H., Kałkowska J. 2011. Organizational Structures In Knowledge Based Enterprises. [In:] Lewandowski J., Jałmużna I., Walaszczyk A. (Ed.). Contemporary and Future Trends In Mangement. Technical University of Lodz., pp 113-131

Perechuda K. (ed.). 2005. Zarządzanie wiedzą w przedsiębiorstwie, PWN, Warszawa

Peters T.J., Waterman R.H. 2000. Poszukiwanie doskonałości w biznesie. Wydawnictwo Medium, Warszawa

Probst G., Raub S., Romhardt K. 2002. Zarzadzanie wiedzą w organizacji. Oficyna Ekonomiczna, Kraków

Oslo Manual. 2006. Podręcznik Oslo. Zasady gromadzenia i interpretacji danych dotyczących innowacji. OECD, Warszawa

Romanowska M. 2001. Kształtowanie wartości firmy w oparciu o kapitał intelektualny. [In:] Borowiecki R., Romanowska M. (Ed.). Systemy informacji strategicznej. Wywiad gospodarczy a konkurencyjność przedsiębiorstwa. Difin. Warszawa

Senge P.M. .2003. Piąta dyscyplina. Teoria i praktyka organizacji uczących się. Oficyna Ekonomiczna, Kraków

Stabryła A. (ed.). 2009. Doskonalenie struktur organizacyjnych przedsiębiorstw w gospodarce opartej na wiedzy, Wydawnictwo C.H. Beck, Warszawa

Trzcieliński St. (ed.). 2005. Nowoczesne przedsiębiorstwo, Monografia wydana przez Instytut Inżynierii Zarządzania, Politechnika Poznańska, Poznań 2005

Wang C.L. and Ahmed P.K. 2003. Structure and Structural Dimensions for Knowledge – Based Organizations. Measuring Business Excellence. Vol. 7 No 1

Zgrzywa-Ziemak A., Kamiński R. 2009. Rozwój zdolności uczenia się przedsiębiorstwa, Difin, Warszawa

CHAPTER 2

Chosen Methods Supporting Management of Enterprise's Agility

Stefan Trzcieliński

Poznan University of Technology
Poznan, Poland
stefan.trzcielinski@put.poznan.pl

ABSTRACT

Agile enterprise is able to use short life time opportunities. This is possible if it is bright, flexible, intelligent and shroud. These four features constitute the agility. To manage agility it is necessary to shape these features. That can be achieved by using proper methods which are suggested by some metaphors of agility. In this paper three metaphors are used as a background to select the methods: bright eye of chameleon, flexibility of car engine, and the contextual sub-theory of intelligence as a metaphor of the shrewdness. Next, the methods that support achieving agility are presented.

Keywords: agility, bright organization, flexible organization, intelligent organization

1 INTRODUCTION

The civilization development is an evolutionary process although from time to time it become essentially accelerated. Such accelerating moments, particular when the technological development is concerned, are caused by scientific discoveries and inventions and following them turning points are usually considered as a revolution. In the technological civilization such turning point was an industrial revolution in the second half of 18th century that initiated the paradigm of craft manufacturing. This paradigm was based on universality of both the workers skills and the equipment and machines they used (Womack, Jones and Roos, 1990; Dove, 2001).

The products were customer tailored and expensive.

The second breakthrough in the industrial manufacturing was caused by mastering and used the production technology of standardized spare parts in 1904 by Cadillac Motor Car Company (Bicheno, 2000). In the connection with achievements in the work study (Taylor, 1911), simplification of construction of products and used of production lines by Ford Motor Company in 1913, all these events resulted in mass production. Mass production was based on the paradigm of narrow and deep specialization leading to the economies of scale. Products were/are highly standardized but essentially cheaper comparing with craft manufacturing.

The next breakthrough was the consequence of assuming by Toyota by the end of 40' in the XX century that the production costs can be cut down without the economies of scale. The assumption was taken in very hard condition that existed after the II World War in Japanese economy. The country was destroyed, peoples were pour, inflation was very high, there was no capital to invest, the inflow of foreign direct investment was strongly reduced, the American administration introduced the work law protecting employees, the businesses were small and used to use rather the craft than mass production technology (Womack, Jones and Roos, 1990). To become competitive on international markets Toyota has implemented a row of organizational inventions known commonly as Toyota production System or Lean production. The philosophy of these inventions was based on paradigm of liquidation and prevention of wastes.

Although lean enterprises are more flexible in terms of offering the customer oriented products, yet they are interested in extending the production batches, to balance the capital-intensive fixed assets in which they are equipped. There is a question if liquidation of wastes is enough to compete effectively in the continuously increasing turbulent and unpredictable environment. The new paradigm of industrial manufacturing bases on the assumption that a such environment can be recognized not as a hostile but as a friendly one, as changes generate opportunities. The enterprise that is short life time opportunity oriented is called the "agile enterprise" (Goldman, Nagel and Preiss, 1995).

Although the essence of enterprise's agility depends on the ability of the enterprise to use the opportunity, yet to shape the ability the term "agility" requires operationalization. One of the proposal includes meaning the agility as a system of such features like brightness, flexibility, intelligence and shrewdness of the enterprise (Trzcieliński, 2007).

2   METHODS SUPPORTING AGILITY OF ENTERPRISE

Opportunities are passing favourable situations appearing in the enterprise's environment. Thus, they exist for a limited time. Looking on the agility from that perspective the following requirements can be expressed about the agile enterprise:

- It should be able to perceive the events appearing in the environment and form them in the situations which are favourable to the enterprise. This feature is called "brightness".

- It should assess its own resources and that which have to be acquired from the environment to be able to respond quickly for variety of customers expectations. This feature is called "flexibility".
- It should be able to learn and adopt itself to the new situations appearing in the environment. This feature is called "intelligence".
- It should be able to manage the currently appearing problems to use the opportunity including reconfiguration of the accessible resources, initiation and modification of necessary projects and controlling of their flow. This features is called "shrewdness".

Skipping grading of the features and limiting only to statement if the enterprise possesses them a fifteen cases of agility can be distinguished. Below are only four cases characterized by lack of the following feature:

- Shrewdness; such enterprise can be categorized as wasting the opportunities because its system of operations management is not efficient.
- Brightness; such enterprise can be categorized as using the opportunities accidently as its management system do not scan the environment.
- Intelligence; such enterprise can be categorized as temporally agile as its management system do not play the role of strategic adjustment of its potential to act in the conditions of discontinuity.
- Flexibility; such enterprise can be categorized as oriented only on long life cycle opportunities as its potential is narrowly specialized and its adoptability is strongly limited.

These examples confirm that agility should be managed.

2.1 Brightness

Some methods to shape the brightness of the enterprise can be withdrawn from the metaphor of bright chameleon's eyes (Trzcielinski, 2010). The reptile perceives the opportunities – the flying insects in the wide visual field. To notice the prey the chameleon moves quickly its one eye knob and scan the environment when at the same time the second eye is turn on long perspective vision. The process is switched in short intervals between both eyes. To use the opportunity i.e. to capture the insect, the chameleon shuts with great precise its long tongue for the distance of the flying insect. Its very precise judgment of distance is achieved by coupled accommodation in both eyes.

Using the metaphor of the chameleon we can say that the methods that the enterprise uses to improve its brightness should enable: segmentation of the environment (sectors observed by the reptile), analyses of the segments (looking for the potential victim), and recognition if the situation is favorable to capture the opportunity (assessment of the distance and the dimension of the flying insect).

Segmentation of the environment
The following methods can be used to segment the environment of the enterprise: PEST, structural analysis of the sector and identification of the stakeholders.

The PEST method is a general approach to the segmentation of the macroenvironment (Morrison, 2006). It divides the environment on political (and legal), economic, social (and demographic), and technological segments. Each of them should be observed in short and long perspective by proper organizational units (departments or specialists).

The structural analysis of the sector was elaborated by Porter (1980). It divides the industry environment on segments that include: the existing competitor, new entrants to the sector, suppliers, customers and substitute products. There are events in each of the segment that individually or together with another events can create the opportunities.

The environment can be segmented also according to the stakeholders of the firm. This term refers to these groups whose help is critical to the existence of the enterprise. The stakeholders can wield the positive or negative influence on the firm creating either opportunities or threads.

Analyses of the segments
These methods of strategic management that are focused on analysis of factors belonging to particular segments of the environment can be useful to search for opportunities. Among them there are: structural analysis of the sector, assessment of the sector attractiveness, white industrial intelligent, SWOT analysis, trends extrapolation, and Delphi method. In the Table 2.1 there are presented factors which can be analyzed with use of the two first methods.

The white industrial intelligence depends on collecting data about industrial environment, particular about competitors, using such sources like: newspapers, journals, industry reports, corporate reports, court announcement about bankruptcy, internet, etc. The data can concern e.g. products, prices, cost of production, technologies, human resources, nets of collaboration, intended movements and strategies of competitors, and others.

SWOT analysis enables to create the existing or future picture of the environment in the segments according to PEST analysis or structural analysis of the industry. The picture presents strengths, weaknesses, opportunities and threats in particular segments.

Extrapolation of the trends can be used to forecast states of these environmental factors which are subject of continuous changes. Examples can be some factors of economic growth like: rate of unemployment, inflation, GDP, or demographical factors like: the birth rate, population of inhabitants in cities, and social structure of unemployment. There are several of details methods of trends extrapolation. They can be categorized into two groups: correlation analysis and regression analysis. Correlation analysis is used in forecasting as a screening process to identify dependent variables that can be used in forecasting model. Example can be looking for correlation between competitor's average price (independent variable) and demand for the company's product (dependent variable). Regression analysis is used to depict a relationship between dependent variable and a single or set of independent variables. The relationship is expressed in mathematical formula that enables to receive the value of dependent variable according to the future value of independent variable(s).

Table 2.1 The states of features of industry environment that can create favorable situations

Industry environment factors connected with:				
suppliers	customers	new entrants	competitors and substitute products	Attractiveness of the sector
Suppliers concentration	Customers concentration	Economy of scale	Number and similarity of firms	Market demand and rate of increase
Substitutes offered to the buyers	Volume of purchases	Access to new technologies	Trends in demand for products	Seasonality and cyclicality
Importance of the buyers	Participation of purchase in the total volume of purchases	Effect of experience	Products differentiation	Profitability
Importance of the products for the buyers	Standardizatio n of the products	Preferences of Brand and customer loyalty	Effectiveness and price of the substitutes	Changeability of the technology
Diversity of the products offered	Switching cost of the suppliers	Capital requirements and accessibility	Switching cost of substitutes	Leader participation in the market
Switching cost of the buyers	Profit of the buyer	Distribution channels accessibility	Diversity of competitors and their strategies	Requirements about New skills
Credible threat of foreword integration	Credible threat of backward integration	Concessions and governmental regulations	Cost of exit from the sector	Natural environment hazard
Economic situation of the supplier	Influence of the purchase on the quality of own product	protectionism	–	–

Delphi method enables forecasting future events and situations not on the base of time series like methods of trends extrapolation but on the base of experts' knowledge. The experts are questioned in this way that the answer can be given in measurable way. They work independently and their answers are collected in questioners. These answers which are out of inter-quartile interval, requires

18

justification and explanations and the explanations are shared with other experts. In this way during four rounds the experts make their stands resemble (Figure 2.1).

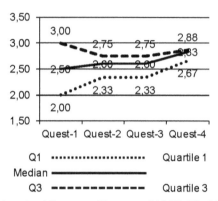

Figure 2.1 Experts opinion about the competitiveness of Middle-West Poland wood processing industry in European Union (Likert scale). (From L. Pacholski, S. Trzcielinski, and M. Wyrwicka. 2011)

Associating events in favorable situations
To model opportunities such events must be taken into consideration which associated together create a favorable situation. A scenario method can be used to support this process. One of the variants to build the scenarios is cross-impact method (Grodon, 1994). It depends on building a probably scenario of connected events that can happen in the future. First the probability of each event is estimated and next the conditional probability of each pair of events. On this base the probably sequence of events that create favorable situation is identified.

2.2   Flexibility

Flexibility of car piston engine can be used as a metaphor of flexibility of enterprise and it delivers an interesting suggestions of methods that the enterprise can use to manage its flexibility. The piston engine flexibility its ability to adjust itself to changeable loads and rotational speed and it depends on its turning moment. In case of the enterprise the turning moment is the analogue of the enterprise's capacities which are determined by its own and outsourced resources (Figure 2.2).

- Bigger the turning moment (enterprise's capacities) bigger the engine's work (production volume) is possible.
- The flexibility of the engine increases when the maximal turning moment moves in the direction of lower rotational speed. This leads to slower wear of the engine. In case of the enterprise the analogue of the scope of the rotational speed is the variety of the accessible production capacities particular these which are determined by the fixed assets. The flexibility of the enterprise increases when it reduces its own capacities and substitutes them by external resources (outsourcing).

- Wither the scope of rotational speed with the maximal turning moment, higher the flexibility of the engine. Then the power of the engine increases almost in linear mode and there is no need to change the gear frequently. In case of the enterprise the analogue is an extension of the repertoire of the assortment of the product that can be delivered by the enterprise not in result of investment in own capacities but by use of the unlimited capacities of small and medium enterprises (in the Figure 2.2 the shadow area is stretched to the left). This leads to the increase of the production power of the enterprise.

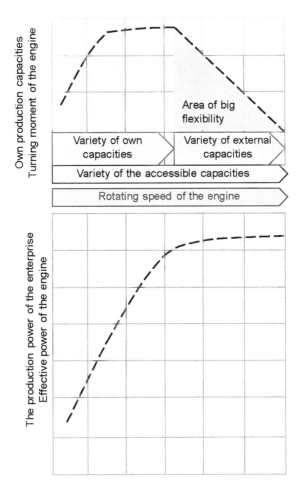

Figure 2.2 The rotating moment of the car piston engine as an analogue of the production capacities

- More rapidly goes the curve of rotating moment down, more easy the engine adjust itself to the external loud. It enables to conclude that the bigger portion of the enterprise capacity is substituted by the external resources, more the enterprise is flexible.

Flexibility refers to the capacity of an organization to move quickly from one task to another one (Goldman, Negal, and Preiss, 1995). Such meaning flexibility encompasses at least two aspects: flexibility of resources end flexibility of organizational structures.

Resource flexibility is determined by the quantitative and qualitative structure of the resources. In the quantitative dimension the flexibility can be increased by the excess of the resources and in the qualitative dimension – by their universality and diversification. Diversification can be achieved by increase the variety of own resources or by outsourcing.

The excess or diversified own resources kept to use them when the opportunity appears increases the risk, that they will not be exhausted in satisfactory level. Thus it generate not justified fixed costs. So the most rational way of increasing the enterprise's flexibility is by using the universal resources and extending the access to the occasionally needed resources.

In contemporary production systems the universality of fixed assets is reached by use of the flexible manufacturing systems. Concerning the human resources the increase of their universality depends on the development of the staff knowledge, competencies, and skills. Thus the concept of learning organization should be implemented.

The access to the external resources can be extended by creating networking company. From one side this includes supply chain partnering – the management method oriented on establishing long time partnership with suppliers and customers like strategic alliances or consortiums, from the second side – temporary relationships with subcontractors that leads to such structures like virtual organization (Trzcielinski, and Wojtkowski, 2007), reconfigurable organization (Galbright, 1997) and chameleon organization (Miller, 1997). The flexibility of enterprise can be also supported by proper structuring of internal resources, mainly the human resources. Example can be structures like matrix, task, project or hybrid structures.

2.3   Shrewdness

Opportunities are passing favorable situations. Their life time depends on changeability of the environment. More changeable is the environment, particular the industry one, shorter lasts the opportunity (Trzcieliński and Trzcielińska, 2011). The brevity of the opportunity makes difficult the use of it, as the enterprise has to configure the necessary resources in relatively shorter time. In such circumstances there is a demand for special managerial skills that depend on practicality of actions. The ability of the enterprise to use quickly the opportunities in beneficial mode is called the shrewdness of the enterprise.

A useful metaphor of company's shrewdness is delivered by the contextual subtheory of Triarchic Theory of human intelligence (Sternberg, 1985). It involves the ability to grasp, understand and deal with everyday tasks. This is the contextual aspect of intelligence and reflects how the individual relates to the external world about him. It combines: (1) adaptation to the environment in order to have goals met, (2) changing the environment in order to have goals met, (3) moving to a new environment in which goals can be met.

These three kinds of behavior directly refer to the organizations. The shrewd enterprise is the one that is able: to adopt to the environment, change the environment, and/or move to another environment to achieve its goals.

The adaptation is a reactive model of the organization adjustment to the changing environment. It is usually used when relatively small changes caused the loss of the organization's effectiveness and efficiency. To regain the balance the enterprise has first to identify the changes using the mentioned above methods of segmentation and analysis of the segments. Next it has to implement some modifications in the organization. A useful method that can be used is a diagnostic analysis of the organization.

The enterprise can take the attempt to change the environment to make it friendly. Changes in the industry environment can be achieved by negotiations with suppliers and customers, elaboration a competitive strategy or relation strategies. More difficult is to implement changes in the macroenvironment. As usually a single enterprise is not able to do this itself, an effective method can be lobbing or representation of the enterprise interest by the chamber of commerce, confederation of the employers or business clubs.

It is not always possible for the enterprise to adopt itself to the environment or change it. Than the option is to move to another environment. For example it can depend on entering a new industry, changing the location of the business or registration the firm in "tax paradise".

3   CONCLUSIONS

Some authors who write about agile enterprise appoint that it uses the same concepts and methods as lean enterprise. For some extend it seems to be obvious, as lean laid the foundations for agility and lean organization and methods are key components for agility (Goldman, Nagel, and Preiss, 1995). Although the enterprise cannot be agile if it is not lean, it does not mean that these concepts are similar one each other. Lean is oriented for using long life time opportunities and therefore uses methods that protect the enterprise against changeable environment. Contrariwise, agile enterprise needs the changeable environment as such environment generates short life time opportunities. However to be effective the organizational system of the enterprise must enable to perceive and next to use the opportunities. There are four features that make the system opportunity oriented. They are: brightness, flexibility, intelligence and shrewdness. Each of these features can be enhanced by a variety of methods. Some of them have been referred in this paper. There is a question who in the organization should be responsible for establishing the agility, that is for brightness, flexibility, intelligence and the shrewdness of the enterprise. In big enterprises the responsibility is allocated to the functionally specialized organizational units. In small and medium enterprises the organizational units are usually widely specialized. Because of this the responsibility is rather connected with functions than with departments. Thus for example representatives of the following roles and functions should be engaged in searching for opportunities in particular segments:

- P – political and legal segment: board of directors, managers and specialists who are involved in B&R, business economics, finances, personal, sale, work and natural environment protection,
- E – economic: personal, business economics, finances, merchandise,
- S – social and demographic: marketing, personal, business economics, natural environment protection,
- T – technological segments: B&R, finances, personal, quality, procurement.

Thus, making the organization agile is a collective effort.

REFERENCES

Bicheno, J. 2000. *The lean toolbox.* PICSIE Books: Buckingham.

Goldman, S., Nagel, R., and K. Preiss. 1995. *Agile competitors and virtual organization. Strategies for enriching the customer.* Van Nostrand Reinhold: New York.

Galbraith J.R. 1997. The reconfigurable organization. In. *The organization of the future*, eds. F. Hesselbein, M. Goldsmith, and R. Beckhard. The Drucker Fundation: New York.

Gordon, T. 1994. *Cross-impact method.* A Publication of United Nations Development Program's. African Futures Projects in collaboration with United Nations University's. Millennium Project Feasibility Study – Phase II.

Miller, D. 1997. The future organization. A chameleon in all its glory. In. *The organization of the future*, eds. F. Hesselbein, M. Goldsmith, and R. Beckhard. The Drucker Fundation: New York.

Pacholski, L., Trzcieliński S., and M. Wyrwicka. 2011. Clustered makroergonomic strucrures. *Human Factore and Ergonomic in Manufacturing and Service Industries*, 21: 147-155.

Sternberg R.J. 1985. *Beyond IQ: A triarchic theory of human intelligence.* Cambridge University Press: Cambridge.

Taylor, F. 1911. *The Principles of Scientific Management.* Harper & Brothers: London.

Trzcieliński, S. (Ed.). 2007. *Agile enterprise. Concepts and some results of research.* IEA Press: Madison.

Trzcielinski, S. 2010. Some metaphors of agile enterprise. In. *Advances in humen factors, ergonomice, and safty in manufacturing and service industries*, eds. W. Karwowski and G. Salvendy. CRS Press: Boca Rsaton.

Trzcieliński S., and W. Wojtkowski. 2007. Toword the measure of organizational virtuality. *Human Factors and Ergonomics in Manufacturing*, 17: 575-586.

Trzcieliński S., Trzcielińska J. (2011). Some elements of theory of opportunities. *Human Factors and Ergonomics in Manufacturing and Service Industries*. Vol. 21, 124-131.

Womack, J.P., Jones D.T., and D. Roos. 1990. *The machine that changed the world.* Rawson Associates: New York.

CHAPTER 3

Workforce Diversity and Ergonomic Challenges for Sustainable Manufacturing Organizations

Amjad Hussain, Russell Marshall, Steve Summerskill, Keith Case
Loughborough University, UK.
A.Hussain@lboro.ac.uk

ABSTRACT

Demographically, it is evident that the composition of the workforce is becoming more diversified and this trend is very significant in most developed countries such as the US, UK, Canada and Australia. Workforce diversity covers a wide range of dimensions like age, gender, culture, ability, background, level of skill, marital status etc. Because of this, workers share different attitudes, working behaviors, needs, desires and values. Workforce diversity management needs the development and management of such an environment where all individuals with these differences can perform at their full potential, so that any organization can draw an optimum benefit from its diversified workforce. Like many others, manufacturing organizations are also facing the issue of workforce diversity where it affects work performance capabilities. Organizational sustainability can only be ensured by workplace safety, employee satisfaction and retention along with health and well-being. In spite of highly automated systems, manufacturing activities like manual assembly tasks with sustained high quality requirements demand highly repetitive movements with high physical demands at the highest level of work pace. Ergonomics plays a vital role in the development of work environments that ensure a healthy, safe, risk-free and productive use of human capital. Yet there has been little investigation of workforce diversity management with reference to ergonomic issues, challenges, opportunities and strategies. This paper reveals the need for an ergonomics-based 'design for all' approach to address the issues of a diversified workforce. This approach is based on the use of a digital human modeling system where an individual's actual working capabilities along with coping strategies are used at a pre-design phase for any design assessment. A database of 100 individuals belonging to different age groups and working capabilities provides an opportunity to assess any workplace, product, and process or environment design at an early design phase. In this way, it provides design solutions that are equally acceptable for a broad range of humans belonging to different backgrounds, age groups and levels of ability to do the work. Current ongoing research is focusing on capturing working strategies of a diversified workforce in the furniture manufacturing industry where workers belonging to different age groups, backgrounds, experience and levels of skill will be analyzed. Subsequently this data will be used in a digital human modeling system called HADRIAN providing designers and ergonomists with the ability to access and address the design needs of a more diversified workforce. This strategy helps in addressing global workforce challenges where organizations can effectively utilize their human capital by providing them with a healthy and safe working environment.

Keywords: Workforce diversity, organizational sustainability, ergonomics, inclusive design

1   INTRODUCTION

Workforce diversity has become a primary concern of many organizations. It is demographically well-evident that future organizations will be facing a challenge of a more diversified workforce and this trend is very prominent in developed countries like US, UK, Canada and Australia. Workforce diversity covers a wide range of dimensions like age, gender, culture, ability, background, level of skill, marital status etc. Because of this, workers share different attitudes, working behaviors, needs, desires and values. This clear global demographic trend demands a working environment where people with different working capabilities, attitudes, behaviors, age and gender can co-exist effectively within the same organization. Diversity management accentuates the development and implementation of specific skills, policies and practices that aim to get the best from every employee. The ultimate objective of these strategies is to win a competitive advantage for the organization by recognizing the importance of each employee. So, a proper understanding of differences that exist among workers belonging to different age groups, levels of skills, working capabilities, gender and ethnic backgrounds becomes vital for the achievement of effectiveness and productivity. There is a need to implement an inclusive design strategy that can overcome workplace design difficulties by promoting design practices where a maximum proportion of the workforce, with their existing differences, is considered at some earlier design stage.

2   WORKFORCE DIVERSITY AND ORGANIZATIONAL PERFORMANCE

Diversity is typically referred to as differences between individuals that may lead to the perception that other persons are different. Diversity mainly focuses on the differences in gender, age, functional capability, ethnic and cultural background, and education (Knippenberg, 2007, Williams, 1998).

Workforce diversity management is not so straight forward as diversity is a double-edged sword which comes with potential benefits and challenges. The literature clearly shows that the relationship between work group diversity and work performance on individual and organizational levels is inconsistent. As mentioned above, diversity has different dimensions like age, race, ethnicity, cultural background, gender, disability etc; so different dimensions of diversity might have positive as well as negative effects (Knippenberg, 2004, Shore, 2009). Evidence suggests that effective workforce diversity management can contribute to organizational performance in terms of improved group performance, friendlier attitudes, better cooperation, innovation and better decision-making as people from different backgrounds, cultures, experiences and knowledge provide a larger pool of novel and diverse problem solutions. Moreover, it also helps organizations in winning desirable work behaviors from the employees which contributes to organizations in achieving their goals. It adds value to the organization and contributes a competitive advantage to firms. Richard (2000) also concluded that a positive impact of diversity management will depend on the context and absence of diversity context may lead to negative outcomes. Diversity management can increase coordination and control costs of the organization (Mamman, 2012, Richard, 2000, Williams, 1998). On the other hand, evidence also shows that failure to manage a diverse workforce can lead to a perception of injustice among the members which may lead to an environment of conflicts, frustration and odd behaviors that can have very serious consequences for the organization. Results show that these experiences ultimately promote behaviors like absenteeism, high turnover and job dissatisfaction, lower work commitment and withdrawal from organizational citizenship behaviors (Shore, 2009, Mamman, 2012).

The above discussion reveals the complexity of diversity management and demands strategies that might foster positive aspects and prevent negative outcomes. It requires the exploration of diversity from a new positive and proactive standpoint. Recently, researchers have already started working on new ideas like diversity climate and inclusiveness (McKey, 2007, Roberson, 2006)

3   ORGANIZATIONAL SUSTAINABILITY

Much has been written on the concept of sustainability in the last few years and debate is still going on. This might be due to the varying conceptual roots of defining the term 'sustainability'. Indeed, the sustainability concept has inherent positive meanings that can appeal to everybody at individual and organizational levels. There are two very common perspectives of sustainability mentioned in the literature. The first concept is based on Brundlandt's definition of sustainability, where sustainability is defined as, "meeting the needs of the present generation without compromising the ability of future generations to meet their needs" (WCED, 1987). Later on, Dyllick and Hockerts (2002) conceptualized the definition again in organizational stakeholder's perspective, when they defined it as, "meeting the needs of firm's direct and indirect stakeholders (such as employees, shareholders, clients, pressure groups, communities etc.) without compromising its ability to meet the needs of future stakeholders as well" (Dyllick, 2002). The second popular concept of sustainability was defined by Elkington (1997), where the triple-P perspective was introduced. The Ps stand for people, planet and profit. An organization might be considered sustainable, if a certain minimum performance can be achieved in these areas. In practical terms, organizational sustainability can be achieved by finding and achieving a balance between financial or economic goals (profit), social goals (people), and ecological or environmental goals (planet) (Elkington, 1997). The core of the organizational sustainability concept lies in the understanding of the fact that multiple stakeholders share different objectives of sustainability as it is directly related to their needs and the extent to which these needs are fulfilled. Moreover, it is a continuous process where the relative needs of different stakeholders might change with the passage of time.

As mentioned previously, the organizational workforce is becoming diversified with every passing day. Here it becomes important for organizations to understand the changing needs of their future diverse workforce, so that they can retain their experienced, skilful and committed workforce. Organizational sustainability can be promoted by achieving a safe, friendly, productive and healthy working environment. As we know, diversity management demands a working environment where people with different backgrounds, races, age, working capabilities, behaviors etc. can co-exist happily in the presence of all these differences. So, the objective of organizational sustainability in workforce diversity management can only be achieved by achieving an environment where differences among the workers are recognized and their job needs are fulfilled according to their capabilities. Workforce dissatisfaction results in higher turnover, lack of interest and absenteeism. Removal of an experienced worker is not simply a loss of a person but it is the drainage of skills, relations and knowledge and regaining these will need resources such as money, time and commitment (Dychtwald, 2004).

4   FUTURE ERGONOMIC CHALLENGES

The International Ergonomics Association (IEA) states "Ergonomics (or human factors) is a scientific discipline concerned with the understanding of interactions among humans and other elements of a system, and the profession that applies theoretical principles, data and methods to design in order to optimize human well-being and overall system performance. It is broadly divided into three main domains; physical, cognitive and organizational ergonomics which shows its multi-disciplinary nature" (IEA). A multi-disciplinary ergonomics approach provides an option to understand differences in human beings that leads to the addressing of the workforce diversity issue for the achievement of organizational sustainability. Ergonomics has contributed well in the recognition of the mismatch between human work capabilities and work demands. In spite of all this, still there are many areas like understanding human differences that directly or indirectly affect work performance and need the urgent attention of ergonomists, planners, managers and designers. Unlike many other workforces, a manufacturing organization's workforce is still supposed to complete their work manually where high work demands with repetitive motions creates many difficulties for the workers. On the other hand, the globally competitive market forces organizations to develop strategies like 'doing more with less' so that they may sustain themselves in the market. For example, it is demographically very clear that the global workforce is ageing and this trend is seen in nearly all parts of the world. United Nations statistics show that there were 378 million people aged 60 or above in 1980 and that figure had approximately doubled to 759 million in 2010. It is further projected that the people in the world aged 60 or over will be increasing to 2 billion by 2050 (U.N.O., 2009). Age affects humans in different ways that directly or indirectly affect human work performance. Functional capacity mainly depends on the musculoskeletal strength of the body, which starts declining after the age of 30 (Wanger, 1994). Moreover, decline in many other functions like joint mobility, balance, visibility, and higher reaction time have been described in the literature (Chung, 2009, Hultsch, 2002, Sue, 2008, Sturnieks, 2008). As age diversity is increasing, such issues are becoming more serious as variations in human capabilities due to age directly influence work performance. For example, most manual assembly tasks require fast and accurate movements of different parts of the body whereas the decline in joint mobility decreases flexibility. Similarly, manual material handling requires muscular strength to safely handle heavy weights but a person 50 years of age is surely less capable of handling such tasks as compared to a 25 year old person. These variations in human capabilities that relate to work performance demand such design solutions that can accommodate a wide range of the worker population. Usually designers and planners target fully capable and young people when they set their organizational goals and ignore these variability issues that create problems for the workers at some later stage.

Like age, skill variation is also an important area that must be considered at a pre-design phase so that skill variability issues might be addressed properly. Human working skills might be influenced by work experience, age, level of education, background etc. An experienced and skillful worker is supposed to perform tasks in the least possible time by adopting less physically strenuous, safe and easy working methods. Conversely, less experienced and younger workers normally go for the strategies that expose them to a number of risk elements.

In future, ergonomists, designers, managers and planners will be facing many problems linked to human variability and its impact on an individual's work performance. More realistic design decisions will be needed to accommodate a diverse workforce so that retention of experienced and skillful workers might be assured. Ergonomics plays a vital role in the designing of workplaces where we can proactively access the suitability of any product, process and environment design. It would be challenging to understand differences due to human variability, their impact on individual's and organizational work performance and adoption of the strategies that can materialize effectively the benefits of workforce diversity.

5   AN INCLUSIVE DESIGN METHOD

Inclusive design is an approach used to address the design needs of the broader range of the population. The inclusive design approach aims to understand

28

and address design requirements proactively at some pre-design phase so that any product, environment, service, equipment or tool can be designed in such a way that it could be used by a broad range of population. It takes notice of human variability in shape, size, age, working capabilities and behaviors and uses this data for the assessment of why some people are excluded from any design and how they can be accommodated. Keeping in view the aim of the inclusive design method, it is proposed that the approach can be used for addressing the challenges faced by a diversified workforce. One established way of evaluating the suitability of any design or environment is to use digital human modeling (DHM) tools together with the CAD model of the product, workplace or environment. As the use of DHM tools allows the designers and planners to evaluate any design at an early stage of design against a variety of potential users, so the problems can be addressed early on, when changes are less costly and easier to implement. However, challenges lie in the understanding of differences that exist among the potential users and transformation of this valuable information into a format where it can be used for design recommendations.

To address these issues discussed above, an inclusive design tool called HADRIAN (Human Anthropometric Data Requirements and Analysis) was developed. It is a software database of 103 people consisting of more realistic information about sizes, shapes, working capabilities like joint range of motion and behaviors that influence task performance. It is integrated with a digital human modeling tool SAMMIE (System for Aiding Man Machine Interaction Evaluation); where a task analysis tool was developed to support designers. Any task can be broken down into basic task elements such as look at the screen, reach to the card slot etc. where its automated evaluation process facilitates the users by providing details of those who experienced difficulties in task performance and what was the reason for that. Then user can explore the individual and get exact information about capabilities and behaviors of the individuals designed out and try new design solutions by modifying the computer model of the products or workplaces (Gyi et al., 2004, Marshall, 2010, Case, 2001). Previously, the HADRIAN design evaluation system has been used for a variety of applications including the use of ATMs by wheelchair users, wheelchair access to trains and road vehicles and task performing strategies in kitchen and transport activities (Figure 1and 2).

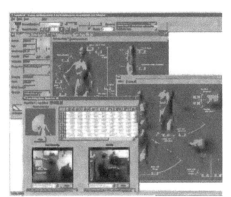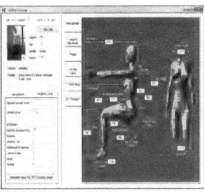

Figure 1. Screenshot of a part of HADRIAN data presentation

Figure 2. Screenshot of HADRIAN task driven evaluation

The above discussion highlights that the HADRIAN inclusive design approach can address human variability issues and a task evaluation system might be useful for this purpose. However, the HADRIAN automated task evaluation strategy still needs data about task performing strategies for a wide range of workers so that designers can recommend workplace design solutions for a diverse workforce. The next section discusses how this inclusive design approach can be used to capture working strategies of workers having different working strategies, behaviors, levels of skill and experience in a manufacturing industry context. Finally, there is a discussion on how this data might be used in the HADRIAN task evaluation system where upcoming challenges that relate to workforce diversity in manufacturing industries can be addressed.

6   MANUFACTURING INDUSTRY PERSPECTIVE

Industrial workforces are becoming more diversified with people from different races, backgrounds, experiences and skills work together for the same organization. Manufacturing industrial tasks, especially manual assembly activities, are greatly influenced by these differences as physical, physiological and cognitive variations among the workers affect task performance. These variations need to be understood by the designers and planners so that organizations can draw optimum benefit from their human capital. Therefore, it is proposed that the HADRIAN inclusive design approach might be equally applicable to workplace design, especially manual assembly tasks. It is believed that through the provision of more applicable and realistic data about task performing strategies of a diverse workforce this can be a good source for the promotion of optimum and effective utilization of organizational workforce diversity.

Recently, research has started where task performing strategies and behaviors of a diverse workforce have been studied for a variety of manual assembly tasks. At present, these strategies have been video recorded and their ergonomic assessment underway. Subsequently, this data will be used to define basic assembly task elements and assessments will be made against the HADRIAN database where individual's specific data about anthropometry, shape, size, joint range of motion etc. will be utilized in a digital human modeling environment. As mentioned, the HADRIAN database is representative of a more diverse population where data about their capabilities can be exploited to prevent any design exclusion.

For example, Figure 3 shows two different activities performed during sofa assembly, where the worker is using different parts of his body for the completion of the two tasks. Both of these tasks show different joint mobility requirements for successful completion (Table 1).

Table 1. Typical task completion requirements in sofa assembly process

Task	Critical joint mobility requirements
Task 1	Upper arm flexion
	Upper arm abduction
Task 2	Upper arm flexion
	Upper arm abduction
	Wrist flexion

Figure 3. Recording individual's capabilities and behaviors performing manual assembly tasks

The HADRIAN database contains joint range of motion data for many older people and evidence shows that age is responsible for a significant decrease in joint range of motion values, especially for arm abduction and wrist flexion. Using these preliminary findings, designers can generate the same kind of scenarios where they can validate these findings by using joint mobility data of older individuals. The HADRIAN design exclusion process can, for example, give feedback that individual 10, aged 55 is unable to perform this assembly task because of joint constraints. The database provides an opportunity for designers to access detailed data about name, nationality, background, age, anthropometry, joint constraints etc. and this helps designers in understanding why this particular individual was excluded and what kind of design changes can allow that individual to use that product, workstation or environment comfortably. In this way, we might understand potential differences among the workers and design a more inclusive work environment where workers from different age groups, levels of ability, background and experience can work together. Future research will be focusing on using working strategies data of a diverse workforce in a manufacturing assembly environment and using this data for an automated task evaluation method within HADRIAN.

7   Conclusion

Demographic changes and economic considerations require the attention for global workforce diversity management so that organizational performance sustainability can be sustained. An inclusive design method is considered useful as its aim of designing products, services, workstations or environment for a broad range of population is well-suited to diversity management issues. A digital human modeling based HADRIAN database is considered helpful where its automated task evaluation approach helps designers and ergonomists to address the design needs of a diverse population. Current ongoing research is focusing on the use of the HADRIAN tool to address the design requirements of a diverse workforce in manufacturing industries, especially manual assembly activities where most of the work is completed physically. Design recommendations achieved through the HADRIAN system can be very helpful in addressing the issues related to workforce diversity. This strategy will ultimately give benefits to the organization by providing safe, healthy, productive and progressive environments for the workers where they might happily co-exist and perform well. Workforce satisfaction leads to an organizational citizenship behavior that positively affects individual and overall organizational performance.

References

CASE, K., PORTER, M., GYI, D., MARSHALL, R., OLIVER, R. (2001) Virtual fitting trials in 'design for all'. *Journal of Materials Processing Technology*, 117, 255-261.
CHUNG, M. J., WANG, M.J. (2009) The effect of age and gender on joint range of motion of worker population in Taiwan. *International Journal of Industrial Ergonomics*, 39, 596-600.
DYCHTWALD, K., ERICKSON, T., MORISON, B. (2004) It's time to retire retirement. *Harvard Business Review*, 82, 48-57.
DYLLICK, T., HOCKERTS, K. (2002) Beyond the business case for corporate sustainability. *Business Strategy and the Environment*, 11, 130-141.
ELKINGTON, J. (1997) *Cannibals with forks: The Triple Bottom Line of 21st century business*, Capstone, Oxford.

32

GYI, D. E., SIMS, R. E., PORTER, J. M., MARSHALL, R., CASE, K. (2004) Representing older and disabled people in virtual user trials: data collection methods. *Applied Ergonomics*, 35, 443-451.

HULTSCH, D. F., MACDONALD, S.W.S., DIXON, R. A. (2002) Variability in reaction time performance of younger and older adults. *Journal Of Gerontology: Psychological Sciences*, 57B, 101-115.

IEA International Ergonomics Association. Accessed on February 27, 2012. http://www.iea.cc/01_what/What%20is%20Ergonomics.html

KNIPPENBERG, D. V., SCHIPPERS, M.C. (2007) Work group diversity. *The Annual Review of Psychology*, 58, 515-541.

KNIPPENBERG, V., D., DE DREU, C. K. W., HOMAN, A.C. (2004) Work group diversity and group performance: An integrative model and research agenda. . *Journal of Applied Psychology*, 89, 1008-1022.

MAMMAN, A., KAMOCHE, K., BAKUWA, R. (2012) Diversity, organizational commitment and organizational citizenship behavior: An organizing framework. *Human Resource Management Review*.

MARSHALL, R., CASE, K., PORTER, J.M., SUMMERSKILL, S.J., GYI, D.E., DAVIS, D., SIMS, R.E. (2010) HADRIAN: a virtual approach to design for all. *Journal of Engineering Design*, 21, 253-273.

MCKEY, P. F., AVERY, D.R., TONIDANDEL, S., MORRIS, M.A., HERNANDEZ, M., HEBL, M.R. (2007) Racial differences in employee retention: are diversity climate perceptions the key? *Personnel Psychology*, 60, 35-62.

RICHARD, O. C. (2000) Racial diversity, business strategy and firm performance: A resource based view. *Academy of Management Journal*, 43, 164-177.

ROBERSON, Q. M. (2006) Disentangling the meanings of diversity and inclusion in the organizations. *Group and Organization Management*, 31, 212-236.

SHORE, L. M., HERRERA, B.G.C., DEAN, M.A., EHRHART, K.H., JUNG, D.I., RANDEL, A.E., SINGH, G. (2009) Diversity in Organizations: Where are we now and where are we going? *Human Resource Management Review*, 19, 117-133.

STURNIEKS, D. L., GEORGE, R.' LORD, S.R. (2008) Balance disorders in the elderly. *Neurophysiology Clinique/Clinical Neurophysiology*, 38, 467-478.

SUE, B. (2008) The association between low vision and function. *Journal of Aging and Health*, 20, 504-525.

U.N.O. (2009) World population prospects, 2008 revision. Accessed on February 27, 2012. http://social.un.org/index/Ageing/DataonOlderPersons/ADemographicsCharts.aspx

WANGER, S. G., PFEIFER, A., CRANFIELD, T.L., CRAIK, R.L. (1994) The effects of ageing on muscle strength and function: A review of the literature. *Physiotherapy Theory and Practice*, 10, 9-16.

WCED (1987) *Towards Sustainable Development. Our Common Future. World Commission on Environment and Development.*, Oxford University Press: Oxford, 43-66.

WILLIAMS, K. Y., O'REILLY, C.A. (1998) Demography and diversity in organizations: A review of 40 years of research. *Research In Organizational Behavior*, 20, 77-140.

CHAPTER 4

Integrative Production Technology for High-Wage Countries - Resolving the Polylemma of Production

Christian Brecher, Wilhelm O. Karmann,
Stefan Kozielski, Cathrin Wesch-Potente

Laboratory for Machine Tools and Production Engineering (WZL)
at RWTH Aachen University
Aachen, Germany
c.wesch@wzl.rwth-aachen.de

ABSTRACT

Competition between producers in high-wage and low-wage countries is typically carried out in two dimensions, the economy of planning (scale vs. scope) and the economy of production (plan vs. value) which together form the "Polylemma of Production". The determination of a company's optimal operating point is carried out depending on the cost of labor and is thus dependent on location. Recent developments show that the key to sustainably produce in high-wage countries cannot lie in achieving a better position within each of the dichotomies but implies the resolution of the contradictions in the fields of planning economy and production economy. Within the Cluster of Excellence "Integrative Production Technology for High-Wage Countries" (CoE), researchers from the areas of production engineering and material sciences at RWTH Aachen University develop tailored methods for the organization and the use of technology in high-wage production sites. The scientific core result of the CoE is the development of a production engineering based theory of production which includes holistic description, explanation and design models for production systems and delivers answers how future production in high-wage countries can turn out both sustainable and successful.

Keywords: Production Technology, Self-Optimization, Polylemma,

34

1 INTRODUCTION

Production has a central role in industrial, developed high-wage country societies, contributing to welfare and social stability. Over 30% of the total workforce in Europe depends on production. Global megatrends that infer economic, ecologic and social challenges increase the pressure on production steadily. In addition production in high-wage countries has been facing a tremendous increase in volatility due to financial systems' crises, a paradigm change in energy supply and the growing scarcity of raw materials within the last five years (Brecher et al. 2011).

For production in high-wage countries it is vital to (i) offer products matching customer and societal demands at competitive prices (market-oriented perspective) and (ii) quickly adapt to market and societal changes while assuring optimized use of resources (resource-oriented perspective). Two fundamental dichotomies that restrict the simultaneous fulfillment of both objectives form the polylemma of production (Brecher et al. 2008, 2012), see Figure 1:

- The market-oriented dichotomy is characterised by the objective to manufacture products at mass production costs (scale), which perfectly match individual customer demands (scope)
- The resource-oriented dichotomy is characterised by optimising the synchronisation of production resources (planning orientation) while simultaneously achieving highest system dynamics (value orientation).

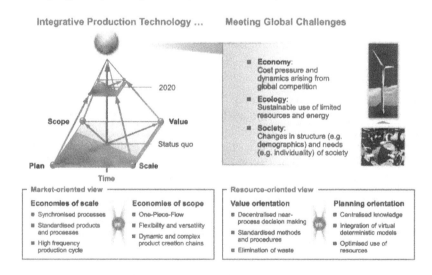

Figure 1 Meeting economic, ecological and social challenges by means of Integrative Production Technology aimed at resolving the polylemma of production (Brecher et al. 2011)

2   IMPORTANCE OF DOMESTIC PRODUCTION FOR HIGH-WAGE COUNTRIES

Global competition and new competitors mainly from low-wage countries ("best-cost countries") force the manufacturing industry in Western Europe to relocate their production. These companies generate their economic advantage mainly based on the low wages and simultaneously improving their technological capabilities. The distribution of the worldwide production quantities in the course of the last centuries reflects this development (Tseng 2003). Until approximately 1930 the percentage of the western industrial nations (Europe and North America) constantly increased. In this period the competitive pressure was mainly generated by these countries, which is also reflected by the high gradient in wealth and economic power during this time. In the following years, this situation slowly started to change as a consequence of the growing global trade and the increasing technological capability (Brecher et al. 2012).

Production accounts for the main fraction of German national economy. In the second quarter of 2010, the industry nearly employed 7.7 million people, which is around one fifth of the country's working population. A distinctive feature inherent to the manufacturing industry is its multiplier effect: intra-industry changes have a significant inter-industrial impact, especially on the other industries value creation and on the level of employment. Contrary to this, services have rarely any inter-industrial effects employment-wise. This is rooted in the fact that its value creation chain only comprises a small number of levels and rarely requires any preliminary products in comparison to the manufacturing industry. Beyond that, services generate no specific after-sales revenue. Industrial production can therefore be regarded as the main driver of Germany's value creation and employment.

In order to respond to this issue and sustainably secure Germany's level of employment, changes in the manufacturing industry must be recognized and corresponding adjustments must be implemented at an early stage. Companies of the manufacturing industry in high-wage countries are constantly put under increasing pressure by international competition. One reason for this development are the manufacturing costs of low-wage countries, which seem to be significantly cheaper in comparison (Brecher et al. 2012).

In addition, new competitors not only take advantage of their advantage in labor costs, but constantly upgrade their technological capabilities. According to a survey carried out by the German chamber of industry commerce DIHK, nearly one in four German industrial companies reacts to this threat by relocating production to foreign countries, mostly to the new members of the EU, South America and Asia (DIHK 2010). Among relocated R&D activities, design, technical development, testing, software development and even fundamental research are the most prominent (Rose and Treiver 2005). As mentioned before, the relocation of production sites is likely to result in a relocation of services and R&D. For high-wage countries like Germany, securing domestic production therefore is a vital part of securing national wealth (Fig. 2).

36

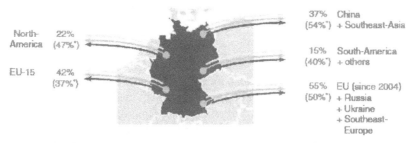

North- 22%
America (47%*)

EU-15 42%
(37%*)

37% China
(54%*) + Southeast-Asia

15% South-America
(40%*) + others

55% EU (since 2004)
(50%*) + Russia
+ Ukraine
+ Southeast-
Europe

* Proportion of companies undertaking investments in the region (multiple answers possible)
** Proportion of total production-related relocation

Figure 2 Foreign investment (DIHK 2010)

With the Cluster of Exrcellence "Integrative Production Technology for High-Wage Countries", RWTH Aachen University responds to this development with concepts to increase competitiveness of western high-wage countries by enabling them to implement a more economically efficient production. Based on the inherent advantages of domestic production sites, new approaches in different fields of science are integrated into a comprehensive approach, in order to meet global competition and strengthen innovation. Main part of the concept is the resolution of the so-called polylemma of production (Brecher et al. 2012).

3   THE POLYLEMMA OF PRODUCTION

Manufacturing companies have to define their position in the field of tension between production and planning profitability, thereby striving towards their specific ideal operating point. In this process, labor costs are of high importance. Hence, the position in the above mentioned area of conflict is mainly location-dependent. In terms of their production efficiency, companies in low-wage countries generally focus solely on mass production (economies of scale) while companies based in high-wage countries need to balance their production system between mass production (economies of scale) and production of customized products (economies of scope) (Brecher et al. 2012). Regarding the second dimension in the area of conflict, the planning efficiency, manufacturers in high-wage countries continuously make efforts to optimize their processes using sophisticated, hence capital-intensive planning tools and production systems. In contrast to that, companies in low-wage countries generally employ simple and robust value stream oriented process chains. Recent developments, however, show that despite a seemingly perfect corporate positioning in the area of conflict, relocation of production from high-wage to low-wage countries still continues (Schuh and Orilski 2007). Thus, the key to strengthen the competitiveness of high-wage countries is the resolution of the contradictions in the fields of production and planning efficiency. These dichotomies, which are specific to the scientific disciplines, form the so-called polylemma of production.

In the context of production economy, low unit costs can be achieved by focusing the production system on the economies of scale. Hereby, increases in efficiency necessary for the utilization of scale effects can be accomplished through process standardization. This includes standardization in terms of both the organization's business processes and its technical and manufacturing processes, which are part of the production system. (Schuh et al. 2012) In mass production this includes interlinked machineries with a high degree of automation. However, the downside of this configuration is that, when put into practice, standardization often leads to restrictions in terms of production flexibility. Beneficial economies of scale like these are gained by a low adaptability of the production system to changing boundary conditions, like e.g. a modified market behavior. Contrary to that, adaptability is of top priority when configuring a production system for the economies of scope. This involves business processes and technical systems, which are designed in a way that they allow a high degree of freedom for variants of producible goods (Schuh and Gottschalk 2008). However, this comes at the price of additional investment or higher proportion of manual work, leading to higher costs per unit compared to a production system optimized for the economies of scale (Schuh et al. 2012).

The dimension of planning economy features a similar dichotomy. A high degree of planning orientation leads to an extensive use of models, simulations and optimization approaches. (Brecher et al. 2012) These support operational processes as well as planning and decision-making processes (e.g. production planning and control, design of production processes and machinery). Since those kinds of activities do not immediately add value, corresponding approaches are contrary to the concepts of lean management, which aims to maximize the value added by employing efficient planning concepts. In the dichotomy of planning economy, this position is represented by a high degree of value orientation.

The resolution of both the dichotomy between the economies of scale and the economies of scope as well as between planning and value orientation therefore is the key to sustainable preservation of production in high-wage countries and thus needs to be in the focus of production research. The solution hypothesis to this objective is to integrate the approaches of different fields of science into a comprehensive strategy to resolve the polylemma of production (Brecher et al. 2008).

4 THE CONTRIBUTION OF SELF-OPTIMIZATION TO THE RESOLUTION OF THE POLYLEMMA

A successful enterprise that connects both planning-oriented and value-oriented approaches uses the advantages of both strategies. In order to achieve this, on the one hand it is necessary to exactly model the production process. On the other hand, by identifying the essential parameters to be influenced, a foundation has to be laid for the ability to autonomously and flexibly make decisions. However, the existing possibilities to optimize the behavior of one element within the whole system can be

38

focused too tightly and bind resources even though in certain situations this could lead to an adverse behavior in other areas. This optimization task within a production system usually cannot be solved analytically.

The solution of this conflict becomes possible if a system is designed that can adjust its goals situatively. While in most cases the optimization of a system is controlled from the outside, e.g. by a person, in many cases the optimization by the technical system itself is a possible option. The developments in automation technology show, however, that even in comparatively simple matters this has not yet been accomplished. Therefore in many cases people still play an important role. The implementation of self-optimizing abilities provides a substantial possibility to reduce the area of tension of planning economy (Schmitt et al. 2012).

Self-optimizing systems are defined by the interaction of contained elements and the recurring execution of the actions continuous analysis of the current situation, determination of targets, and adaptation of the system's behavior to achieve these targets (Adelt et al. 2009). The superordinate objective to design self-optimising production systems ranging from single machines over manufacturing respective assembly cells, up to the factory level that are able to autonomously define, reach and sustain optimal operating points within a socio-technical production network (see Figure 3).

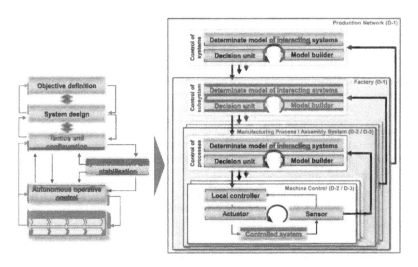

Figure 3 Self-similar cybernetic model of self-optimizing production systems (Brecher et al. 2011)

The research thus has to address strategies, models and methods for the cognition-enhanced self-optimisation of production systems on different levels of abstraction. For the purpose of finding self-similar structures and arranging a basis for self-optimising production systems, a cybernetic based reference model for self-optimization scaling from loosely coupled processes to highly integrated socio-technical production networks needs to be designed. To control the complex interplay of heterogeneous structures, it provides interfaces for both product manufacturing and assembly systems. To enable self-optimisation on the process level the rigging, monitoring and control strategies need to be improved significantly and made available for optimisation on the production system or network level. Furthermore the assembly of small batch products of arbitrary complexity within a production network poses a substantial challenge for product design, assembly modelling and planning. Hence, novel strategies and methods for self-optimisation of assembly systems for varying parts and orders need to be developed (Schmitt et al. 2012).

5   SUMMARY AND OUTLOOK

Regarding highly automated manufacturing systems that shall produce customer-specific products, an increase in conventional automation will not necessarily lead to a significant increase in productivity. Novel concepts towards proactive, agile and versatile manufacturing systems have to be developed to solve the polylemma of production (Schuh and Orilski, 2007). E.g., cognitive automation is a promising approach to improve proactivity and agility. Despite this novel automation approach the experienced machining operator will always play a key architectural role as a solver for complex planning and diagnosis problems. Moreover, he/she is supported by cognitive simulation models which can solve algorithmic problems on a rule-based level of cognitive control (Rasmussen, 1986) quickly, efficiently and reliably and take over dull and dangerous tasks.

Within the Cluster of Excellence "Integrative Production Technology for High-Wage Countries" (CoE), researchers from the areas of production engineering and material sciences at RWTH Aachen University draw possible scenarios for the future of production, see figure 4. An area of focus regards methods, concepts and theories of self-optimization.

Figure 4 Scenario of self-optimizing production systems

The scientific core result of the CoE is the development of a production engineering based theory of production which includes holistic description, explanation and design models for production systems and delivers answers how future production in high-wage countries can turn out both sustainable and successful.

ACKNOWLEDGMENTS

The authors would like to thank the German Research Foundation DFG for the kind support within the Cluster of Excellence "Integrative Production Technology for High-Wage Countries".

REFERENCES

Adelt, P., et al. 2009 Selbstoptimierende Systeme des Maschinenbaus. In. *HNI-Verlagsschriftenreihe, vol 234*, eds. J. Gausemeier, et al. Westfalia Druck GmbH, Paderborn

Brecher, C., et al. 2008 Integrative Produktionstechnik für Hochlohnländer. In. *Wettbewerbsfaktor Produktionstechnik,* eds. C. Brecher, et al Aachener Perspektiven, Apprimus Verlag, Aachen

Brecher, C., et al. 2011 Integrative Production Technology for High-Wage Countries, Renewal Proposal for a Cluster of Excellence

Brecher, C., et al. 2012 Integrative Production Technology for High-Wage Countries. In. *Integrative Production Technology for High-Wage Countries*, eds. C. Brecher, et al, Springer, Berlin

Deutsche Industrie- und Handelskammer (DIHK) eds 2010, Auslandsinvestitionen in der Industrie. Ergebnisse der DIHK-Umfrage bei den Industrie- und Handelskammern, Deutsche Industrie- und Handelskammer, DIHK, Berlin

Rasmussen, J. 1986 Information Processing and human-machine interaction. An approach to cognitive engineering. North-Holland, New York

Rose, G., V. Treiver, 2005 Offshoring of R&D – examination of Germany's attraktiveness as a place to conduct research. DIHK, Berlin

Schmitt, R., et al. 2012 Self-optimising Production Systems. In. *Integrative Production Technology for High-Wage Countries*, eds. C. Brecher, et al, Springer, Berlin

Schuh, G., S. Orilski, 2007, Roadmapping for competitiveness of high wage countries, in: *Proceedings of the XVIII ISPIM Annual Conference*, Warsaw, Poland

Schuh, G., S. Gottschalk, 2008 Production engineering for self-organizing complex systems, Prod Eng Res Dev 2: 431-435

Schuh, G., et al. 2011 Integrative Standardisation – Theoretical Model and Empirical Investigation of German Toolmaking Firms. In. *ICE Conference Proceedings 2011*, Vol. 17 (2011)

Schuh, G., et al. 2012 Individualised Production. In. *Integrative Production Technology for High-Wage Countries*, eds. C. Brecher, et al, Springer, Berlin

CHAPTER 5

Management Audit as Part of an Ergonomics Management in Production Systems

Max Bierwirth, Ralph Bruder

Institute of Ergonomics
Technical University Darmstadt
Darmstadt, Germany
sek@iad.tu-darmstadt.de

ABSTRACT

Shorter cycle times, continuous improvement activities and the demographic change have drawn more attention to the design of manual work and the prevention of work related risks through ergonomics. As a consequence, activities to improve ergonomics in production systems have multiplied. To achieve effective risk reduction, however, ergonomics need to go beyond isolated improvements on the individual work station level. For this purpose, a management model for integrating systematic ergonomics in the processes of work design along with an audit to control all ergonomics integration activities has been developed. The audit assesses the current status of the integration of systematic ergonomics in the processes of work design and evaluates their effectiveness in terms of a systematic reduction of work related risks. Audit results from 11 manufacturing companies in Germany show that the audit does not only stand the test in practice but reveals that several companies have already implemented a sound ergonomics management in their production system.

Keywords: Ergonomics-Management, Audit, Mangement-Model

1   ERGONOMICS AND PRODUCTION SYSTEMS

Modern production systems are based on continuous improvement processes and standardization to increase productivity. Due to global competition this is crucial for all production facilities. In manufacturing, improvement activities typically focus on a more efficient use of the human work force, mostly resulting in shorter cycle times and higher workload for the workers.

Parallel to the strive for more productivity, the demographic change compels companies in industrialized countries to devote more attention to occupational health and safety, especially to the prevention of work related health risks for the workers. As a consequence, more and more companies implement OHS management systems such as the BS OHSAS 18001:2007 and provide health promotion programs. Also, ergonomics activities are initiated to improve workstations. To reduce or even eliminate work related risks effectively through ergonomics, however, activities must not only focus on the improvement of individual workstations in production systems.

Instead, a systematic approach is needed which not only considers all workstations but is also integrated into the design processes of the work stations and work organization. This comprises the design process of new work stations as well as all activities in the context of continuous improvement processes. Only then, sub-optimal improvements which solely focus on (short-term) efficiency, can be avoided and ergonomic design deficits of new work stations can effectively be reduced (NIOSH, 2007). The full potential of ergonomic work design can only be achieved if and when the abilities of the actual work force are taken into account in all risk assessments already at the level of work design and organization. Such a comprehensive approach becomes more and more important as the variance of abilities grows with the age of workers, which is bound to raise the heterogeneity of the performance of the work force.

2   INTEGRATION OF ERGONOMICS INTO PRODUCTION SYSTEMS MANAGEMENT

2.1   Management Model

To implement such a comprehensive approach, the processes of work design and organization need to be adapted and several departments of the organization have to be involved. This makes the integration of ergonomics into a production system a complex task (Siemieniuch & Sinclair, 2002). A management model provides the vision of a comprehensive ergonomics approach and can serve at the same time as the basis for the controlling of all integration activities (Zink, Steimle & Schröder, 2008). For this purpose, Bierwirth, Bruder and Schaub (2010) have developed the Total-Ergonomics-Management-Model (see Figure 1).

This Management-Model divides the necessary implementation activities into four consecutive modules meant to achieve results in terms of a risk reduction on four different levels. The implementation of ergonomic risk assessment tools, e.g. the European-Assembly-Worksheet (Schaub et al., 2010) allow both to identify risks objectively and to realize effective improvements of individual work stations. A consistent application of risk assessments creates transparency about the risks at all work stations. Through this transparency isolated, unilateral improvements can be avoided and a comprehensive optimization of the existing work stations can be initiated. The integration of risk assessments into the process of work design, in particular, offers a maximum of possibilities to realize ergonomic work design (Dul & Neumann, 2009). Within work organization and in the design of new work stations the consideration of (physical) abilities of workers helps to lower work related risks for the individual worker even more. A more detailed description of the model is given in Bierwirth, Bruder and Schaub (2010).

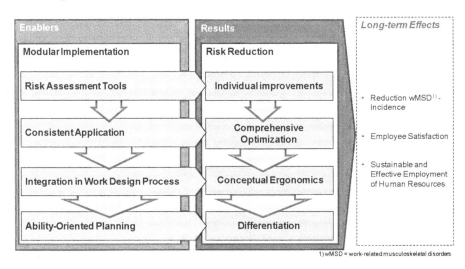

1) wMSD = work-related musculoskeletal disorders

Figure 1: The Total-Ergonomics-Management-Model (adapted from Bierwirth, Bruder and Schaub 2010, p. 114.)

The Total-Ergonomics-Management-Model (TEM-Model) aims at organizational changes, i.e. the systematic integration of ergonomics considerations into processes of work design and organization to enable a company to systematically reduce work related risks in the production system. The separation into "enablers" and "results" is important as "enablers" are preconditions for the desired "results" and the implementation of process changes need time to be effectual (Zink, Steimle & Schröder, 2008). Good results not based on systematic processes and instruments (implemented enablers), by contrast, risk to be random occurrences only. That is why also quality management models such as the EFQM Model for Excellence (EFQM, 2009) differentiate between "enablers" and "results". The TEM-Model's separation into 4 modules describes logical steps of an iterative implementation: Ability-Oriented Planning (module 4) can be fully realized only if an appropriate risk assessment tool is implemented (module 1), if assessments have been conducted for a significant fraction of all work systems (module 2) and if ergonomics are systematically integrated into the work design process (module 3). This also supports the definition of realistic aims for the implementation process.

2.2 Operationalisation through Audit

To offer an effective controlling-tool for the integration of ergonomics into production system management in industrial practice, a suitable audit tool has been developed. In general, audits assess whether an organization functions in conformity with the requirements of a defined management system, e.g. a quality management system (ISO 9000:2005). Similarly, the Total-Ergonomics-Management-Audit allows to assess the degree of integrity, consistency and effectiveness of the current ergonomics management within a company.

The audit questionnaire contains criteria for both enablers and results for each module. In total, it contains 60 items. All items are rated on a four-point scale (see Figure 2). For some items further information, e.g. about involved persons or known risk factors is gathered for the analysis but is not directly rated.

		0 Pts.	2 Pts.	4 Pts.	6 Pts.
B2.1	What percentage of Work stations has already been analyzed?	<20% ☐	20-50% ☐	50-80% ☐	>80% ☐
B2.2	When are ergonomic risk-assessments conducted?	never ☐	only on individual complaints ☐	systematically, where high work load is obvious ☐	systematically, all work stations ☐
B2.3 *	Who conducts ergonomics risk-assessments?	none ☐	external experts, on demand ☐	internal expert, on demand ☐	internal expert, as regular task ☐
B2.4	Which tool is to be used for which workstations is clearly defined?	No regulation ☐	for some tools or areas ☐	mostly ☐	Clearly defined for all tools & areas ☐
B2.5 *	A problem-solving process with clear responsibilities is defined for ergonomics issues?	does not exist ☐	exists, but not defined ☐	exists, some process gaps ☐	fully defined ☐
B2.6	Workers and health and safety experts are involved in the problem-solving process?	never ☐	seldom ☐	regularly ☐	always ☐

Figure 2: Ergonomics-Management Audit Questionnaire (excerpt implementation module 2).

The audit can be used as a self-audit or as an external audit. It is recommended to involve a group of experts from industrial engineering, health and safety and production management in the auditing process.

To get the audit results, all directly rated items within each module are summed up separated in implementation (enablers) and process outcomes (results). Based on the maximal possible score, individual scores are then transferred into a spider-web-chart (see Figure 3). This gives a condensed overview about the current status of the ergonomics management within a company. Improvement potentials are indicated not only on the level of individual ratings but also with regard to gaps between implementation and process outcomes.

The results in the example in Figure 3 suggest, that one (or more) risk assessment tool(s) are well established in the company, yet their effectiveness is clearly limited. A look into the detailed analysis reveals that the risk assessment tools are only used as an expert tool and therefore achieve only little ergonomic awareness within the company. As a consequence, the consistent application is also limited. In this case, a better communication of the benefits of such assessments and a broader user group may overcome these shortcomings and increase the frequency and fields of use.

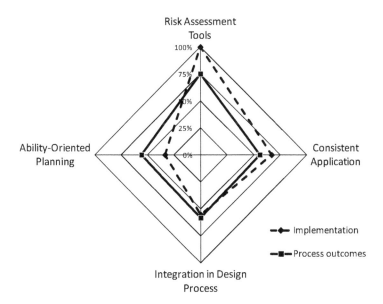

Figure 3: Audit Results Company H.

In company H only little is done concerning the integration of ergonomics into the work design process. Work related risks are systematically considered at a very late stadium of the design process only. This allows only minor changes to reduce risks. A systematic ergonomics lessons-learned and risk assessments at earlier design stages should be defined. Likewise, the abilities of workers are systematically considered in cases of reintegration or disability management only. The higher rating of the process outcomes of module 4 show that currently the assignment of workers to the work systems does not pose great problems. Nevertheless, this finding is not based on systematic processes and is only true for the current situation. For the near future, an efficient and sustainable worker-work system allocation is, therefore, seen as critical. Only a systematic assessment of worker abilities and comparison with the work demands of the current and future work systems can prevent a systematic mismatch that induces higher risks and lower productivity in the production system.

Thus, the audit results clearly show strengths and weaknesses of the current ergonomics management. Based on the structure of the management model and the individual scores of each module, specific aims can be defined and appropriate organizational changes can be initiated.

3 TOTAL-ERGONOMICS-MANAGEMENT IN GERMAN MANUFACTURING COMPANIES

The ergonomics management audit was applied in 11 manufacturing companies in Germany. The sample contained large car manufacturers as well as smaller suppliers. Figure 4 gives an overview of the current implementation status of the 4 modules in the assessed companies. The majority has implemented risk assessment tools on a high level. This means that the tools are well established in the company and that they can assess most of the risks present within the workstations. Lower scores often result from limited ability to assess all types of risks and from limited acceptance of the tools and their assessment results within the company.

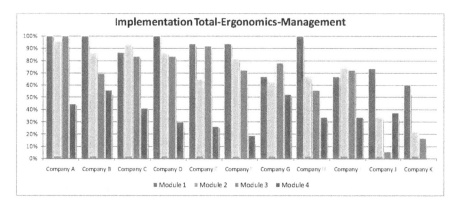

Figure 4: Implementation of the Total-Ergonomics-Management in 11 manufacturing companies.

Concerning the consistent application (Module 2) bigger differences between individual companies are found. Especially the institutionalization of a standardized ergonomic problem solving process with clear responsibilities and strict management controlling in terms of a PDCA-Cycle (Deming, 1986) is often lacking. Although most ergonomic risk assessment tools provide quantified data, the systematic integration of these results into decisions of work design and organization is not yet common practice.

Several companies have initiated activities to integrate ergonomics into the process of work design (Module 3). In some companies ergonomics has become part of quality-gate reviews. However, only few companies have established so far a systematic consideration and integration of risk assessments providing process-oriented information for ergonomic work designs. Therefore, future research in the field of human factors and ergonomics is needed to provide appropriate risk assessment tools and guidelines.

Ability-Oriented Planning (Module 4) is not yet realized in any of the assessed companies. However, some companies have already established IT-based matching routines that could be used for a general ability-oriented worker assignment. Yet, the available information about workers' abilities is scarce. Aspects of future development of workers' abilities and possible consequences for the work design are considered in only two companies of this sample.

4 CONCLUSIONS

The audit results from 11 manufacturing companies show that steps to a more systematic ergonomics management are undertaken. Interviews with experts from the audited companies gave a very positive feedback concerning the TEM-Model and audit. The comprehensive approach of TEM is strongly supported. However, a complete realization of module 4 (Ability-Oriented Planning), in particular the assessment of workers' abilities, is seen critical due to necessary efforts and legal privacy policies. The TEM-Model itself is highly valued as a vision and strategy-roadmap of an ergonomics management. It helps to create a common understanding on a strategic level. As several departments are involved, this aspect is of paramount importance. The similarity to established quality-management systems.

The TEM-audit was in all companies the first time that ergonomics activities and processes were systematically assessed. With the audit specific strengths and weaknesses were identified in each company. The involved experts saw two positive outcomes of the audit: First, the objective and comprehensive identification of strengths and weaknesses. From the analysis specific aims can easily be defined and appropriate actions to improve the ergonomics management can be chosen.

Second, the possibility to measure and quantify their ergonomics management activities offers the possibility to prove the case and get more management attention. The intuitive visualization in a spider-web was seen as very helpful in this context. Benchmarking within a company or with external organizations is also possible and allows to identify best-practices (Jarrar & Zairi, 2000).

REFERENCES

British Standards Institution (BSI), *Occupational Health and Safety - Specifications*. BS OHSAS 18001:2007.

Dul, J & Neumann, WP 2009. 'Ergonomics Contributions to Company Strategy', *Applied Ergonomics* 40: 745–752.

EFQM 2009, *EFQM-Model 2010*. EFQM, Brussels.

International Organization for Standardization (ISO), *Quality management systems – Fundamentals and vocabulary*. ISO 9000:2005.

Jarrar, YF & Zairi, M 2000. 'Best practice transfer for future competitiveness: A study of best practices. Total Quality Management', *Total Quality Management* 11: 734–740.

National Institute for Occupational Safety and Health (NIOSH) 2010, *Prevention through Design. Plan for the national initiative*, National Institute for Occupational Safety and Health. Available from: http://www.cdc.gov/niosh/docs/2011-121/pdfs/2011-121.pdf [15 February 2011].

Schaub, K, Caragnano, G, Britzke, B & Bruder, R 2010. The European Assembly Worksheet. In. *Proceedings of the Eighth International Conference on Occupational Risk Prevention 2010*, eds. P Mondelo, W Karwowski, KL Saarela, P Swuste & E Occhipinti, Valencia, Spain.

Siemieniuch, CE & Sinclair, MA 2002, 'On complexity, process ownership and organisational learning in manufacturing organisations, from an ergonomics perspective', *Applied Ergonomics* 33: 449–462.

Zink, KJ, Steimle, U & Schröder, D 2008, 'Comprehensive change management concepts: Development of a participatory approach', *Applied Ergonomics* 39: 527–538.

CHAPTER 6

Ergonomic Intervention Plan for Machinery Operators

Milena Drzewiecka, Beata Mrugalska, Leszek Pacholski

Poznan University of Technology
Poznan, Poland
milena.drzewiecka@gmail.com, beata.mrugalska@put.poznan.pl,
leszek.pacholski@put.poznan.pl

ABSTRACT

In this article an overview of ergonomic aspects of work environment such as anthropometric, physiological, biomechanical, psychophysical, hygiene requirements is presented as they are the basis for effective and safety cooperation between an operator and means of work. Furthermore, workplace ergonomic intervention research is provided and its benefits are highlighted. Then, the ergonomic intervention plan for machine operators is developed on the basis of the analysis of nine job positions. For this purpose, the risks for machine operators in different areas of ergonomic requirements together with their sources and possible consequences are presented. On the basis of them, actions, which the employer should take to ensure workers safe and ergonomic working conditions, are specified. In the last part of the paper the practical benefits thanks to development of the intervention plan for machine operators are discussed.

Keywords: ergonomic aspects, ergonomic intervention plan, machinery operators

1   INTRODUCTION

Human being is the subject in the process of work, often described as an operator who is responsible for installing, use, adjustment, maintenance, cleaning, repairs as well as relocation of machines. To secure effective and safe cooperation of the operator with the means of work rules of ergonomics and technical safety requirements should be followed during formation of working means (EN ISO 12100:2010; EN 614-1:2006; Mrugalska, and Kawecka-Endler, 2011; Pacholski, 2009). For this purpose a detailed analysis of ergonomic aspects of work environment formation is necessary and it includes the following:

- anthropometric requirements that describe the relationship between the operator and machine and/or elements of work environment. These requirements relate to technical objects and determine their adjustment to dimensions and weight of human body or his parts in static and dynamic configuration to secure for example rational position at work,
- physiological requirements that concern the adjustment of technical objects to physiological human characteristics. These requirements relate both to the process of work and technical objects and determine their adjustment to the physiological capabilities of human body through the determination of the load value of muscles, bone system, respiratory system, cardiovascular system, joints and limbs,
- biomechanical requirements that result from the adjustment of technical objects to the capabilities of motion organ, first of all to the musculoskeletal system and peripheral nervous system,
- psychophysical requirements that tackle the problem of the adjustment of technical objects to the functioning of human senses: sight, hearing, smell, taste and touch,
- hygiene requirements that result from the adjustment of the environment to the human being i.e. lighting, acoustic environment, vibrations, microclimate and electromagnetic field to reduce harmful environmental factors (Horst, 2006; Wykowska, 2009; Olszewski, 1997).

The appropriate adjustment of these ergonomic aspects decides about safety and psychophysical comfort. Moreover, proper conditions of working environment are one of indispensable factors influencing creativity of employees work and thus creativity of the process realized in a given company (Dul, and Ceylan, 2011; Mrugalska, and Kawecka-Endler, 2012; Pacholski, 2003).

2   WORKPLACE ERGONOMIC INTERVENTIONS

New trends in the work organization, such as flexible and lean production technologies, flatter management structures and nontraditional employment practices have aroused concerns about their negative influence on worker health and safety (Murphy, and Sauter, 2004). It is noticed that in order to counter their consequences effective work organization interventions are required (Cooper, 2004). Thus, nowadays more and more workplace ergonomic programmes are introduced to show how to ensure employee health and safety and optimize business performance. In such programmes a significant attention is paid to ergonomics in workplace settings (Tompa et. al., 2010). There are differentiated two distinguished approaches to ergonomics. The first one is based on the prevention of health and the other focuses on promotion of human performance (Koningsveld, 2009). It is worth to emphasis that these two directions coexist very well. When economy is in depression most of businesses introduce cuts of cost. It contributes to paying attention to the costs of poor working conditions and occupational injuries and accidents at work. Such activities are favourable to creation of preventive projects on market (Dul, and Neumann, 2007; Koningsveld, 2009). The review of literature shows that primary interventions take place at the level of legislative/policy, employer/organization, job/task, and individual/job interface (Cooper, 2004; Murphy and Sauter, 2004). However, they are still infrequent in spite of its positive influence (Cooper, 2004; Gray, and Scholz, 1993; Guastello, 1993). Moreover, in the published literature the strongest support of the economic values of ergonomic interventions is noticeable in the manufacturing and warehousing jobs for individuals working with machinery. Much worth effects are observed in the administration, service and health care sectors and limited one in the transportation (DeRango et al., 2003; Lewis et al., 2002; Li et al., 2004; Reese, 1998; Richardson, 2002; Tompa et. al., 2010).

3   ERGONOMIC INTERVENTION PLAN

Employers are often of the opinion that providing ergonomic conditions of work environment is a complicated process and limit themselves to compliance with the rules aiming at provision of safe and healthy working conditions only. However, the basic points that should be executed in the frame of ergonomic programmes are workplace design, job design, work rates and selection of tools taking consideration all groups of requirements (Hägg, 2003; Robertson, 1999). Such an ergonomic intervention plan should include the following core set of elements (Sharan, 2004):

1. Worker involvement.
2. Management commitment.
3. Identification of risk factors.
4. Development of solutions.
5. Implementation of solutions.
6. Ongoing review for effectiveness.
7. Training and education for workers.
8. Early reporting of symptoms.
9. Appropriate medical management.

As it can be seen it should start from an engagement of workers and their all managers and supervisors. The basic action of such a plan should be evaluation of risk connected with ergonomic aspects. Such evaluation should contain not only indication and estimation of the risk of threats but also gives reasons and possible results of those threats. Then it is necessary to indicate what protection means should be undertaken to prevent the occurrence of the threat (eliminate it) or reduce the risk of its occurrence. In the next step, adequate changes consistent with the principles of ergonomics as well as accomplishment of re-evaluation to check if really taken actions contribute to reduction of risk, should be introduced. These actions may provide necessary minimum of ergonomic working conditions. In the final part of ergonomic intervention programme the attention should be paid to training and education of workers. Early reporting of symptoms and appropriate medical management should be also done.

An ergonomic intervention programme is perceived as a powerful mean of organizational effectiveness (Robertson, 1999). It is said that good ergonomics programmes are always cost-effective. What is more, they can save more than they cost as it is presented in Figure 1.

Figure 1 Benefits of ergonomic interventions (Adapted from (Koningsveld, 2005))

As it is shown, ergonomists should focus on the profits of their work in terms of core business values. It means that the effects on the system's performance should be studied. However, the confirmation of effectiveness is a rather new field of interest in the profession of ergonomics (Koningsveld, 2009).

1   INVESTIGATION OF CHOSEN WORKSTANDS FOR CREATION OF AN EXEMPLARY ERGONOMIC INTERVENTION PLAN

To elaborate ergonomic intervention plan of machine operators an analysis of such work places as fork lift truck operator, gardener– gardening machines operator,

53

injection moulding machine operator, baler operator, electro-hollowing machine operator, dielectric welder operator, compressor, pump and other equipment operator in wastewater treatment plant, eccentric press operator and excavator operator was done (Romanowska-Słomka, 2002, 2003a, 2003b, 2003c, 2004a, 2004b, 2005, 2009, 2010). Repeatability of certain group of threats was observed in Figure 2.

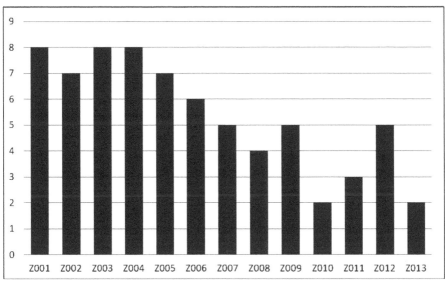

Legend (ID of threats): Z001 – hitting by moving objects, Z002 – stumbling and falling, Z003 – hitting, squashing, Z004 – electric shock, Z005 – fire, Z006 – explosion, Z007 – cuts, Z008 – intoxication, Z009 – polluting, Z010 – vibrations, Z011 – noise, Z012 – forced body position, Z013 – electromagnetic radiation

Figure 2 Characteristic threats at machine operators working place

As shown in Figure 2 there is a certain group of threats typical for machine operators working places. For example threads of hitting by moving objects, hits, squashing and electric shock concern eight considered working places. Thus, it is necessary to elaborate a universal plan of ergonomic intervention for machine operators.

4.1. Evaluation of machine operators risks

In order to evaluate machine operators risks the ergonomic intervention plan should be elaborated. To begin with all threats should be identified as well as their reasons and possible consequences. Characteristic threats together with sources and consequences for machine operators are presented in Table 1.

54

Table 1 Identification of characteristic threats in work places of machine operators (Adapted from (Romanowska-Słomka, 2002, 2003a, 2003b, 2003c, 2004a, 2004b, 2005, 2009, 2010))

Threat	Source	Possible consequence
Z001	Objects which are equipment of work place	Death, disability, bruise
Z002	Way of access to the machine, slippery and uneven surfaces, blocked passages, leakage of liquids and lubricants	Disability, limb fracture
Z003	Machines and other elements which are equipment of work place	Death, disability, bruise, hands injuries
Z004	Equipment powered by electricity, charging of batteries	Death or other consequences of electric shock
Z005	Calling the fire – inflammable materials, short circuit in electric installation	Death, burns
Z006	Charging of battery, gas installations damages, solvents used	Death, disability, burns
Z007	Sharp, rough surfaces	Hand wounds, minor injuries
Z008	Exhaust fumes, chemicals	Headaches, malaise (bad feeling), impairment of internal organs, acute intoxication, death
Z009	Powdery materials, dust generated during operation	Respiratory disease
Z010	Machines and equipment generating vibrations	Vibration disease, damage of the motion system (musculoskeletal system)
Z011	Machines generating oversized noise	Possibility of weakening and damage of hearing
Z012	Work in forced body position (sitting, standing, monotype movements), manual transporting work	Diseases of motion system (musculoskeletal system)
Z013	Equipment operation	Influence on functioning of cardio – vascular system, nervous system, headaches, quick tiredness, memory impairment (disorder)

As the presented threats are basic ones for the machine operators, they will significantly improve the process of providing ergonomic and safe working conditions to the employers who consider these issues as too complicated. Additionally the indication of possible reasons as well as their consequences will facilitate undertaking eventual preventive actions. However, the proposed threats are universal ones and it is also necessary to consider others which are characteristic only for particular operators.

4.2. Proposal of protection means

The aim of the next stage of ergonomic intervention plan should propose preventive actions. It is essential to undertake such actions that first of all will lead to the elimination of a specific threat and only then, when it is impossible to do it, its limitation should be introduced. In Table 2 protection means for identified threads in machine operators jobs were presented.

Table 2. Protection means necessary in machine operators jobs (Adapted from (Romanowska-Słomka, 2002, 2003a, 2003b, 2003c, 2004a, 2004b, 2005, 2009, 2010))

Threat	Protection means proposed
Z001	Increased attention, observance of regulations and instructions of execution of works
Z002	Maintaining order, appropriate shoes, increased attention
Z003	Observance of procedures and instructions, increased attention, personal protection means
Z004	Observance of procedures and instructions, proper technical condition of installation, overhauls and measurements of insulation resistance, maintenance of requirements of fire protection
Z005	Observance of procedures and instructions, smoke detectors, increased attention
Z006	Observance of procedures and instructions, increased attention, site recognition
Z007	Personal protection means, increased attention
Z008	Room ventilation, observance of procedure and instructions, personal protection means
Z009	Room ventilation, durable packaging of powdery materials, personal protection means (protective glasses, masks)
Z010	Ergonomic seat protected against vibrations transmission when given work is executed in sitting position on the machine, limited working time according to the instruction, breaks in work

Z011   Use of hearing protectors, limited exposure

Z012   Proper organization of work place, possibility of rests

Z013   Electromagnetic protection (voltage filtering, shielding), observance of procedures and instructions, control measurements, trainings, proper technical condition of installation

Moreover, the structure of ergonomic action programme should also include the evaluation of risk connected to ergonomic aspects, evaluation of introduced changes from the point of view of principles of ergonomics, workshops on ergonomics problems, input of ergonomics to the engineering design as well as training and formation of skills of ergonomic techniques (Słowikowski, 2004).

It should be underlined that there are many methods in reference to risk assessment. Predominantly used is Risk Score method or classic one (Berkowska, and Drzewiecka 2011).

5   CONCLUSIONS

Employers often see ergonomics as a complicated activity, time consuming and redundant. The proposed plan of ergonomic intervention for the machine operators is an indication how in simple way ergonomic and safe working conditions may be assured.

The intervention plan for machine operators can bring the following practical benefits:

- reduction of accidents,
- increase of safety and comfort machine operators,
- promotion of ergonomic approach to the development of ergonomic working conditions among employers,
- practical indication of what actions should be taken to ensure the operators safe and ergonomic working conditions.

REFERENCES

Berkowska, A., and M. Drzewiecka. 2011. The choice of optimal risk assessment method for an exemplary post. In. *Safety of the system: Human-technical object-environment. Work health and safety in production, operation and maintenance*, ed. Sz. Salomon. Częstochowa, Poland: 55–74.

DeRango, K., B.C.III Amick, M.M. Robertson, T. Rooney, A. Moore, and L. Bazzani. 2003. *The productivity consequences of two ergonomic interventions. Working paper #222.* Toronto: Institute for Work & Health.

Dul, J., and C. Ceylan. 2011. Work environments for employee creativity. *Ergonomics* 54(1): 12–20.

Dul, J., and W.P. Neumann. 2007. The strategic business value of ergonomics. In. *Meeting Diversity in Ergonomics*, eds. R.N. Pikaar, E.A.P. Koningsveld, and P.J.M Settels. Oxford, UK: 17–28.

EN ISO 12100:2010 Safety of machinery. General principles for design. Risk assessment and risk reduction.

EN 614-1:2006 Safety of machinery. Ergonomic design principles. Part 1: Terminology and general principles.

Guastello, S.J. 1993. Do we really know how well our occupational accident prevention programs work? *Safety Science* 16(3-4): 445–463.

Gray, W.B., and J.T. Scholz. 1993. Does regulatory enforcement work? A panel analysis of OSHA enforcement. *Law & Soc Rev* 27: 177–213.

Hägg, G.M. 2003. Corporate initiatives in ergonomics – an introduction. *Applied Ergonomics* 34(1): 3–15.

Horst, W. 2006. *Ergonomia z elementami bezpieczeństwa pracy. Przewodnik do ćwiczeń laboratoryjnych.* Poznań: Wydawnictwo Politechniki Poznańskiej.

Koningsveld, E.A.P. 2005. Participation for understanding: an interactive method. *Journal of Safety Research – ECON Proceedings* 36: 231–236.

Koningsveld, E.A.P. 2009. The impact of ergonomics. In. *Industrial Engineering and Ergonomics*, ed. C.M. Schlick. Berlin Heidelberg, Springer-Verlag: 177–195.

Lewis, R., M. Krawiec, E. Conifer, D.R. Agopsowicz, and E. Crandall. 2002. Musculoskeletal disorder worker compensation costs and injuries before and after an office ergonomic program. *International Journal of Industrial Ergonomics* 29(2): 95–99.

Li, J., L. Wolf, and B. Evanoff. 2004. Use of mechanical patient lifts decreased musculoskeletal symptoms and injuries among health care workers. *Injury Prevention* 10: 212–16.

Mrugalska, B. 2011. Incorporation of ergonomics into machinery design In. *Human potential management in accompany. Knowledge increase.* ed. S. Borkowski and J. Sujanová.Trnava: Publisher AlumniPress: 103-116.

Mrugalska, B., and A. Kawecka-Endler. 2012. Practical application of product design method robust to disturbances. *Human Factors and Ergonomics in Manufacturing & Service Industries* 22(2): 121-129.

Mrugalska, B., and A. Kawecka-Endler. 2011. Machinery design for construction safety in practice. *Lecture Notes in Computer Science.* Springer. 6767(III): 388-397.

Murphy, L.R., and L.S. Sauter. 2004. Work organization interventions: state of knowledge and future directions. *Social and Preventive Medicine* 49(2): 79–86.

Olszewski, J. 1997. *Podstawy ergonomii i fizjologii pracy.* Poznań: Wydawnictwo Akademii Ekonomicznej.

Pacholski, L. 2009. Macroeconomic premises for errors in institutionalised treatment processes. *Ergonomia.* 31(3-4): 141-142.

Pacholski, L. 2003. Macroergonomic circumstances of the manufacturing company development. In. *Ergonomics in the digital age*: Proceedings of the XVth Triennial Congress of the International Ergonomics Association and the 7th Joint Conference of Ergonomics Society of Korea/Japan Ergonomics Society, Seoul, Korea, August 24 - 29, 2003. - Seoul, Korea: The Ergonomics Society of Korea. Vol. 6: Proceedings: Safety and Health Miscellaneous Topics: 556-559.

Reese, S. 1998. Helping workers stretch away strain: a semiconductor equipment manufacturer slashes its on-the-job injury rate with a low-tech program and 16 min. a day. *Business Health* 16: 21–23. **ZACYTOWANE JEST REESE 2002 a nie 1998 ale tego 1998 nie moglem znaleźć w internecine - numeróᵉ stron**

Richardson, D. 2002. Ergonomics & retention. *Caring* 21: 6–9.

58

Robertson, M.M. 1999. Office ergonomic interventions. In. *Human-computer Interaction: Ergonomics and user interfaces*, eds. H.J. Bulinger, and J. Ziegler. New York: Lawrence Eelbaum Associates, Inc. 1: 205–212.

Romanowska-Słomka, I. 2002. Operator sprężarek, pomp i innych urządzeń w oczyszczalni ścieków. Ocena ryzyka zawodowego. *Atest*, 5: 41–43.

Romanowska-Słomka, I. 2003a. Operator koparki. Ocena ryzyka zawodowego. *Atest*, 5: 41–42.

Romanowska-Słomka, I. 2003b. Operator prasy mimośrodowej. Ocena ryzyka zawodowego. *Atest* 9: 42–43.

Romanowska-Słomka, I. 2003c. Operator wózków widłowych. Ocena ryzyka zawodowego. *Atest* 11: 42–43.

Romanowska-Słomka, I. 2004a. Operator belownicy. Ocena ryzyka zawodowego. *Atest* 11: 42–43.

Romanowska-Słomka, I. 2004b. Operator elektrodrążarki. Ocena ryzyka zawodowego. *Atest* 10: 42–43.

Romanowska-Słomka, I. 2005. Ogrodnik-operator maszyn ogrodniczych. Ocena ryzyka zawodowego. *Atest* 5: 50–52.

Romanowska-Słomka, I. 2009. Operator zgrzewarki dielektrycznej. Ocena ryzyka zawodowego. *Atest* 7: 62–63.

Romanowska-Słomka, I. 2010. Operator wtryskarki. Ocena ryzyka zawodowego. *Atest* 11: 60–62.

Sharan, D. 2003. "RSI Prevention Programme" Accessed February 27, 2012, http://www.deepaksharan.com/cri_prevention.html

Słowikowski, J. 2004. Zastosowanie zasad ergonomii w przedsiębiorstwie – przegląd rozwiązań. *Bezpieczeństwo Pracy* 4: 24–26.

Tompa, E., R. Dolinschi, and C. de Oliveira. 2006. Practice and potential of economic evaluation of workplace-based interventions for occupational health and safety. *Journal Occupational Rehabilitation* 16: 375–400.

Tompa, E., H. Scott-Marshall, M. Fang, and C. Mustard. 2010. *Comparative benefits adequacy and equity of three Canadian workers' compensation programs for long-term disability. Working Paper # 350*. Toronto: Institute for Work & Health.

Wykowska, M. 2009. *Ergonomia jako nauka stosowana*. Kraków: Uczelniane Wydawnictwo Naukowo-Dydaktyczne: 287-330.

CHAPTER 7

Human-Machine-Systems for Future Smart Factories

Detlef Zuehlke

German Research Center for Artificial Intelligence (DFKI)
Innovative Factory Systems (IFS)
Kaiserslautern, Germany
zuehlke@dfki.de

ABSTRACT

Our daily life depends more and more on the advantages of information and communication technologies. Information and communication is available anywhere, anytime with any content for any user using any device and any access. No doubt, we are entering the era of ubiquitous computing where computers are becoming smaller, more powerful, invisible and permanently networked. But in our factories the old traditions are still hold: a lot of paperwork, many cables and traditional interaction devices. But as modern ICT technologies are penetrating our homes and lives they will penetrate our factories in future as well. In 2004, a group of German companies and researchers met to discuss the impact of these technologies on future factories. As a result the idea of the smart factory was conceived and made reality. The factory built in 2006 is based on a real production setting in a downsized but representative factory environment. Such a large-scale, real-life production process is the perfect testbed for new methods, concepts and technologies in and for the industrial field. This paper will present the changes and challenges we are facing in future human machine systems and will deduce the technologies and design paradigms from the gathered experiences.

Keywords: Advanced Manufacturing, HMI, ICT, SmartFactory, model-based design

1 INTRODUCTION

Over the last decades the interaction between machines and the human operator has not seen too many technological leaps forward. We started with black-and-white text-only screens and keyboards in the 80th which were slowly replaced in the 90th by color screens and block graphics. The only real technological step forward may be seen when PC technology and WINDOWS made their way into the factories. We moved on to fully graphical screens and also introduced mouse devices for screen interaction. Since then the screens got bigger, the controllers faster and the systems more powerful but cheaper.

But today it seems that we are ready to enter a new era of industrial interaction. And this is mainly driven by the advances in information and communication technologies (ICT). In our daily life everyone appreciates these many technologies which help us to communicate anywhere, anytime with any content for any user using any device and any access (Zuehlke, 2010). We have powerful smartphones with high resolution screens, multi-touch gestures, photo and video capabilities, location sensing (GPS), entertainment functions and many other things. But the most important enabler for new applications must be seen in the widespread use of wireless networks which allow us to share all information in real-time around the globe. By this the user becomes nomadic and can work anywhere.

And as these technologies have reached a well accepted level of maturity, it is not a surprise that other application fields are interested in a fruitful use for the benefit of the users. And this has a deep impact also on the industrial field. On the first run useful applications are seen for those applications where expanding machines are operated in quite clean environments with a high degree of agility. A good example is the packaging industry. They typically operate large machines which are build in a modular way, with dedicated controllers and a high degree of flexibility and reconfigurability. Instead of installing many small operator panels at each module the use of smartphones or tablets is very promising. The user can enjoy these wearable devices exactly where he needs the interaction.

Another application area which shows great promise for the use of smart technologies is in plant maintenance. For cost reasons maintenance is more and more centralized and performed by specially trained staff. On the other side, an effective maintenance is crucial for an efficient plant operation. If a production device is broken down the maintenance staff must be informed quickly, guided to the defective device reliably, be enabled to immediately identify the cause of malfunction, optimally supported in identifying spare parts and repair procedures and finally be enabled to restart the production. This should be described in the following scenario.

A machine sends out an error message into the factory network. The maintenance department receives it and starts a first situation and cause analysis. A maintenance order is generated and send to the smartphone of a staff member. The maintenance engineer is then guided by the smartphone-GPS and indoor location systems to the place of malfunction. There he uses the smartphone camera to identify the device by shooting a picture of the manufacturer identification plate. This automatically invokes the current version of the technical documentation over the network including audiovisual repair procedures. Using pattern matching techniques the parts are identified and a spare part request generated online. After receiving the spare part the crew can watch a short video on the replacement procedure and change the device. Then the staff can use their smartpads via a direct M2M connection to check the correct version and parameters and restart production. After this successful maintenance procedure on their way back to the office they can also do the administrative work of closing the maintenance order also using their various smart devices.

2   WHERE ARE WE TODAY

The described scenario sounds very inspiring but it is still far from reality. We have powerful nomadic devices and also the necessary wireless communication infrastructure. But there are still many problems to be solved.

First, today´s smart devices were not made for the harsh factory environment. The footprint is often too small for workers hands, the touchscreens will often not allow for glove operation and furthermore get dirty in short time.

Second, these devices need reliable wireless network connections. But what is stable in home applications often shows problems in industrial applications. These consumer devices communicate in the open ISP-Bands (mostly 2.4GHz) which get more and more occupied by many devices (WLAN, Bluetooth, wireless cameras, toys). For a stable operation in industrial environments the companies must care for a restrictive radio band use in their factories, e.g. by prohibiting the use of private devices and registering each device in a radio band use list. It is also recommended to install monitoring systems which track the wireless traffic and help to identify radio problems.

Thirdly, wireless communication is vulnerable to unintended or –even worst-targeted jamming. Provisions must be made to detect such problems as the basic requirement for counter measures.

Fourthly, one advantage of today´s smart devices must be seen in their universal usability. But when using it in a specific interaction scenario it must adapt to the required machine context. Wired operator consoles are preloaded with the specific HMI software and the control system knows the location and implicitly also the context by design. With wireless consoles the context specific HMI software must be downloaded context dependent and this requires the knowledge on the place of interaction. For outdoor use the build-in GPS can be used but for indoor applications we still need convincing localizing solutions which also work in harsh industrial environments.

Fifthly, the equipment used in the consumer world and in the industrial world widely diverges in life cycle. Industrial equipment has a typical life cycle of 15 to 30 years whereas the smart devices from the consumer world are mostly obsolete after 12 months. Spare parts will be available for perhaps another year but not for 15 or 20. And also the user wants to benefit from newer devices much earlier pushing the supplier to update the HMI in shorter periods. As no industrial supplier can afford to reprogram the HMI systems once a year the only solution must be seen in hardware abstraction. We can no longer design HMI systems on the WYSIWYG-level like today but must instead define and apply a (standardized) markup language which describes separately functionality and appearance. Instead of designing a push button at design time we should define an abstract representation for the action which can be mapped to either a push button or e.g. a speech input at runtime depending on the hardware device and the users preferences.

3   TARGETING SOLUTIONS

As we learned from the CIM era in the late 80ies the goal of fully-automated factories with no human workers is a fallacy. Instead our factories must be designed to optimally support the humans. In Germany this led to the foundation of a demonstration and research testbed for smart information and communication technologies: the *SmartFactory*[KL] (see figure 1). Supported by many industrial partners and research institutions a downsized but realistic factory was set up in Kaiserslautern which serves as a testbed for smart solutions in industrial environments ("SmartFactory - Towards a Factory-of-Things", 2012).

Figure 1   The living lab *SmartFactory*[KL] in Kaiserslautern (Germany) represents a realistic test and demonstration facility for ICT in industrial environments.

Besides other topics like smart control architectures, also smart interaction techniques are developed and tested. E.g. Smartphones and smart tablets are used here as preferred interaction devices. These are linked to the controllers by wireless connections such as Bluetooth, WLAN, ZigBee. By this the user gains mobility and can benefit from all build-in technologies of today´s devices. He can shoot and send photos, can read barcodes or RFID´s, communicate with his supervisors and maintenance staff. But this new world will also bring up new design and work challenges: *Who is interacting with the machine? Where is this person located? Is he allowed to perform the function?* and many more.

3.1   Communication setup

In a future world of smart devices the communication between the machines and the operating devices cannot be predesigned. Instead a user can appear somewhere around a machine wanting to communicate with it or even a single subsystem. A still very useful model in spite of its age for approaching this task, is the ISO-OSI 7-layer model (Zimmermann, 1980), which is outlined in figure 2.

First, it must be clearly defined which devices should communicate. The user may select on his HMI screen the right device from a list, he can use the camera and a pattern recognition technique to "point-and-shoot" the right device, or he can use RFID or mostly better NFC wireless identification to ensure the desired communication link. Then the physical connection must be established on layer 1. Here the devices will scan the wireless bands for the right communication partner and then establish the protocols on the transport layers 2-4. Mostly, the well known wireless technologies like WLAN or Bluetooth offer already the associated standards on these layers. In the next step, the authentication and authorization must be checked. After that the communication on the application layer 7 must be installed e.g. by downloading the HMI model into the mobile device.

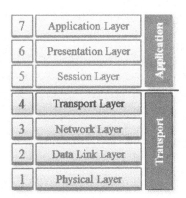

Figure 2 Scheme of the *Open Systems Interconnection* (ISO OSI) seven layers model.

But besides the protocol and authentication questions several other problems have to be solved. How do we define an active communication? Will we force it to an inactive state after a predefined inactivity? We also have to deal with parallel or integral communications where several users with different access rights may operate systems. How can we easily switch between several applications? And last but not least, from the wireless systems – with the exception of NFC – we normally will not be able to extract the place of interaction. A user may be in front of the device or many meters away from it. At least in safety critical applications a solution is required. Therefore, we need location sensing systems in addition.

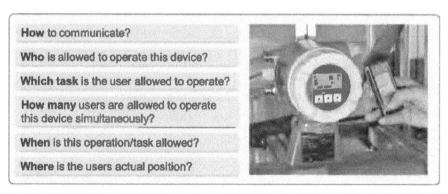

Figure 3   Open questions concerning the future communication and interaction between the machine and the operating devices on the shop floor.

3.2   Location based services

For location sensing many solutions are already available. Without additional effort we can extract location information from the wireless systems. This works with WLAN and Bluetooth, but the achieved accuracy especially under industrial conditions will not exceed 3-4 meters. Commercial location sensing systems like the Ubisense system ("Real-time Location Systems", 2012) offer much higher accuracy ranging down to 0.5m under industrial conditions. But as they are add-on solutions, cost and applicability must be taken into account. Today, we are still missing convincing location sensing systems for in-factory use and a simplicity comparable to the GPS. Such novel solutions are also developed and tested in the *SmartFactory*[KL]. In addition, a location information can not be reduced to local locations but instead must be seen as a universal data set for worldwide use. As shown in the scenario above location information will be a central backbone for future systems (Stephan, and Heck, 2010).

Location again is a prerequisite for identifying general context information which will be important for supplying the right HMI information. In general, we see location sensing as an enabler for location based services.

3.3   Augmented service support

Augmented reality is an established field of research since many years. But the technological requirements are high and have been a brake shoe ever since. But today's smart devices already offer many of the needed technologies at reasonable cost and with good performance. We have high resolution cameras and screens, pattern recognition software, gyro and radio sensors and many more. So promising solutions are enabled. In the *SmartFactory*^KL we operate a demonstration system using smartpads which guide users to the right place, visually identify devices and overlay service information on the screen in real time to support the user in various training and maintenance actions.

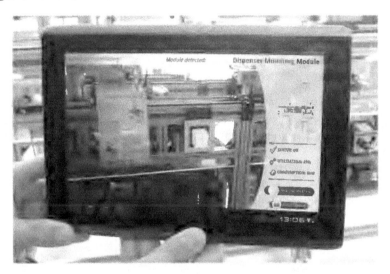

Figure 4 The demonstration system *Augmented SmartFactory*^KL provides relevant information in real-time directly at the point-of-action.

3.4 Model-Based User Interface Development (MBUID)

As stated earlier, the user interface development process of the future must be hardware independent to comply with the short lifecycles of the smart devices. One very promising solution can be seen in a model-based approach (Meixner, Paternó, and Vanderdonckt, 2011).

"The purpose of Model-Based User Interface Development (MBUID) is to identify high-level models, which allow developers to specify and analyze interactive software applications from a more semantic oriented level rather than starting immediately to address the implementation level. This allows them to concentrate on more important aspects without being immediately confused by many implementation details and then to have tools which update the implementation in order to be consistent with high-level choices" (Cantera Fonseca, González Calleros, Meixner, et al., 2010). In MBUID the user interface is therefore split up into several distinct models like e.g., task model (describes which activities need to be performed in order to reach a specific users' goal), dialog model (describes dialog state changes of a user interface), presentation model (describes how the interaction objects or widgets appear in different dialog states and on different hardware platforms) accompanied by a function or domain model (describes the data binding to the functional core of the software system) and a context model (describes context factors like e.g., user groups, device profiles for adaptation of the user interface). These models (often described by an XML-compliant language) are related by each other at different levels of abstraction.

The Cameleon Reference Framework (Calvary, Coutaz, and Thevenin et al, 2003) describes a framework with different levels of abstraction (composed of different models) that serves as a reference for classifying user interfaces that support multiple targets, or multiple contexts of use on the basis of a model-based approach. Between these levels there are different relationships e.g., Reification covers the inference process from high-level abstract descriptions to runtime code. It is not needed to go through all steps: one could start at any level of abstraction and reify or abstract depending on the project.

Such a model-based approach is currently being developed and tested at the *SmartFactory*[KL] (Hussmann, Meixner and Zuehlke, 2011).

3.5   Design for usability

Whatever we design finally must optimally support the user. So our development processes must be driven by user needs and not by technologies. This leads to promoting a new appreciation by the system suppliers. New interactive systems must therefore be methodically developed with a user-centered development process as described in ISO 9241-210.

4   WHAT IS STILL TO COME

The described technologies and methods have demonstrated their potential in supporting the human worker in our *SmartFactory*[KL]. Though we are still years away from an industry wide application, the innovation speed is increasing mostly driven by the advances in the consumer world. The workers today are much more open to new technologies than ever before and the manufacturers have also recognized the benefits. Over the next years we are facing new challenges in HMS.

Closely linked to the success of smart devices is the APP concept. So it's somehow understandable that innovative manufacturers are experimenting with it already. If a customer buys a smart sensor or actor in future, he can download "APP's" for internal device control as well as for operating it via smart handhelds. By this basic innovation both manufacturer and user will not only benefit from more up-to-date software and easier distribution but also enter a new business model where users are more willing to pay for new gadgets over the lifetime of a device.

Second, all future devices will get smart, i.e. they will offer smart server capabilities into networks. This will not only ease the system integration but also lead to new control architectures. We will slowly replace the old hierarchical control structures with a strict horizontal integration by network structures with a distinct vertical integration.

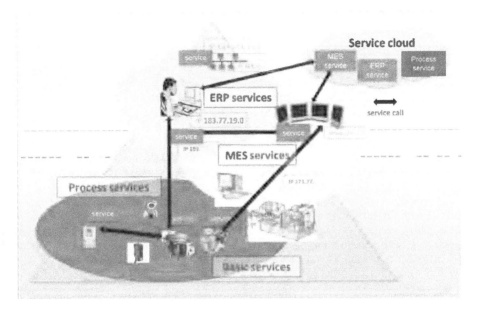

Figure 6 The schema shows how hierarchical control structures ("automation pyramid") will be replaced by network structures with distinct vertical integration.

By the initiative of several high-ranking people from german industry, academia and politics a new paradigm was created in 2011: Industry 4.0 – as the symbol for the 4th industrial revolution, which is characterized by the pervasive use of information and communication technologies in industry (Kagermann, Lukas, and Wahlster, 2011).

As we have learned from the first three revolutions (steam engine > power generation, mass production and micro-electronics) technologies must lead to a better life for the individual. The human must benefit from it and recognize the advantages. This will also be true for the postulated 4th revolution: systems must be designed for the user instead of forcing the user to work with technological dinosaurs.

REFERENCES

Calvary, G., J. Coutaz, and D. Thevenin, et al. 2003. A Unifying Reference Framework for multi-target user interfaces. Interacting with Computers, 15(3): 289-308.

Cantera Fonseca, J. M., J. M. González Calleros, and G. Meixner, et al. 2010. Model-Based UI XG Final Report, W3C Incubator Group Report, Accessed February 14, 2012, http://www.w3.org/2005/Incubator/model-based-ui/XGR-mbui-20100504.

Hussmann, H., G. Meixner, and D. Zuehlke, 2011. Model-Driven Development of Advanced User Interfaces. Heidelberg: Springer.

Kagermann H., W.-D. Lukas, and W. Wahlster, 2011. Industrie 4.0: Mit dem Internet der Dinge auf dem Weg zur 4. industriellen Revolution, In: VDI Nachrichten, April 1, 2011, VDI Verlag.

Meixner, G., F. Paternó, and J. Vanderdonckt. 2011. Past, Present, and Future of Model-Based User Interface Development. i-com, 10(3): 2-11.

"Real-time Location Systems", Accessed February 14, 2012, http://www.ubisense.net.

"SmartFactory - Towards a Factory-of-Things", Accessed February 14, 2012, http://www.youtube.com/watch?v=EUnnKAFcpuE.

Stephan P. and I. Heck. 2010. Using Spatial Context Information for the Optimization of Manufacturing Processes in an Exemplary Maintenance Scenario. Proceedings of the 10th IFAC Workshop on Intelligent Manufacturing Systems. Lisbon, Portugal.

Zimmermann, H. 1980. OSI Reference Model — The ISO Model of Architecture for Open Systems Interconnection. IEEE Transactions on Communications, Volume 28, No. 4: 425 – 432.

Zuehlke, D. 2010. SmartFactory — Towards a factory-of-things. In: IFAC Annual Reviews in Control, Volume 34, Issue 1: 129-138, ISSN 1367-5788, Elsevier Science Ltd.

CHAPTER 8

The Role of Ergonomic Factors in the Modeling of Project Management Organization

Paweł Pietras

Technical University of Lodz
Poland
ppietras@konto.pl

ABSTRACT

In this paper, experiments in creating and modeling organizational structures of companies on different levels, specified above, are presented. Knowledge and application of behavioral factors, especially those culturally determined, play a crucial role in this process. For instance, one of the major factors hindering a smooth transition from ad hoc level to operational management is culturally determined tendency for individualism. Yet, by planning the transition from operational to strategic level, a commonly applied model is a combination of a stiff Prince2 methodology oriented on competences distribution with agile methodologies, even when the projects are not in an IT field.

In the concluding part of a paper, a model of Project management in a dispersed or network organization will be presented. This is an original model combining features necessary to supervise the project by supervisory bodies (eg. sponsor of the project) and enabling a flexible completion of project tasks by their performers. From the technical side, it is available in the form of Java technology based Internet system. Thanks to web applications, joint project work of teams from different organizations is possible and no software, nor IT infrastructure is required.

1. INTRODUCTION

An increasing number of actions taken by enterprises is being completed through projects. It results mainly from the progressive specialization of particular companies, organizational units and people who, when faced with the necessity to complete increasingly more complex tasks, organize themselves into teams which use the cooperation and experience exchange synergy. Team-working, as the more and more frequent form of cooperation, poses a challenge to science and makers of tools which facilitate cooperation between humans, with a particular focus on cooperation of teams consisting of employees of different entities (organizational and formal scattering in network organizations). In these cases using traditional project management supporting tools (such as, for instance, IT systems based on one-user or server software versions) is not very effective (because of the high cost of acquiring rights) or is even technically not feasible.

In such cases the solution is to apply one of the numerous on-line tools which support project management. However, a thorough analysis of these solutions market demonstrates that there are not too many solutions which would offer at the same time popular project planning and project control tools, tools which help supervise the course of work by units superior to the project team and/or work flow type tools that are additionally equipped with ergonomic mechanisms encouraging their everyday use.

Having observed such a gap in the market, the author has compiled the methodology of project management based on best practices known, such as Prince2 technology and IPMA and PMI standards. It has been computerized and made available on-line. The further part contains its basic assumptions and an IT tool created for it.

2. METHODOLOGICAL ASSUMPTIONS

When compiling methodology it was assumed that it would be addressed at enterprises executing projects of the following characteristics:
- short execution time, most often one quarter
- the cost of the project is relatively low and consists of organizational costs mostly
- the risk of the project execution is low and easy to revise
- the organization executes a few dozens of projects per year
- the organization addressees the projects on a routine basis
- the execution of the project uses the processes already existing in the organization

With such projects and at this type of organizations project management poses a huge number of managing problems which on their own are not that complex, yet combined cause significant delays, budget overrunning and in many cases end without completing targeted goals.

The method proposed divides the course of project management into several basic processes. Each of them covers a set of actions to be undertaken and supported by suitably matched tools. Standards of documents, which help record and disseminate information about the executed project, have been compiled for all the processes.

The methodology distinguishes the following processes:

1. Initialization. This process consists of the following elements:
   - project description with technical specification,
   - defining project goals,
   - launching a project team,
   - defining project milestones,
   - identifying the stakeholders of the project,
   - establishing the requirements regarding project results,
   - preliminary risk assessment,
   - creating a project charter.

2. Planning. This process consists of the following elements:
   - defining the project deliverables,
   - defining the task list,
   - structuring of the project (creating the structure of work breakdown),
   - time and deadlines for completing tasks,
   - estimating costs of the tasks,
   - planning resources and describing their roles and responsibilities.

3. Steering. This process consists of the following elements:
   - estimating the work progress,
   - announcing the end of the package,
   - change management,
   - establishing the earned value,
   - data reporting,
   - data analysis of work progress trends and schedule progress.

4. Risk management. This process consists of the following elements:
   - risk identification,
   - threats classification,
   - risk control.

5. Closing. This process consists of the following elements:
   - formal disbanding of the project team,
   - summarizing the results and comparing them to the estimated values.

6. Communication. This process is used for disseminating information about the project progress. It contains the following issues:
   - preparing meetings of the project group,
   - maintaining ongoing information exchange among the project participants.

It is assumed that the project is executed within a certain organizational structure whose fixed units are the steering committee and the project manager. A steering committee is a unit formed by at least one person – the head of the steering committee, for whom the executed project is a tool for completing business

72

objectives. Apart from the head on the steering committee there may be other persons from the organization who have the real impact on its course, for instance the direct superiors of the project executors or future users of its products. A project manager is a person who assumes responsibility for the efficient project execution.

Each project is a consequence of the appearance of a more or less identified need. Its appearance starts the process which results in formal launching of the project. The process is called **Initialization**. The time of its course depends mostly on the awareness level regarding particular needs and technical and organizational possibilities of fulfilling them.

The material effect of the process is a project charter which contains basic information about the project environment, reasons for its initialization, stakeholders of the project and which mostly points at the directions of development. An important element of the project charter is establishing its milestones. Pieces of the project which lead to achieving the consecutive milestones mark out the iteration cycle – the recurrence of the planning and steering processes (see figure 1). The project charter should be approved by the steering committee and accepted for execution by the project manager.

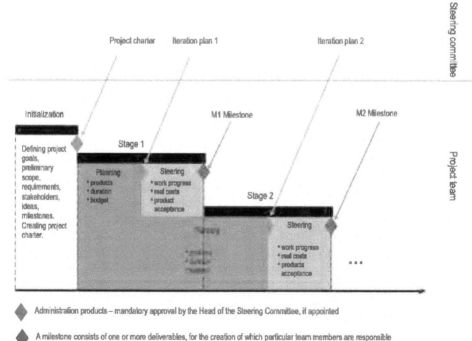

Administration products – mandatory approval by the Head of the Steering Committee, if appointed

A milestone consists of one or more deliverables, for the creation of which particular team members are responsible

Phase model of the project course according to the compiled methodology.

The **planning** process is launched repeatedly, each time when a consecutive iteration begins at the latest. The most popular planning methods and techniques may be used to prepare the plan.

1. The planning of the scope of the project is done by establishing the deliverables and building the structure of work breakdown (Work Breakdown Structure). For greater transparency the following dependance between work packages contained in the work breakdown structure and the tasks is enforced: work packages are connected with particular deliverables and there are functional, qualitative and other requirements described for them; and with the tasks are connected pieces of information about the time of their duration, order of actions and costs.
2. Time planning is based on network techniques and the work schedule itself is visualized as a Gantt chart.
3. Planning project resources is executed by roles and responsibilities matrix. Particular project team members assume responsibility according to the following key:
    - task executor (E) – a person physically executing entrusted task,
    - responsibility for execution (R) - a person who assumes responsibility for executing the entrusted task, the role may be combined with the executor's role, yet it is not a rule,
    - responsibility for defining requirements (R) – a person fulfilling this role establishes how a particular product should be made and/or what quality requirements it should fulfill; then this person must conduct acceptance tests for the product and ultimately approves the fact of producing a deliverable,
    - cooperation (C) - a person who cooperates with the project executor, i.e. does not do work for them, but provides technical/organizational/substantial or office support, etc.
4. Planning the budget of the project is based on the technique of planning costs of particular tasks, which form costs of particular work packages and groups of tasks. Since each deliverable is connected with a particular work package, costs of gaining particular deliverables become known.

The basic task of the **Steering** process is monitoring the work progress and providing tools for managing potential changes in the project. The process uses the following work techniques.

1. Defining the monitoring cycle – it occurs separately for each iteration. This results from the fact of different work intensity during particular phases of the project and the varied planned length of their duration.
2. Analysis of the product trends (milestones). It is possible to define the planned date of achievement in relation to each deliverable which is connected with a particular work package. During the project execution persons responsible for particular deliverables are supposed to estimate at the end of each monitoring cycle the currently planned date of submitting a particular deliverable. Due to this fact the project manager and the steering committee have the possibility to familiarize themselves with the schedule progress in relation to particular deliverables and also the final product of the project.
3. Analysis of the project using the earned value method. The earned value

method allows for establishing how the real course of the project deviated from the schedule and the planned budget and also for estimating the final cost of the project.

In case when changes occur in the project, they are reported to the steering committee, which decides whether or not to introduce a particular change. In case when the decision about implementing a change is made, the plan of the given cycle is changed, i.e. deliverables are added or removed and tasks and relations between them are changed.

When it comes to communication the methodology envisages cyclic project meetings taking place within the monitoring cycle dates as a tool of information exchange about the work progress and the general course of the project.

3. ERGONOMIC FACTORS SHAPING FUNCTIONALITY OF THE PROJECT MANAGEMENT SYSTEM

3.1. Access to the system

Accessing the application requires merely the usage of an internet browser and it is not necessary to instal any client software. This approach makes the program much more attractive and accessible (the program may be used for example by employees who do not have the right to install the software on their computers) and also hugely simplifies the up-date process, as the application is developed (re-installing software on many computers is not necessary, an upgrade is conducted only once, on a server).

The IT system has been installed on an internet portal under the tools4pm.com address. At the moment the portal works in the Polish language, and by the end of June 2012 the service in English, German and French will have been provided. The system uses all tools and techniques listed in the previous paragraph and also connections between particular processes and the flow of documents among the project participants. For the user using these functions does not require the knowledge of methods or techniques and the interpretation of results achieved is facilitated by a support system.

3.2. Using the methodology

Taking into consideration the organizations and projects for whom the methodology has been designed, the IT system has been prepared in such a way to make it possible for the user to 'switch off' the methodology (i.e. enforcing the flow of information and decisions) and use only the project planning and steering tools and communication tools.

The user who registers a new project becomes automatically its administrator. As an administrator he can add new users to the project, assigning appropriate roles to them: the role of a project manager, member of the project team, member of the steering committee, head of the steering committee and the observer.

When the system administrator does not define the head of the steering committee, the IT system will not 'enforce' the document circulation, so the methodology as a system of correlation of decisions and documents will not work. This facilitates using the system by organizations which are less developed when it comes to organizational issues, at the same time not limiting the technical quality of managing the project itself.

3.3. System of decision making

Most IT systems which support management make their resources available by logging in to a system and forcing the users to have a computer with software and to familiarize themselves with the system. For users who actively participate in project management (planning the project, inputting data about the project, etc.) this is obvious. However, as the office reality proves, we do not like to do that. So it becomes very difficult for the project manager to 'enforce' on the decision-makers from the steering committee familiarizing with the system, logging in to it and choosing particular tools. In the presented system the decision-making process has been significantly simplified.

The head of the steering committee and its members as a rule do not have access to the planning tools (i.e. they can view the data, yet cannot introduce their own). Their role is to supervise the project process, so they make decisions about the key elements of the project (initialization, approval of the project plan or stage, approval of the project or stage execution). Members of the steering committee and its head receive through their email boxes information about the state of the project in form of ready project documents saved in PDF format. Naturally, they have an opportunity to log in to the system at any time and view the data in the system instead of a PDF file, however the principle when forming the portal's functionality was maximum simplification of tools usage for the project decision-makers. Following this rule, it is enough for a decision to be made, when a decision-maker clicks on a suitable link in an email message (e.g. 'agree', 'not agree', 'please provide me with more information'). Therefore to make a decision (on the technical side, of course) merely a telephone with an email service is required.

3.4. Work automation

From the level of all implemented tools one can print the results of work to a PDF file or export numerical results in MS Excel format. Additionally at any stage and in each process one can attach to the project external documents, for instance, attachments to project documentation, files with data, figures, schemes, etc. They are kept in a repository of project documents, which is a system of catalogs assigned to processes and stages of the project.

An additional facility for project managers is a project closing tool. It produces an automatic report which summarizes the project, comparing the planned data (e.g. duration time of particular work packages and their costs) against the actual data achieved. Naturally, the head of the project (or persons from the team assigned by him/her) has a possibility to add their content and comments. At the end of the project the system automatically generates a file which contains compressed data generated by the system during the project and kept on the portal in the document repositories.

3.5. Communicating in a virtual team

Next to substantial tools (for planning and steering) the portal provides communication tools. They are supposed to facilitate information exchange among the project team members, especially when the teams are geographically scattered. The portal provides such tools as an internet forum, chat room or video connections. The system has been equipped with a mechanism which supports conducting project meetings. In such case a project meeting takes place on the project internet forum and all threads are automatically saved and attached to the report from the project meeting.

4. SUMMARY

The author has made a decision to compile a simple and clear methodology of project management having observed a few dozens of projects, after numerous interviews with their participants and having executed many projects by himself. The conclusions of this research point unequivocally at two basic barriers in using any methodological approach in project management. The first one is accessibility of the project management methodologies. Naturally, there are many methodologies, however the most popular ones (e.g. Prince2, guidelines contained in PMBoK or IPMA guidelines) are very expanded and to a certain extent they 'deter' potential users. The second barrier is the accessibility of IT tools. Of course there are many of them on the market, yet they can be divided into two categories: 1) complicated, comprehensive and usually very costly software which supports the whole process of project management (e.g. Primavera) or provides basic tools (e.g. MS Project); 2) simple, cheap (sometimes free) software providing a chosen tool and not offering support during the whole process of project execution.

The compiled methodology uses the most frequently applied planning and steering tools and principles of work organization, responsibility and decision breakdown in projects. It is comprehensive since it includes the whole process of preparing a project for its formal existence in an organization and planning, execution and also the decision-making process in the project. The IT side of it provides ergonomic and functionally attractive tools, simple in usage and at the same time providing the freedom of application, regardless of the stationary software used by the particular organization or work place. The functionalities applied, especially pertaining decision-making do not require logging in to the system and familiarizing with it by the decision-makers, which hopefully means that they will be willingly used or recommended by company management. Since, as the practice proves, the more complicated the tool, the less willing the users are to use it.

REFERENCES

Balser T., Evolution of Enterprise – from Executing Projects to Business Orientation, in: Project Management – Efficient Business Ventures Management. Wydawnictwo WEKA, Warsaw 2000

Pietras P., Wasiela-Jaroszewicz J., Internal Communication in Network Organizations, Academic Conference 'Improvement of Management Systems in Information Society. Methodology and Experiences of Corporations and Public Sector Units', University of Economics in Katowice 2006

Pietras P., Pietras A., Wasiela-Jaroszewicz J., Virtual Teams in Network Oranizations, in: Penc J. (ed.) Problems of Human Resource Management in a 21st Century Organization, Wyd. Wojskowa Drukarnia, Łódź 2007

Pietras P., Wasiela-Jaroszewicz J., Interpersonal Communication within Project Teams, in: Organizations Management in Knowledge Based Economy, Nicolaus Copernicus University in Toruń, Toruń 2008

Pietras P., Network Organization Management Based on Web 2.0 Idea, in: Management of Organizations Development, Cracow University of Economics, Cracow 2008

Pietras P., Pietras A., Virtual Team Management – Case Analysis, in: Enterprise Management in Circumstances of High Technologies Development, Technical University of Łódź Publishing House, 2008

Pietras P., Project Management in Manufacturing Enterprise – Oryginal Method, 11th Academic and Technical Conference 'Management in Enetrprise. Production and Technologies in Metallurgy' Częstochowa University of Technology, Ustroń-Jaszowiec 2003.

Pietras P., Characteristics of Management Condition in Polish Enterprises, 8th International Academic Conference 'Modern Management of Enterprises', University of Zielona Góra, Łagów 2003.

Pietras P., Szmit M., Project Management. Selected Methods and Techniques, Wydawnictwo Horyzont, Łódź 2003.

CHAPTER 9

User-centered Design of a Game-based, Virtual Training System

Dominic Gorecky[1], Glyn Lawson[2], Katharina Mura[1], Setia Hermawati[2], Mikkel Lucas Overby[3]

[1]German Research Center for Artificial Intelligence (DFKI)
Kaiserslautern, Germany
Dominic.Gorecky@dfki.de, Katharina.Mura@dfki.de

[2]Human Factors Research Group
Faculty of Engineering
University of Nottingham
Nottingham, England
Glyn.Lawson@nottingham.ac.uk, Setia.Hermawati@nottingham.ac.uk

[3]Serious Games Interactive
Copenhagen, Denmark
mlo@seriousgames.dk

ABSTRACT

Virtual training systems show a high potential to complement or even replace physical setups for training of assembly processes. A major precondition for the breakthrough of virtual training is that it overcomes the problems of former approaches, especially the poor user acceptance. In this paper we describe different design factors, which have or might have an effect on the user acceptance of virtual training. Recommendations are given, how these factors can be addressed and measured though a user-centered design process. The paper presents the design approach taken during the development of a game-based, virtual training system in the EU-FP7 project VISTRA.

Keywords: Assembly Processes, Game-based Training, Serious Game, User-centered Design, Virtual Reality

1   MOTIVATION AND BACKGROUND

The current developments in the production domain are characterized and driven by two major challenges:

i)   globalization and the pressure to innovate force manufacturers to shorter innovation cycles and market introduction times while simultaneously reducing the development and production costs.

ii)   the increased variety and complexity of products places high demands on the skills and knowledge of the employees.

One phase of the product life cycle, where these challenges become apparent, is the transition between product and production planning and the physical start of production (compare figure 1). Up to now, the planning and training of manual manufacturing processes are still carried out in physical stages, which is expensive and often ineffective. A better knowledge transfer between both phases could help i) to reduce times and costs for training, and ii) to improve the skills and competencies of the workers. This applies especially to the automotive industry, where the launch of a new vehicle necessitates a long planning and training period. Traditionally, the operators train the assembly on dedicated hardware prototypes for several weeks in advance. To set up the training, the parts which are involved in the training must first be obtained. Afterwards the worker can perform the training by assembling these parts according to the correct build sequence. After each training loop, the parts must be removed again to prepare the next training. This training method exhibits certain disadvantages:

- high parts costs due to prototype or preproduction status of the parts,
- only a few operations and variations can be trained,
- late start of training due to hardware availability,
- part condition will decrease within each training loop,
- limited access to the few prototypes for a high number of operators, and
- time and effort needed to disassemble the parts, which has no benefit to training activities.

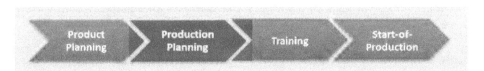

Figure 1 Extract of a simplified product lifecycle showing the training phases.

Compared to this, virtual assembly training has a high potential to complement or even replace current training methods, and thus to reduce the time, costs and efforts for testing on physical prototypes. What is most crucial for the breakthrough of virtual training of assembly processes is that it overcomes the major problem of insufficient user integration. In the following, we will examine which factors, might have an influence on the user acceptance and how a user-centered design process can measure and address them.

2 VIRTUAL TRAINING SYSTEMS

In this section we describe different design factors, which have or might have an effect on the user acceptance of virtual training systems. Each sub-chapter concludes with a recommendation, how we intent to improve poor user acceptance in the VISTRA system.

2.1 Hardware setup

In recent years, many virtual training systems have been developed for different application domains. They can basically be distinguished in two groups related the hardware setup:

i) highly instrumented virtual reality (VR) environments (e.g. CAVE, head-mounted displays, haptic feedback devices) using advanced interaction devices (e.g. static or body-mounted motion capturing systems), and

ii) desktop-based concepts using traditional interaction devices (e.g. mouse, keyboard, joystick).

The systems of type i) enable an immersive user experience, and are used e.g. in construction ("Catsimulator", 2012), medicine or military ("Cubic", 2012). The high initial outlay as well as the immobility and inflexibility of the hardware restrict the training on a large-scale e.g. in automotive industry. In manufacturing, such systems are rather used for the support of decision-making in product and production planning than for actual worker training. The systems of type ii) are available at low cost, but they are less or non-immersive. For assembly training, there are a number of commercial systems available on the market ("Cortona3d", 2012, "Ngrain", 2012, "Vizendo", 2012). However, they are described as unintuitive and are often rejected by the trainees. Furthermore, concerns exist over the transfer of training from non-immersive systems to tasks conducted in "real" space.

Main criteria, influencing the decision of the hardware setup, are costs and scalability, ease of use as well as the required immersion. For VISTRA, we propose a system, which combines the advantages of both categories described above – i) *immersive user experience* and ii) *large-scale training at reasonable costs* – by using immersive, but out-of-the-shelf hardware setups (e.g. standard-monitor for visualization, *Microsoft Kinect* in combination with a *Nintendo-Wii* device for interaction) in order to allow a natural and more realistic interaction with the virtual objects and environment (see figure 3).

Figure 3 Interaction Concept for the Virtual Assembly Training.

2.2   Learning concept

A framework for a training concepts was shown in (Malmsköld, Örtengren, Carlson, et al., 2007a), discussing on the fundamental learning theories and influencing parameters on the learning effect. Several studies have proven that computer-based, virtual training has a positive effect on the procedural learning in manual, industrial tasks (Malmsköld, Örtengren, Carlson, et al., 2007b; Adams, Klowden, and Hannaford, 2001). However, the motor skill development through virtual training is very limited, due to the lack of user interactivity and immersion (Dawei, Bhatti, Nahavandi, 2009).

The central issue related to the learning concept is if cognitive procedural knowledge and/or skills should be addressed. This question is basically related to the goal and content of the training. The VISTRA system should focus only on training of procedural knowledge about how a task should be performed. The most relevant knowledge components are clustered in different semantic classes (compare figure 2). The training of fine motor skills should not be addressed.

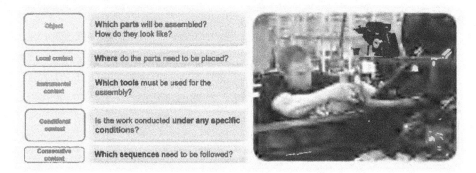

Figure 2 Overview about relevant semantic classes of training content for operator training in the automotive industry.

2.3   Game-based training concept

In recent years, the increase of serious games and games-based training in companies and the educational sector, in hospitals and health-care institutions, within the military, and several other areas, presents a clear indication of the novel potential of games besides being mere entertainment (Bergeron, 2006). Despite this development, until now, only little attention has been drawn to the design of game-based training for the factory-floor level. Due to an absence of typical gaming elements, the beforehand described training approaches show a more or less sterile character of a simulation.

Our hypothesis is that gaming characteristics (such as fun, challenge, play, outcomes and feedback) increase the user acceptance of training systems and lead to a more enjoyable learning experience. The motivational strength of games lies in the ability to keep the user interested through interactivity (Egenfeldt-Nielsen, 2007). In order to create an engaging user experience and stimulate the motivation of the users, we are choosing for the virtual training a serious games approach (that is, using gaming for a serious purpose, other than solely entertainment, yet incorporating the characteristics described above). In contrast to pure simulation, game-based virtual training requires additional components, such as a storyboard, entertaining elements and a reward system.

Can a game that represents the daily work be motivating and entertaining? Firstly, intrinsic and not extrinsic motivation is the focus of the game design. As it is known from many conventional games the playing itself constitutes a reward, e.g. through the gamer's experience of mastering levels with increasing difficulty. Sometimes, extrinsic incentives can be even counterproductive and lead to less motivation and acceptance, known as overjustification effect (Deci, Koestner, and Ryan, 1999). Though, possible reward can be given within the virtual environment through game internal features as through the incentive of collecting skills and properties for the personal avatar. In addition, outside the virtual training system reward can be realized through monetary, e.g. gift certificates, or non-monetary measures, e.g. public acknowledgment as "trainee of the month" or the like. In this case best performance in competition with other trainees or trainee groups is rewarded. Yet, personal improvement or achievement of a certain performance criterion may be acknowledged as well.

Besides this, the design of a game for the industrial context raises the question about the general acceptance for games. *How is operators' perception of usefulness on virtual training with game based concept? Will operators accept and deal with new technologies and game-based concepts?* Initial analyses in the automobile industry have shown that acceptance differs among operators – depending on age, education and personal interest.

2.4   Fidelity

Fidelity is the degree of realism and the depth of detail presented in a virtual environment. It is supposed to have a high effect on user acceptance. According to (Stone, 2008), we must distinguish between physical fidelity, which relates to how the virtual environment and its component objects mimic the appearance of their real-world counterparts and the psychological fidelity, which relates to the degree to which simulated tasks reproduce behaviors that are required for the actual, real-world task. Psychological fidelity has proven to be more closely associated with positive transfer of training than physical fidelity.

Which degree of realism is necessary to achieve the best learning effects and which abstraction is needed to create an engaging game-like virtual environment? The question of the appropriated degree of fidelity is not easy to answer. The decision must primarily be based on the training concept (motor skill training or procedural learning) and the complexity of the task and its context. In VISTRA, we try to achieve a high psychological fidelity, whereas the physical fidelity is subordinate.

3   A USER-CENTERED DESIGN PROCESS

(Stone, 2008) reports that developments in virtual reality and game-design often are only concentrating on technological issues, as opposed to meeting the needs of the end user. In response to this, we proposed a user-centered design (UCD) approach (Norman and Draper, 1986) to address the design questions raised in section 2. UCD is a common term, encompasses a philosophy and a variety of methods, which refers to how end-users influence a design through their involvement in the design processes. UCD has been shown to have positive effects on various aspects, i.e. the quality or speed of the design process; the match to the end users' needs or preferences; and end user satisfaction (Kujala, 2003). It has also been shown contributing to the acceptance and success of products (Preece, Rogers and Sharp, 2002). In the following we will propose several types of analyses that will assist providing answers to the design questions in section 2 whilst ensuring that end users' needs are accommodated. The proposition for the analyses was based on the guidance provided by (Leonard, Jacko, Yi et al., 2006) in which factors such as the purpose of the study (explanation driven or implementation driven) and availability of resources (time, experience, funding and previous research) were taken into account in choosing the analysis method. In the following, we describe the aim and background of these analyses in detail.

3.1   Stakeholder Analyses

The stakeholder analysis identifies and examines users that are likely to be somewhat affected by a product. (Eason, 1987) identified three types of users: i) primary, users that will use the product; ii) secondary, users that will use the product occasionally or through an intermediary; iii) tertiary, users that will be affected by or make decision to purchase the product. It is paramount that stakeholder analysis investigates the effect of a product for these three types of users. Stakeholder analyses have been recognized as an important part to ensure the effectiveness of information system design and development in healthcare (Atkinson, Eldabi, Paul et al., 2001). Stakeholder analysis has also been applied successfully to support product development (Amiri, Dezfooli and Mortezaei, 2012), space planning (Brooks, 1998), integration of ergonomics into manufacturing sector (Neuman, 2004) and job design and potential training (Neary and Sinclair, 1998). (Thun, Größler and Miczka, 2004) reported that, as the demographic of workers at manufacturing companies shifted towards an ageing, the adoption of new technologies could be hampered. Thus, it is important that a clear understanding of users' background, their general skill level and their acceptance regarding computer games and interactive learning tools are investigated.

3.2   Task Analyses

According to (Kirwan and Ainsworth, 1992), task analysis is a method to study actions and cognitive processes that are required by an operator to do in order to complete the system goals. One of commonly user task analyses is *Hierarchical Task Analysis* (HTA). HTA involves breaking down the task under analysis into a nested hierarchy of goals, operations and plans. HTA can be applied to a variety of applications, e.g. interface design and evaluation, training, job aid design (Stanton, 2004). With regard to (ISO 13407, 1999), HTA will complement stakeholder analyses in providing the context of use and user/organizational requirements for the intended virtual training system. The results from HTA will particularly inform details of operation involved in manual assembly in automotive manufacturing processes. Through HTA, the type and extent of the training content (cognition vs. motor skills) could be defined further.

3.2   Thematic analysis

Thematic analysis provides a descriptive presentation of qualitative data by identifying common themes and then categorizing the qualitative data under suitable theme (Franzosi, 2004). Thematic analysis helps to select, focus and transform qualitative data into manageable information segments to show its patterns. Thematic analyses have been shown to successfully support UCD of technological products. For instance, thematic analysis is utilized to inform design guidelines related to usability, ease of navigation in 3D spaces and information design on websites. Because thematic analysis is guided by specific research question or sub questions by making inferences of an intended population (Lapan and Quartaroli, and Riemer, 2012), it can be used to provide answers for design questions that could not be directly addressed by other proposed analyses (e.g. degree of physical fidelity) and complement the results from other proposed analyses (e.g. training content).

4 APPLYING USER-CENTERED DESIGN TO THE DESIGN QUESTIONS

In this section, method(s) of measurement for each design question is proposed (see Table 1). It is important that answers to the design questions shown in Table 1 are based on the input from carefully chosen of users representatives. (Damodaran, 1996) suggested that choice of users' representative should be aimed to ascertain that "those appointed are genuinely representative of the user population and possess the necessary personal attributes". In addition to this, users need to be given a clear brief on their role in the design process of the intended system in order to ensure the success of users' involvement in UCD (Damodaran, 1992). Last but not least, it is also essential that once the answers of the design questions were established, implementation of these findings should be continually assessed by asking questions such as: *"what is the consequences for users if we do not implement a specific requirement?, how important are the different requirements for the users?"* etc. (Rexfelt and Rosenbald, 2006).

Table 1 Design questions, which has an influence on the user-acceptance with the related method for measurement.

Design Question	Decision- making-factor	Method(s) for measurement
Hardware setup	Costs and scalability	Stakeholder analysis, Thematic analysis
	Ease of use	Stakeholder analysis, Thematic analysis
	Required immersion	Task analysis, Thematic analysis
Training Concept	Type and content of training	Task analysis, Thematic analysis
Game-based approach	Acceptance for games	Stakeholder analysis, Thematic analysis
Fidelity	Type of training (procedural knowledge or motor skills); Complexity and context of the task	Thematic analysis

5 CONCLUSIONS

The lesson learned from former developments is that ensuring a high user acceptance plays a key role in designing new virtual training systems. We showed some important influencing factors on the user-acceptance and described how to analyze them through user-centered design. In particular, the game-based approach for industrial assembly training represents a novelty, which requires further attention during the design process. Potential pitfalls related to user involvements are also identified in order to raise awareness and ensure the fitness of the intended virtual training system to the users.

ACKNOWLEDGMENTS

The VISTRA project is an industry-driven project with the aim to minimize the number of physical prototypes in the automobile industry by developing a training platform on the basis of available product and process data ("VISTRA EU-FP7-Project", 2012). The project is funded in the 7th Framework Programme of the European Union (project number: ICT-285176).

REFERENCES

Adams, R. J., D. Klowden, and Hannaford, B., 2001. Virtual training for a manual task. Haptics-e, Vol. 2, no. 2.

Amiri, Dezfooli, Mortezaei, 2012. Designing and ergonomics backpack for student aged 7-9 with user centred design approach. Work: A Journal of Prevention, Assessment and Rehabilitation, 41 (1), pp. 1193-1201.

Atkinson, C., Eldabi, T., Paulm R. J., Pouloudi, A., 2001. Investigating integrated socio-technical approaches to health informatics. Proceedings of the 34th Hawaii International Conference on System Sciences.

Bergeron, B. 2006: Developing Serious Games. Hingham, MA: Charles River Media Inc.

Brooks, A., 1998. Ergonomics approaches to office layout and space planning. Facilities, 16(3/4), pp. 73-78.

"Catsimulator", Accessed February 15, 2012, https://www.catsimulators.com/c-7-simulators.aspx.

"Cortona3d", Accessed February 15, 2012, http://www.cortona3d.com/.

"Cubic", Accessed February 15, 2012. http://www.cubic.com/Solutions/Defense-Systems/Training-Systems/Virtual-and-Immersive-Training-Systems.

Dawei, J., A. Bhatti, S. Nahavandi, S. 2009. Design and evaluation of a haptically enable virtual environment for object assembly training. Haptic Audio visual Environments and Games. ISBN: 978-1-4244-4217-1

Damodaran, L., 1992. Integrating human factors principles into a version of SSADM: the practice and the problem. Invited paper to DTI seminar on Human Factors in the System Design Life Cycle, London, June, HUSAT Memo 592.

Damodaran, L., 1996. User involvement in the systems design process-a practical guide for users. Behaviour and Information Tehnology, 15, 6, pp. 363-377.

Deci, E. L., Koestner, R., Ryan, R. M., 1999. A meta-analytic review of experiments examining the effects of extrinsic rewards on intrinsic motivation. Psychological Bulletin, 125, 6, pp. 627-668.

Eason, K., 1987. Information Technology and Organizational Change. Taylor and Francis, London.

Egenfeldt-Nielsen, S. 2007. Educational Potential of Computer Games. Continuum International Publishing Group

Franzosi, R., 2004. Content analysis. In M S Lewis-Beck, A Bryman and T F Liao (Eds), The Sage Encyclopaedia of Social Science Research Methods, Thousand Oaks, CA: Vol. 1: 186-189.

"ISO 13407", 1999. ISO – International Organization for Standardization.

Kirwan, B., and L. K. Ainsworth. 1992. A Guide to Task Analysis. London, UK: Taylor & Francis.

Kujala, S., 2003. User involvement: a review of the benefits and challenges. Behaviour and Information Technology, 22(1): 1-17.

Lapan, S.D. and Quartaroli, M. T., 2012. Qualitative Research: An Introduction to Methods and Designs,

Leonard, V. K., Jacko, J. A., Yi, J. S., and Saintfort, F., 2006. Human Factors and Ergonomics Methods. In: Salvendy, G. (Eds.), Handbook of Human Factors and Ergonomics, John Wiley & Sons, Inc.

Malmsköld, L., R. Örtengren, B.E. Carlson, and al., 2007a. Virtual Training – Towards a Design Framework. Proceedings of the World Conference on E-Learning in Corporate, Government, Healthcare, and Higher Education: 6299-6307

Malmsköld L., 2007b. Preparatory virtual training of assembly operators – an explorative study of different learning models. Preprints, The Swedish Production Symposium 2007

Neary and Sinclair, 1998, A case study of job design in a steel plant. In: Hanson, M. A (Ed), Contemporary Ergonomics, Taylor and Francis, pp. 218-222.

Neuman, W. P., 2004. Production Ergonomics: Identifying and managing risks in the design of high performance work system, Doctoral Thesis, Lund Technical University, Lund.

"Ngrain", Accessed February 15, 2012, http://www.ngrain.com/solutions/virtual-task-trainer/.

Norman, D. A. and Draper, S. W, 1986. User-Centred System Design: New Perspective on Human-Computer Interaction, Lawrence Earlbaum Associate, Hillsdale, NJ.

Preece, J., Rogers, Y and Sharp, H., 2002. Interaction Design: Beyond Human-Computer Interaction, New York, NY: John Wiley and Sons.

Rexfelt, O., and Rosenblad, E., 2006. The progress of user requirements through a software development project. International Journal of Industrial Ergonomics, 36, pp 73-81.

Stanton, N. A., 2004. The psychology of task analysis today. In: Diaper, D. and Stanton, N. A. (Eds), The Handbook of Task Analysis for Human-Computer Interaction, Mahwah, NJ: Lawrence Erlbaum Associates: 569-584.

Stone R. J., 2008. Human Factors Guidelines for Interactive 3D Game-Based Training Systems Design, University of Birmingham. Accessed February 23, 2012 http://www.hfidtc.com/

Thun, J-H., Größler, A., Miczka, S., 2007. The impact of the demographic transition on manufacturing: Effects of an ageing workforce in German industrial firms. Journal of Manufacturing Technology Management, 18, 8, pp.985 – 999

"VISTRA EU-FP7-Project", Accessed February 15, 2012, http://www.vistra-project.eu.

"Vizendo", Accessed February 15, 2012, http://www.vizendo.se/.

CHAPTER 10

Age-appropriate Workplace Engineering with the Aid of Cardboard Engineering

Gerrit Meyer, Peter Nyhuis

Institute of Production Systems and Logistics
Leibniz University of Hannover, Germany
meyer@ifa.uni-hannover.de

ABSTRACT

Companies in Germany are facing the challenges of demographic change. The changing efficiency of an aging workforce calls for the development of new concepts for age-appropriate job engineering. To increase the acceptance of these concepts and use the broad implicit know-how of the workforce, it is recommendable to involve the active employees in the development process (Dombrowski and Riechel, 2010). Cardboard engineering constitutes a participatory method for the design of age-appropriate working systems (Nyhuis and Zoleko, 2010).

In cardboard engineering, workplaces are modeled by using cardboard boxes and other modeling material to simulate and analyze the production flow (Nyhuis and Zoleko, 2010). By integrating the know-how of employees of different hierarchy levels into the development process, this method links expert and workforce knowledge. The advantages are the enhancement of creativity and willingness to change as well as promoting self-initiative on the part of the employees (Schuh, Kampker, Franzkoch, Wesch-Potente, and Swist, 2010).

Primarily, cardboard engineering is suitable for the implementation of new production systems or efficiency improvement projects, but it is also applicable for the development of age-appropriate workplace systems (Nyhuis and Zoleko, 2010).

The Institute of Production Systems and Logistics has developed a concept for age-appropriate work engineering, which uses the method of cardboard engineering. In this case, the focus is placed on an age-appropriate ergonomic job. With this concept, workplaces of different areas can be designed age-appropriately. To validate the concept the workplaces in the installation area of a company were designed in an age-appropriate way with the help of cardboard engineering. This gives the company the opportunity to meet the challenges of demographic change.

The article describes the developed concept for age-appropriate work engineering and opportunities to improve workplaces and work processes under consideration of critical-aging criteria in an easy and inexpensive way.

CONSEQUENCES OF THE DEMOGRAPHIC CHANGE
DEMOGRAFISCHER WANDEL UND DEREN AUSWIRKUNGEN

The consequences of the demographic change increasingly pose challenges for politics and companies in Germany and Europe (Langhoff, 2009).

Within a short to medium term, the current mortality and fertility behavior as well as the reduced immigration will result in a population decrease and a change of the age structure of the population (Veen, 2008).

As a consequence of these demographic developments, it will come to a descent of the working population rate in the next decades. In addition, the current age structure will abet the trend. Due to the age structure and political measures (e. g. the increase of the retirement age), the quote of older employees (> 50 years) of the working population and average age of the employees as well as the age heterogeneity will rise concurrently (Ries, Diestel, Wegge, and Schmidt, 2010).

Age heterogeneity describes the differences in the age peculiarity of single group members and can have both positive and negative effects (Ries, Diestel, Wegge, and Schmidt, 2010). In age-heterogeneous groups, the probability of group formation, dissatisfaction, communication and integration difficulties increases. All these effects can finally lead to a reduced productivity and an increased conflict potential. Otherwise, age-heterogeneous groups have several advantages such as high flexibility and creativity due to their diversity, which affect problem identification and solution finding positively by means of the force of innovation. Furthermore, the decisive advantage is the transfer of knowledge and experience. In this case, the transfer of implicit know-how is especially relevant (Veen, 2008). Studies have shown that skills are differently distinctive depending on age and subject to changes in the working life.

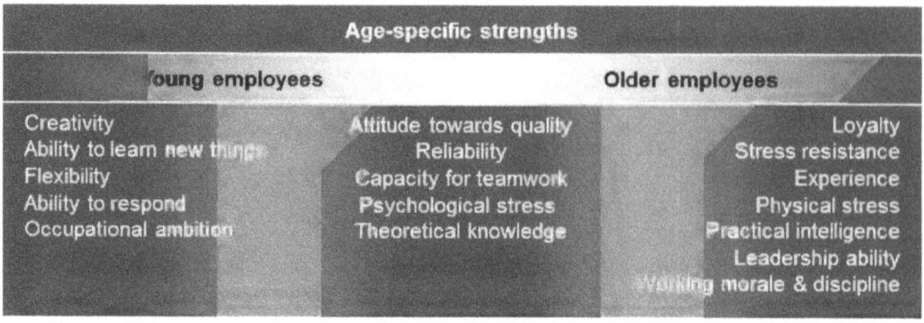

Age-specific strengths		
Young employees		Older employees
Creativity Ability to learn new things Flexibility Ability to respond Occupational ambition	Attitude towards quality Reliability Capacity for teamwork Psychological stress Theoretical knowledge	Loyalty Stress resistance Experience Physical stress Practical intelligence Leadership ability Working morale & discipline

Figure 1: Age-specific strengths (Buck, Kistler, and Mendius, 2002; Reindl, Feller, Morschhäuser, and Huber 2004)

In comparison, older employees do not perform worse than younger employees as it is often assumed. The results they achieve are comparable (Buck, Kistler, and Mendius, 2002). The age-specific strengths are listed in figure 1.

Companies rank the older employees' attitude towards quality, reliability and leadership ability higher than that of younger employees. If you contemplate the theoretical knowledge, the ability to work in a team or the psychological ability to work under pressure, younger and older employees are on the same level. On the other hand, the creativity of younger employees is highly appreciated in comparison to that of older employees. Analyses have also shown that the following performance potentials are higher with younger staff: willingness to acquire new skills, flexibility, ability to respond and occupational ambition. Particularly with regard to the physical ability to work under pressure, studies have demonstrated that younger employees are able to work under pressure to a higher degree than older employees (Buck, Kistler, and Mendius, 2002).

Companies in Germany and Europe must adjust themselves to an aging workforce, however studies show that deficits do exist. Most companies, for example, currently lack a sufficient number of age-adapted workplaces. Contemporaneous concepts lack the transfer of know-how between the working age groups and the involvement of older employees in the innovation process so as to use their age-specific competences in a more efficient way (Nienhüser, 2000).

The risk for companies consists in a sinking performance and innovation ability, especially in the case of older employees, if tasks are not organized beneficially for learning, development and health. Age influences the performance and innovation ability and their development only on a secondary level, but the working conditions on a primary level (Ulich, 2005). The consequence is that the working conditions must be improved first and adapted to the changing general requirements.

MEASURES FOR THE ACCOMPLISHMENT OF THE DEMOGRAPHIC CHANGE IN PRODUCTION

In the light of the aging workforce in nearly all European countries, the topic of how to find managing strategies for the accomplishment of the demographic change is highly relevant in many companies. which cooperate with the Institute of Production Systems and Logistics. In most cases. the relevant question is how to preserve the performance and innovation ability resp. preferably enhance them.

In this context. particularly the factors health. competence and motivation have to be contemplated and strategies to improve these factors have to be developed.

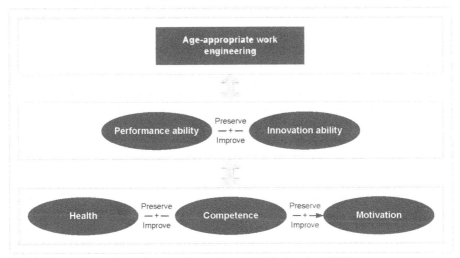

Figure 2: Age-appropriate work engineering (Gerlmaier. 2010)

The objective target of companies must be to develop concepts for age-appropriate work engineering. which preserve and improve health and competence and also motivation.

CONCEPTS FOR AGE-APPROPRIATE WORK ENGINEERING

To meet the requirements of the industry. especially of small and medium-sized enterprises for concepts for age-appropriate workplace and work systems engineering. the Institute of Production Systems and Logistics has developed an integrated concept for age-appropriate work engineering. which is upgraded with the latest scientific findings (Nyhuis and Zoleko. 2010). within the framework of a research project.

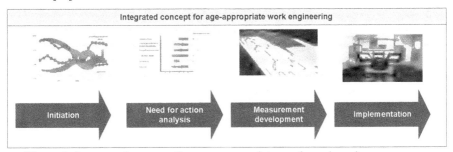

Figure 3: Integrated concept for age-appropriate work engineering

The concept is divided into the steps initiation, analysis of the need for action, development of measures and implementation, which are based on each other in form and content. The user has different tools (e. g. workshops, interviews) at his disposal for each step. The step of initiation includes the sensitization of the workforce for the topic of demographic change and the resultant impact as well as the deduction consequences for the future of their own company. Within the framework of the analysis of the need for action, the need for action regarding an age-appropriate engineering of work system is identified. In the following step, adequate concepts are developed based on miscellaneous tools for age-appropriate work engineering. Practical experience has shown that the use of cardboard engineering, which will be described in the following abstracts in more detail, is especially suitable to achieve these goals. The developed concepts are realized in practice in the final step of the implementation.

Because of the fact that the topic of demographic change is increasingly moving into the focus of the public view, awareness for the problem and the necessary age-appropriate work engineering is given in many companies, especially in big concerns. Furthermore, practical examples illustrate that in some companies an analysis of the need for action has already taken place. But many companies have difficulties with the development of measures.

For this purpose, the following possible causes are assumed:

- The individuality of single workplaces requires individual measures and is therefore very cost and time-consuming
- Complexity of the measures development
- Lack of competence on the field of age-appropriate work engineering

One possibility to solve this problem is the use of cardboard engineering. Cardboard engineering enables the development of individual measures for an individual workplace. Simultaneously, the method has the attribute of describing complex combination in a simple way. By consulting an external moderator, it is possible to integrate the missing competence on the field of age-appropriate work engineering into the process. Furthermore, many companies possess the necessary know-how in form of implicit knowledge, but it is not being considered. However, cardboard engineering supports the use of this implicit knowledge of the workers on age-appropriate work engineering.

FUNDAMENTALS OF CARDBOARD ENGINEERING

Cardboard engineering was developed in the context of the Toyota Production system and is a method for participative workplace engineering. By using this method, the workers are integrated into the planning of the assembly or others areas. The goal of cardboard engineering is to increase the planning and implementing quality of improvement projects by involving the employees and using their implicit knowledge. To allow the use of the workers' implicit knowledge, their participation in the development process of creative and organizational improvement measures is necessary.

Scientific findings show that the participation of the workers in improvement processes of their own work is gaining importance (Zoleko and Nyhuis, 2009). Employee participation is especially recommendable, if there is a high share of manual processes.

With the aid of employee participation, planning mistakes and improvement potentials can be identified because the workers' know-how influences the improvement process positively. Often it becomes evident that employees have ideas for improving their own work process, but these ideas are not communicated. If employees are animated to reflect on the engineering of their workplace and question it, subconscious ideas for improvement come up. Simultaneously, the acceptance of the worked out concepts rises because they were developed by the employees themselves. In many companies, the development and analysis of new processes mostly takes place digitally so as to use time and cost advantages (Dietrich and Schirra, 2006).

However, the operational employees are mostly embedded into the improvement process only insufficiently. The reasons for this are the missing physical models and the active debate with these models (Schuh, Kampker, Franzkoch, Wesch-Potente, and Swist, 2010). The method cardboard engineering applies at this point. With cardboard engineering, complete workplaces and assembly lines up to products are replicated with cardboard, styrofoam and other model making materials, work actions are simulated and work processes are analyzed (Nyhuis and Zoleko, 2010). With the exact replica, reality is visualized and can be tested under real conditions.

In workshops of three to five days in the real environment, up to 8 employees from different units (assembly, logistics, planning, etc.) and at least one workshop moderator work together to check the workplace and the process for potentials and develop suggestions for improvement (Schuh, Kampker, Franzkoch, Wesch-Potente, and Swist, 2010; Nyhuis and Zoleko, 2010). In large projects, the accomplishment of workshops of this duration is recommended. The method connects expert knowledge with employee knowledge by integrating the employees of different hierarchy and experience levels and their know-how into the improvement process.

94

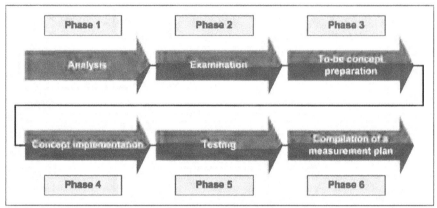

Figure 4: Course of action of cardboard engineering

The course of action of cardboard engineering is divided into 6 steps, which is illustrated in figure 4 (Nyhuis and Zoleko, 2010).

The first step is the definition of analysis criteria. Furthermore, the areas to be improved are viewed and the employees interviewed. In addition, identified problems are noted down by the workshop members. During the evaluation, the identified problems are collected, listed and summarized. Finally, the prioritization of the identified problems is carried out in this step. In the third step, the basic modifications are defined and alternatives drafted. The step concept definition includes the testing of the previously detected alternatives with cardboard and other model-making materials.

Possibly, different alternatives must be combined and tested.

Afterwards, the developed concepts are tested by the employees and adequate improvements are jointly made. In the last step, the tasks, responsibilities and deadlines for the implementation of the concept are determined (Nyhuis and Zoleko, 2010).

The results of the method application are well-engineered assembly concepts, ergonomically improved workplaces, workstations planned in detail and solutions for material supply (Schuh, Kampker, Franzkoch, Wesch-Potente, and Swist 2010).

The use of cardboard engineering offers several advantages. The pro-active involvement of the employees in the improvement process at the workplaces enhances the acceptance of the method and the reorganized workplace. At the same time, the work motivation rises as a result of the participation in the planning and improvement process (Nyhuis and Zoleko, 2010). On the other hand, the rising motivation provides the basis for improving the performance (Schuh, Kampker, Franzkoch, Wesch-Potente, and Swist 2010). Cardboard engineering is also realizable with proportionally low costs and low effort. Furthermore, the method is flexibly applicable, adjustable and deployable in many situations.

For example, cardboard engineering is applicable in plant engineering and construction, continuous assembly lines or prototype manufacturing with job-shop production and various other areas. Yet concurrently improvements of complete assembly and logistics processes, separate workplaces and assembly oriented product engineering are also realizable.

However, cardboard engineering cannot replace the software-oriented planning of the assembly (Schuh, Kampker, Franzkoch, Wesch-Potente, and Swist 2010).

Working in teams also makes a short and direct communication possible. In addition, the increased creativity and willingness for change as well as the promotion of one's own initiative from the staff side are also advantages. The benefit of cardboard engineering is an exact description of the working process and an easy and cost-saving variant of process improvements. Another advantage is the easy validation of the developed concepts and the preparation of the final implementation (Schuh, Kampker, Franzkoch, Wesch-Potente, and Swist 2010).

The production planners involved also benefit. With the help of the model they detect which of their planning results have so far been unsuitable in practice and thus collect valuable suggestions for future planning projects. With cardboard engineering, a major part of cost-intensive modifications during the production after the starting equipment operation is omitted. Cardboard engineering especially helps employees of time management to describe the processes exactly.

If it is not possible to interrupt the current day-to-day process, cardboard engineering offers the opportunity to test new ideas autarkically without influencing the day-to-day business.

CARDBOARD ENGINEERING FOR AGE-APPROPRIATE WORK ENGINEERING

Primarily, cardboard engineering is used for the introduction of new production systems or for efficiency enhancement projects in working systems by reorganizing application. However, cardboard engineering is also suitable for the development of age-appropriate workplace systems (Nyhuis and Zoleko, 2010).

In the context of the research project, the potential of the method cardboard engineering for the improvement of workplaces under the aspect of age-appropriation was identified and enhanced. While in the traditional use of cardboard engineering the different kinds of waste serve as the analysis criteria, the Institute of Production Systems and Logistics has identified and applied age-critical job requirements during the identification of improvement potentials at workplaces within the research project. The focus was laid especially on age-appropriate and ergonomic workplace engineering. Therefore, it was necessary to include particularly the older employees into the improvement process and identify deficits by means of their active collaboration. In the context of the research project, several age-critical job attributes were compiled as analysis criteria for cardboard engineering.

Thus, the workplace has to be engineered on the basis of the following age-critical job requirements (Nyhuis and Zoleko, 2010; Sommer, 2003):

- Avoidance of physically strenuous tasks
- Minimization of stresses and strains in the work environment
- Reduction of high and inflexible performance standards
- Cutback on shift and night work
- Decrease of high psychical pressure

For the improvement of the workplaces, attention has to be paid to the avoidance of physically exhausting tasks such as lifting and transporting heavy loads, one-sidedly tiring or shortly-periodic jobs with constrained postures. In addition, the impact of the work environment should be minimized (e. g. heat, cold, noise and dazzling).

In addition, time pressure, the assignment in clock-dependent assembly systems, shift work and night work should be avoided for older employees, because this kind of work leads to high psychical stresses. If these structural recommendations are disregarded, the consequences are symptoms of fatigue, increase of recovery time and accident hazard, demotivation and loss of competences. The findings of the research project have been validated with an industrial partner to improve the application of the method cardboard engineering for age-appropriate work engineering. At this, special attention was paid to the modification of the method in such a way that it is designed to enhance competence and motivation and thus innovation.

As afore-mentioned, this method increases motivation in the improvement process because of the participative approach and active participation of the employees. For this reason, it is of primary importance to avoid demotivating effects (e. g. social marginalization of single employees, excessive demands) during the execution of the workshop. By means of organizational measures during the workshop planning, it is also possible to influence the competence advancement in a positive way.

Studies have shown that positive effects on the competence advancement are the result of the deployment of age-heterogeneous teams or tutorials in the workshop. At the same time, the common development of new concepts for age-appropriate work engineering in age-heterogeneous teams enhances the ability for innovation. On the other hand, the moderator has to work towards the prevention of the negative effects in age-heterogeneous teams so as to enable increases in motivation.

PRACTICAL APPLICATION

Within the framework of the research project, the workplaces in the assembly area of a consortial partner were designed in an age-appropriate way with the aid of cardboard engineering so as to validate the approach. The target was to enable the company to handle the challenges of demographic change.

In a three-day cardboard engineering workshop, employees of different age, varied areas and hierarchy levels and a member of the workers' council compiled various measures for the design of age-appropriate workplaces. The proceeding was generally carried out according to the standard cardboard engineering process.

For this purpose, the following steps were executed:

- Analysis
- Examination need for action
- Concept development
- Concept implementation

- Testing
- Measurement plan

At first, the workplaces and work process were analyzed considering age-critical attributes and the concrete need for action was identified. For this task it was necessary to impart the relevant knowledge about age-appropriate work engineering to the participants of the workshop and also point out the identified essential age-critical working attributes in this context. Therefore, the workshop participants were able to find deficiencies with regard to age-appropriate work engineering independently during the identification of the need for action. In this context, it was possible to compile deficits from various fields (layout, mode of operation, etc.). In the next step, the workplace is modeled true to scale with cardboard boxes and a concept for age-appropriate improvements of the workplace is developed and implemented.

The main improvement proposal was the reorganization of the work place to a seated/standing work position because young employees prefer to work in a standing work position whereas older employees prefer to work in a seated work position. Furthermore, other miscellaneous opportunities for improvements were developed. For example, with regard to the dimensional engineering of the assembly workplaces, the workshop participants developed footrests and height-adjustable desks and chairs and improved the use of the horizontal reaching distance as improvement suggestions. They also compiled a proposal for the realignment of materials, tools and boxes as well measures for the improvement of the layout.

The mode of operation was to be adapted to the new situation of an aging workforce by means of a standardization of the work process order. In addition, several potentials for improvements such as material supply, stock and lighting were identified.

After the modeling, employees of different ages tested the newly engineered workplaces and evaluated these workplaces on the basis of age-specific criteria.

The feedback on the re-engineering was very positive.

Finally, an action plan was jointly compiled in the cardboard engineering workshop, which included the most important milestones of the implementation of the designed workplaces in practice as well as the responsible persons (Nyhuis and Zoleko, 2010).

CONCLUSION

Cardboard engineering can help to improve in a participative manner. In addition, cardboard engineering is structured in a sufficiently flexible way, which allows the focus on age-specific criteria. By using cardboard engineering, a high improvement demand can be identified by pooling employee and expert know-how in combination with the use of synergy effects between employees of different areas and hierarchies as well as planning authorities. In addition, corresponding concepts for the solution of the problem can be developed.

The use of cardboard engineering for age-appropriate work system engineering offers the opportunity to improve different workplaces and processes with regard to age-critical points in an easy and cost-effective way. The active participation of the employees in the improvement processes and the additional scope of action additionally enhance the employees' creativity. Concurrently, it was detected in the context of the research project that the employees increasingly developed their own initiative, held more intensive group discussions and showed an increased willingness for change.

Fundamental findings of the research project were that the use of cardboard engineering is especially suitable for small and medium-sized enterprises because of their financial and personnel restrictions. On the other hand, the method needs little time and effort for preparation because it is transparent and easy to understand. In addition, it was discovered in the research project that cardboard engineering is ideal for the use in age-heterogenic teams and supports the communication between young and old employees. At the same time, the workforce is motivated to show a higher performance by participating in the measures development and the acceptance for the developed concept increases. In addition to the preservation of health, cardboard engineering also enables the preservation and increase of competences by procurement of competences in the workshop.

Cardboard engineering is a suitable method to develop measures which preserve and support health, competence and motivation. In this way, the capacity for performance and innovativeness of older employees is protected and the company is able to manage the demographic change.

To enhance the effectiveness of the method for age-appropriate work engineering it has to be investigated in more detail in what context cardboard engineering can contribute to the preservation and increase of competences and how the capacity for innovativeness can be sustainably influenced with the aid of this method.

ACKNOWLEDGEMENTS

The authors wish to acknowledge the financial support of the European Regional Development Fund for the research project "Altersgerechte Gestaltung klein- und mittelständischer Unternehmen" (W2-8 80025660).

REFERENCES

Dietrich, L. and W. Schirra, 2006. *Innovationen durch IT - Erfolgsbeispiele aus der Praxis.* Berlin Heidelberg, Germany

Dombrowski, U. and C. Riechel. 2010. Entwicklung eines Multitouch-Planungstischs zur Unterstützung der partizipativen Layoutplanung. *Zeitschrift für wirtschaftlichen Fabrikbetrieb* 12: 1091-1095

Gerlmaier, A. 2010 Innovationsarbeit alternsgerecht gestalten. *Wirtschaftspsychologie* 3/2010: 38-48

Hagen, C. 2009. Cardboard-Engineering und MTM, *MTMaktuell* 04: 10-12

Langhoff, T. 2009. *Den demographischen Wandel im Unternehmen erfolgreich gestalten.* Berlin, Heidelberg, Germany

Nienhüser, W. 2000. Personalwirtschaftliche Wirkungen unausgewogener betrieblicher Altersstrukturen. *Generationenaustausch im Unternehmen*: 55–70. Munich, Germany

Nyhuis, P. and F. Zoleko. 2010. *Leitfaden für die Umsetzung eines Konzeptes zur alternsgerechten Gestaltung der Produktion in klein- und mittelständischen Unternehmen KMU.* Institute of Production Systems and Logistics. Garbsen, Germany

Reindl, J., Feller, C., Morschhäuser, M., and A. Huber. 2004. *Für immer jung? Wie Unternehmen des Maschinenbaus dem demografischen Wandel.* Frankfurt am Main, Germany

Ries, B., Diestel, S., Wegge, J., and K. Schmidt. 2010. Altersheterogenität und Gruppeneffektivität – Die moderierende Rolle des Teamklimas. *Zeitschrift für Arbeitswissenschaft (64) 2010/3*: 137-146

Schuh, G., A. Kampker, B. Franzkoch, C. Wesch-Potente, and M. Swist. 2010. Praxisnahe Montagegestaltung mit Cardboard-Engineering, *wt Werkstattstechnik online* 9: 659-664

Sommer, C. 2003. *"Arbeit-Alter-Qualifizierung. Workshop Leistungsfähigkeit erhalten".* Praxistagung der Industrie und Handelskammer. Stuttgart, Germany

Ulich, E. 2005: *Arbeitspsychologie.* Stuttgart, Germany

Veen, S. 2008. *Demographischer Wandel, alternde Belegschaften und Betriebsproduktivität.* Munich, Germany

Zoleko, F. and P. Nyhuis. 2009. Partizipative Arbeitssystemgestaltung, ja aber wie? – Ein Ansatz für eine systematische Vorgehensweise. *Herbstkonferenz der Gesellschaft für Arbeitswissenschaft:* 309-318. Stuttgart, Germany

Buck, H., Kistler, E. and H. Mendius. 2002. *Demografischer Wandel in der Arbeitswelt – Chancen für eine innovative Arbeitsgestaltung.* Stuttgart, Germany

AUTHORS

Dipl.-Wirt.-Ing. Gerrit Meyer holds a Master's Degree in Industrial Engineering. He studied at the Technical University of Dortmund (Germany). Currently, he is a research associate at the Institute of Production Systems and Logistics at the Leibniz University of Hanover (Germany) involved in the study of labor science, where he is pursuing a PhD.

Prof. Dr.-Ing. habil. Peter Nyhuis heads the Institute of Production Systems and Logistics at the Leibniz University of Hanover and is a member of the board of management of the Institute of Integrated Production, Hanover. He studied mechanical engineering at the Leibniz University of Hanover and subsequently worked as a research associate at the Institute of Production Systems and Logistics. After receiving his doctorate in mechanical engineering he completed his habilitation thesis.

CHAPTER 11

A Flexible Intelligent Algorithm for Identification of Optimum Mix of Demographic Variables for Integrated HSEE-ISO Systems: The Case of A Gas Transmission Refinery

A. Azadeh 1, Z. Jiryaei 2, B. Ashjari 2 and M. Saberi 3

1 Department of Industrial Engineering, College of Engineering, University of Tehran, Iran
2 Department of Industrial Engineering, College of Engineering, University of Tafresh, Iran
3 Institute for Digital Ecosystems & Business Intelligence, Curtin University of Technology, Perth, Australia
aazadeh@ut.ac.ir, z.jiryaei@yahoo.com

ABSTRACT

This study proposes a flexible intelligent algorithm for assessment and optimization of demographic features on integrated health, safety, environment and ergonomics (HSEE)-ISO systems among operators of a gas transmission refinery. To achieve the objectives of this study, standard questionnaires with respect to HSEE and ISO standards are completed by 80 operators. Demographic features include age, education, gender, weight, stature, marital status, and work type. The average results for each category of HSEE are used as inputs and effectiveness of ISO systems (ISO 18000, ISO 14000 and ISO 9000) are used as output for the intelligent algorithm. Artificial Neural Networks (ANN) and Adaptive Neuro Fuzzy Inference System (ANFIS) in addition to conventional regression are used in this paper. Result shows the applicability and superiority of the flexible intelligent algorithm over conventional methods through mean absolute percentage error (MAPE). Computational results show that the proposed ANN performs better than ANFIS and conventional regressions based on its relative error. Finally, the optimum mix of demographic variables from viewpoint of HSEE and ISO are identified. This is the first study that proposes a flexible intelligent algorithm for assessment of optimum mix of demographic features for HSEE and ISO systems in a complex system such as a gas transmission refinery.

Keywords: Health, Safety, Environment and Ergonomics (HSEE); Demographic Variables; Optimization; Artificial Neural Networks (ANN); Adaptive Neuro Fuzzy Inference System (ANFIS)

1   INTRODUCTION

HSE at the operational level will strive to eliminate or decrease injuries, adverse health influences and hurt to the environment. Effective application of ergonomics in work system design can cause a balance between worker characteristics and task requirements. This can increase worker productivity, create improved worker safety (physical and mental) and job satisfaction. Various studies have shown positive influences of applying ergonomic rules to the workplace including machine, job and environmental design (Azadeh et al., 2008a; Shikdar and Sawaqed, 2004). Studies in ergonomics have also produced data and instructions for industrial applications (Blanning, 1984; Bryden and Hudson, 2005). However, there is still a low level of acceptance and few applications in industry. The main concern of work system design in context of ergonomics is improvement of machines and tools. Lack of utilization of the ergonomic rules could bring inefficiency to the workplace. Moreover, an ergonomically deficient workplace can cause physical and emotional stress, low productivity and poor quality of work conditions (Azadeh et al., 2008a). It is believed that ergonomic defects in industry are main cause of health hazards in workplaces, low levels of safety and decreased workers' productivity (Champoux and Brun, 2003). There are close relationship between health, safety, environment and ergonomics factors. Basically, ergonomics is concerned with all those factors that can affect people and their behavior (Azadeh et al., 2008a). Inappropriate design between man and machine could lead to decreased safety. Management error and harmful factors related to work environment could cause human error. HSE has defined human factors and ergonomics as the environmental, organizational and job factors, and human and individual characteristics which influence work behavior. Exact consideration of human factors improves health and safety by reducing the number of injures and unsafe behaviors at work. It also provides considerable benefits by decreasing the costs associated with work injuries and enhancing efficiency (Bellamy et al., 2008; Nachreiner et al., 2006).

Azadeh et al. (2006a) described an integrated macroergonomics model for operation and maintenance of power plants. Torp and Moen (2006) presented the effects of implementing and improving occupational health and safety management system in small and medium-sized companies. Azadeh et.al, (2008b) presented a framework for development of integrated intelligent human engineering environment in complex systems. Moreover, health, safety, environment and ergonomics (HSEE) were developed and introduced. Duijm et al. (2008) showed that HSE management would benefit greatly from existing management systems and also from the further development of meaningful safety performance indicators that identify the conditions prior to accidents and incidents. Mohammad Fam et al. (2008) used behavior sampling technique to evaluate the workers safety behavior in a gas treatment company. Azadeh et al. (2009) implemented a study in a gas treatment company to show the superiority of HSEE over conventional HSE. HSEE integrated the structure of human and organizational systems with a conventional HSE system. It resulted in enhanced reliability, availability, maintainability and safety.

2   METHODOLOGY

A flexible intelligent algorithm is proposed for assessment and optimization of demographic features for integrated HSEE-ISO systems with fuzzy and uncertain variables. To achieve the objective of this study a standard questionnaire are distributed among employees of gas transmission refinery. It is composed of the following 13 steps. Figure 1 shows the flexible intelligent algorithm for assessment and optimization of demographic features on integrated HSEE-ISO systems.

Step 1.   Determine reliability of the questionnaire. In this step Cronbach's Alpha is used for all questions of the questionnaire to check the reliability of the questionnaire.

Step 2.   Determine the validity of the questionnaire. Factor analysis has been applied for determining the validity of questionnaire.

Step 3.   Demographic features include age, education, gender, weight, stature, marital status, and work type.

Step 4.   All mix of demographic features is specified.

Step 5.   Assume that there are n operators to be evaluated. For each combination of demographic features, each operator is placed in one of mixes on respective combination.

Step 6.   Determination of input(S) and output (P) variables of the model.

Step 7.   Divide all data (different statuses for each combination of demographic features) into two subsets: train (S_1) and test (S_2) data.

Step 8.   For each combination of demographic features three method (ANN, ANFIS, conventional regression) are used to estimate relation input(s) and output(s) between that are explained as follow:

8.1. Use ANN method to estimate relation between input(s) and output(s)

- Select architecture and training parameters.
- Train the model by using the train data (S1).
- Evaluate the model by using the test data (S2).
- Repeat these steps by using different architectures.
- Determine the relative error (MAPE: Mean Absolute Percentage Error) of the learned ANN.
- Select the best network architecture (ANN*) with minimum MAPE on the test data set.

8.2. Use ANFIS method to estimate relation between input(s) and output(s)
- Select type and number of membership functions.
- Train the model by using the train data (S1).
- Evaluate the model by using the test data (S2).
- Repeat these steps by using different architectures and training parameters.
- Determine the relative error (MAPE) of the learned ANFIS.
- Repeat these steps using different type and number of membership functions.

8.3. Use conventional regression methods (linear, log-linear and quadratic) to estimate relation between input(s) and output(s)

Select the model with Min of MAPE and run for all possible mixtures in each of demographic feature combination

Step 9. Select the model with Min of MAPE and run for all mixtures in each of demographic feature combination.

Step 10. Calculate efficiency scores by use of best method. For obtain to efficiency scores follow this process:

1. Calculate weigh of (possible mixture)$_j$ (j=1,...,n)

$$V_j = P_{real\ (j)}/Ave(P_{real\ (1)},\ldots,P_{real\ (j-1)},P_{real\ (j+1)},\ldots,P_{real\ (n)});$$ (1)

$$W_j = V_j/sum(V_j);$$

2. Calculate the error between the real output $P_{real(i)}$ and model output $P_{model\ (j)}$ in the period: (j=1,..., n)

$$E_j = P_{real(j)} - P_{model\ (j)}$$ (2)

3. Shift frontier function from model for obtaining the effect of the largest positive error which is one of the unique features of this algorithm:

$$E'_0 = E_j/W_j$$ (3)

This option consists of not considering the largest error, but calculates by noting the DMU scale (Constant Returns to Scale (CRS)). To this end find:

The largest E'_0 indicate the DMU with the best performance. Suppose that DMU_k have the Largest E'_0 and we have:

$$E'_k = max(E'_j)$$ (4)

So, the value of the shift for each of the DMUs is different and is calculated by

$$Sh_j = E_k * W_j/W_k$$ (5)

104

In this approach in spite of the previous studies (Athanassopoulos and Curram (1996) called this measure "standardized efficiency") the effect of the scale of DMUs on its efficiency is considered and the unit used for the correction is selected by notice of its scale (CRS). (Azadeh et al., 2007; Delgado, 2005).

4. The efficiency scores take values between 0 and 1. This maximum score is assigned to the unit used for the correction (Azadeh et al, 2007).

$$F_j = P_j / (P_{(model^*)j} + Sh_j) \qquad (6)$$

Step 11. By use of efficiency scores in previous step, rank mixtures with respect to HSEE and effectiveness of ISO systems in each combination.

Step 12. Determine the best mixtures in each combination of demographic feature.

3   EXPERIMENT

The applicability of the proposed intelligent algorithm is experimented in an actual gas transmission refinery. It is shown how each step is applied for assessment and optimization of demographic features on integrated HSEE-ISO systems in a gas transmission refinery.

A detailed and standard questionnaire containing valuable information related to health, safety, environmental and ISO systems are developed and presented to operators. The questionnaires were distributed to 80 operators of this refinery. The questionnaire is composed of 61 questions. The score (weight) to each question was assigned between 1 and 10. For each question value 1 is shown worst Condition and value 10 is shown best Condition. The input indicators are divided to four main categories which are health, safety, environment, and ergonomic and effectiveness of ISO systems is defined as outputs. Furthermore, the average score for each indicator is computed as a final score of that indicator. Questions of the questionnaire are divided to 13 overall groups covering physical factors of work place, personal protective equipment, Work place safety actions, on-the-job training, monitors and displays, environment, management systems and control, muscular and skeletal disorders, anthropometric features and issues (ergonomics), job characteristics, job satisfactions, general questions related to HSE, focus on work .

4   COMPUTATIONAL RESULTS AND ANALYSIS

Steps 1 and 2: To prove the reliability and validity of the questionnaire, Cronbach's Alpha and factor analysis have been used. The reliability analysis of a questionnaire determines its ability to yield the same results on different occasions and validity refers to the measurement of what the questionnaire is supposed to measure (Cooper and Schindler 2003). In order to assess the reliability of instrument, Cronbach's alpha for all criteria of research variables [HSEE] have been calculated. Reliability index satisfied to an acceptable level over 60%. It is now time to assess the validity of instrument. Construct validity, content validity and predictive validity were analyzed to ensure the validity of the instruments (Nunnally and Bernstein 1994). Construct validity shows the extent to which measures of a criterion are indicative of the direction and size of that criterion (Flynn et al. 1994). It also shows that the measures do not interfere with measures of the other criteria (Flynn et al. 1994). Construct validity of measurement instrument is analyzed through factor analysis. In this study each measurement criterion is considered as a distinct construct. The most common decision-making technique to obtain factors is to consider factors with Eigen value of over one as significant and the factors should set on one component (Olson et al. 2005).

Step 3. In this article we are consider 3 combinations of demographic features. First combination includes age, education and gender, second combination includes weight, stature and marital status and tertiary combination includes age, gender and work type.

Step 4. In continuance of indicate combinations of demographic features, we are considered all possible mixtures in each combination of demographic features.

Step 5. There are 80 operators to be evaluated. In each combination of demographic features all operator is placed in one of possible mixtures. The results show that in combination 1 from 48 possible mixtures only 20 mixture, in combination 2 from 48 possible mixtures only 22 mixture and in combination 3 from 48 possible mixtures only 9 mixtures observed.

Step 6. The next step is determination of input(s) and output (p) variables of the model. We have chosen 4 main categories (health, safety, environment, ergonomic) as input variables and analyze questions that were related with those categories. Then, for each category, the average score is used as final score to be used in the proposed ANN algorithm. In addition, 3 main questions separately are chosen among our questionnaires as the three outputs data, namely, ISO systems. The output questions are stated as follows:

1) Do you think ISO 18000 has been helpful to increase your efficiency?
2) Do you think ISO 14000 has been helpful to increase your efficiency?
3) Do you think ISO 9000 has been helpful to increase your efficiency?

we used the average score as the final scores to be used in the intelligent algorithm.

Step 7. In each combination all data divided into two training (S1) and test (S2) subsets. In combination 1, 16 rows of data for train and 4 rows of data for test are considered. In combination 2, 18 rows of data for train and 4 rows of data for test are considered. In combination 3, 7 rows of data for train and 2 rows of data for test are considered.

Step 8.1. Estimated relationships between the four inputs and three outputs by using ANN model as one of intelligent models. To find the optimum structure for each combination, 26 distinct ANN models are tested. Maximum number of neurons in the first hidden layer is set to 100. Each model is replicated 100 times to take care of possible bias or noise. The optimum architecture of the ANN-MLP models and their MAPE values for 3 combinations of demographic features are shown in table 1.

Table 1: Optimum architecture of the ANN-MLP models and their associated relative error (MAPE)

MAPE	Second transfer function	First transfer function	Number of neurons in first hidden layer	Learning method	Outputs	
0.0356	tansig	purelin	37	BFG	Effectiveness of ISO 18000	Com. 1
0.0740	tansig	purelin	10	GDA	Effectiveness of ISO 14000	
0.0101	Logsig	Logsig	55	SCG	Effectiveness of ISO 9000	
0.0314	Logsig	Logsig	14	CGB	Effectiveness of ISO 18000	Com. 2
0.0417	Logsig	Logsig	75	CGF	Effectiveness of ISO 14000	
0.0222	purelin	tansig	4	GDX	Effectiveness of ISO 9000	
0.0087	Logsig	Logsig	75	B	Effectiveness of ISO 18000	Com. 3
0.0226	purelin	tansig	20	LM	Effectiveness of ISO 14000	
0.0043	Logsig	Logsig	69	BR	Effectiveness of ISO 9000	

Step 8.2. Estimated relationships between the four inputs and outputs 1, 2 and 3 separately by using optimum ANFIS structure for three combinations are shown in table 2. This table presents the ANFIS architecture by four inputs and a single output.

Table 2: Optimum architecture of the ANFIS models and their associated relative error (MAPE)

Outputs		Input Layer		Output Layer	MAPE
		Number of MFs	MF Type	MF Type	
Com. 1	Effectiveness of ISO 18000	3,3,3,3	Triangular-shaped	Constant	21.21%
	Effectiveness of ISO 14000	4,4,4,4	Triangular-shaped	Constant	39.83%
	Effectiveness of ISO 9000	3,3,3,3	Triangular-shaped	Linear	42.96%
Com. 2	Effectiveness of ISO 18000	3,3,3,3	Triangular-shaped	Constant	12.05%
	Effectiveness of ISO 14000	3,3,3,3	Triangular-shaped	Constant	13.61%
	Effectiveness of ISO 9000	3,3,3,3	Triangular-shaped	Constant	31.15%
Com. 3	Effectiveness of ISO 18000	2,2,2,2	Triangular-shaped	Constant	7.79%
	Effectiveness of ISO 14000	3,3,3,3	Triangular-shaped	Constant	19.80%
	Effectiveness of ISO 9000	2,2,2,2	Triangular-shaped	Constant	20.54%

Step 8.3. Estimated relationships between the four inputs and three outputs by using linear, log-linear and quadratic for three combinations are shown in table 3. This table presents the results of MAPE values for these methods.

Table 3: The MAPE values for 3 combinations by use of conventional regressions

Outputs		Conventional Regression		
		Linear	Log-linear	Quadratic
Com. 1	1	30.63%	30.63%	67.26%
	2	24.04%	28.07%	45.32%
	3	171.01%	31.14%	45.81%
Com. 2	1	12.71%	83.25%	65.58%
	2	17.07%	17.07%	38.57%
	3	10.60%	10.60%	21.57%
Com. 3	1	31.24%	31.24%	78.91%
	2	14.96%	14.96%	24.40%
	3	8.00%	8.00%	15.14%

Step 8. Similar to ANN model in combination 1, 16 mixtures are used to find the best regression and ANFIS model and ISO systems efficiency is estimated for the remaining 4 mixtures. Also in combination 2, 18 mixtures are used to find the best regression and ANFIS model and ISO systems efficiency is estimated for the remaining 4 mixtures and in combination 3, 7 mixtures are used to find the best regression and ANFIS model and ISO systems efficiency is estimated for the remaining 2 mixtures. The results are compared with respect to mean absolute percentage error (MAPE). Clearly the ANN model provides better solutions (lower relative error) than conventional regression and ANFIS approaches.

Step 9. Therefore, the optimums ANN-MLP for every combination of demographic features and every output are selected for estimating the performance assessment of possible mixtures. To identify ANN outputs for each possible mixture, the preferred model is tested for all possible mixtures.

Steps 10 and 11: For each combination of demographic feature and each output calculate efficiency scores by use of best ANN method (ANN*). The results of these processes are shown in table 4. This table show the efficiency scores for all possible mixtures by the proposed algorithm. Moreover, all possible mixtures are ranked according to their efficiency.

Table 4: Efficiency scores for all possible mixtures by the proposed algorithm

Possible mixtures	Com. 1		Com. 2		Com. 3	
	Average of efficiency	Ranking	Average of efficiency	Ranking	Average of efficiency	Ranking
1	1	1	1	1	1	1
2	0.801	14	1	1	0.827	8
3	0.785	16	0.888	7	0.944	2
4	0.887	7	0.856	9	0.867	6
5	1	1	0.894	6	0.909	4
6	0.914	5	0.838	11	0.884	5
7	0.952	3	0.689	22	0.924	3
8	0.745	18	0.803	17	0.815	9
9	0.839	10	0.755	19	0.839	7
10	0.822	12	0.776	18		
11	0.598	20	0.863	8		
12	0.905	6	0.836	12		
13	0.919	4	0.734	20		
14	0.861	8	0.937	3		
15	0.725	19	0.934	4		
16	0.77	17	0.821	15		
17	0.817	13	0.898	5		
18	0.799	15	0.722	21		
19	0.845	9	0.825	14		
20	0.837	11	0.811	16		
21			0.834956	13		
22			0.847863	10		

108

Step 10. By regard to above tables we obtain following result:

- Combination 1: mixtures 1 and 5 that include operators by specification of age 18 to 25 years and education diploma for male and female gender are best performance between all possible mixtures of demographic features.
- Combination 2: mixtures 1 and 2 that include operators by specification of weight 45 to 55 kg and stature 140 to 170 cm for single marital status are best performance between all possible mixtures of demographic features.
- Combination 3: mixtures 1 that include operators by specification of age 18 to 25 years and work type fixed day for male gender are best performance between all possible mixtures of demographic features.

5   CONCLUSION

A highly unique flexible ANN algorithm was proposed to measure and rank the mixtures demographic features scores with respect to integrated HSEE-ISO systems. The proposed algorithm is ideal because of nonlinearity of ANN in addition to its universal approximations of functions and its derivates, which makes them highly flexible. Moreover, HSEE factors were considered as input variables and effectiveness of ISO systems was considered as output variables. The proposed algorithm is composed of 10 distinct steps. To show its applicability and superiority it was applied to employers of different shifts in a gas transmission refinery. The efficiency score between 0 and 1 was devised to show effectiveness of ISO systems with respect to the performance of HSEE programs between different mixture demographic features. The mixture demographic features were then ranked by the proposed ANN approach which is capable of handling data complexity because it is composed neural network simulation approach. For 3 combination of demographic feature are obtained best mixture with efficiency scores 1. Selections of combination demographic features are optional. Managers for recruitment of employers can select different combination of demographic feature and by use of this algorithm select employers with best efficiency ground integrated HSEE-ISO systems.

Acknowledgement: The authors would like to acknowledge the financial support of Gas Transmission Refinery (Iran) and aids of Mr. Seyfi for this research.

REFERENCES

Athanassopoulos, A.D. and Curram, S.P. (1996). A comparison of data envelopment analysis and artificial neural networks as tool for assessing the efficiency of decision-making units, Journal of the Operational Research Society, 47 (8), 1000–1016.

Azadeh, A., Mohammad Fam, I. and Azadeh, M.A. (2009). Integrated HSEE Management Systems for Industry: A Case Study in Gas Refinery, Journal of the Chinese Institute of Engineers, 32 (2).

Azadeh, A., Fam, I.M., Khoshnoud, M. and Nikafrouz, M. (2008a). Design and implementation of a fuzzyexpert system for performance assessment of an integrated health, safety, environment (HSE) and ergonomics system: The case of a gas refinery, Information Sciences, 178, 4280–4300.

Azadeh, A., Mohammad Fam, I. and Nouri, M.A. (2008b). Integrated health, safety, environment and ergonomics management system (HSE-MS): An efficient substitution for conventional HSE-MS, Journal of Scientific and Industrial Research, 67, 403-411.

Azadeh, A., Ghaderi, S.F., Anvari, M. and Saberi, M. (2007). Performance assessment of electric power generations using an adaptive neural network algorithm, Energy Policy, 35, 3155–3166.

Azadeh, A., Keramati, A., Mohammad Fam, I. and Jamshidnejad, B. (2006a). Enhancing the availability and reliability of power plants through macroergonomics approach, Journal of Scientific and Industrial research, 65, 873-878.

Azadeh, A., Nouri, J. and Mohammad Fam, I. (2005). The impacts of macroergonomics on environmental protection and human performance in power plants, Iranian j env health sci eng, 2 (1), 60-66.

Bellamy, L.J., Geyer, T.A.W. and Wilkinson, J. (2008). Development of a functional model which integrates human factors, safety management systems and wider organisational issues, Safety Science, 46 (3), 461-492.

Blanning, R.W. (1984). Management applications of expert systems, Information and Management, 7, 311–316.

Bryden, R. and Hudson, P.T.W. (2005). Because we want, Safety and Health Practitioner, 23, 51–54.

Champoux, D. and Brun, J.J. (2003). Occupational health and safety management in small size enterprises: an overview of the situation and avenues for intervention and research, Safety Science, 41, 301–318.

Delgado, F.J. (2005). Measuring efficiency with neural networks, an application to the public sector, Economics Bulletin, 3 (15), 1–10.

Duijm, N.J., Fiévez, C., écile Gerbec, M., Hauptmanns, U. and Konstandinidou, M. (2008). Management of health, safety and environment in process industry, Safety Science, 46 (6), 908-920.

Flynn, B.B, Schroeder, R.G. and Sakakibara, S. (1994). A framework for quality management research and an associated measurement instrument. J. Op. Mgmt,11, 339-366.

Mohammad Fam, I., Azadeh, A., Faridan, A.M. and Mahjoub, H. (2008). Safety Behaviours Assessment in Process Industry: A Case in Gas Refinery, of the Chinese Institute of Industrial Engineers, 25 (4), 298-305.

Nachreiner, F., Nickel, P. and Meyer, I. (2006). Human factors in process control systems: The design of human–machine interfaces, Safety Science, 44 (1), 5-26.

Nunnally, J.C. and Bernstein, I.H. (1994).Psychometric Theory, 3, 214–286, (McGraw Hill: New York).

Olson, E.M., Slater, S.F. and Hult, G.T.M. (2005). The performance implication of fit among business strategy, marketing organization structure, and strategic behavior. J. Marketing,69, 49–65.

Shikdar, A.A. and Sawaqed, M.N. (2004). Occupational health and safety in the oil industry: a managers' response, Computers and Industrial Engineering, Ergonomics, 47 ,223–232.

Torp, S. and Moen, B.E. (2006). The effects of occupational health and safety management on work environment and health: A prospective study, Applied Ergonomics, 37 (6), 775-783.

CHAPTER 12

Hybrid Meta-heuristic Based Occupational Health Management System for Indian Workers Exposed to Risk of Heat Stress

Yogesh K Anand[1], Sanjay Srivastava[2], Kamal Srivastava[3]

[1,2]Department of Mechanical Engineering, Dayalbagh Educational Institute, India
[3]Department of Mathematics, Dayalbagh Educational Institute, India
[1]ykanand@dei.ac.in, [2]ssrivastava@dei.ac.in, [3]kamal.sri@dei.ac.in

ABSTRACT

The present work is carried out in a brick manufacturing (BM) unit near Hathras, India. BM is labor intensive in general and comprises the following major jobs – molding, loading, stacking, covering, firing and unloading. Firing, the most severe job, involves undue exposure of workers to excessive heat. Moreover, they subject themselves to extreme work conditions by working extra hours to maximize their earnings due to economic reasons, and hence are exposed to greater risk of heat stress. To manage the risk of heat stress, we implement a job-combination approach wherein firing workers do another job (molding in this work) along with firing job. We measure the risk of heat stress in terms of composite discomfort score (CDS), computed using factor rating, a method popularly used in location planning decisions. CDS is a direct indicative of perceived discomfort level of workers. Further, we employ hybrid meta-heuristic (HMH), an evolutionary multi-objective optimization (EMO) technique, to search for optimal CDS-earning trade-off (CET) solutions with two conflicting objectives, viz. minimization of CDS, and maximization of earnings.

Keywords: Occupational health hazard, Evolutionary multi-objective optimization, Brick manufacturing, Hybrid meta-heuristic

1   INTRODUCTION

An industrial job is considered having a risk of occupational health hazard (OHH) if job-demands exceed human capabilities. Greater departures from the capabilities are not uncommon in a developing country wherein labor is available at lesser cost. Workers in BM units, in general, aim to maximize their earnings by subjecting themselves to extreme work conditions due to economic reasons. The proposed work introduces an occupational health management system to reduce the risk of heat stress, a well known OHH, in a brick manufacturing unit. Firing, the most severe job among the other jobs, involves undue exposure of workers to excessive heat. Combining jobs is found to be a way of reducing OHH and yet maintaining the good earnings (Srivastava et al., 2010 a). We, therefore, implement a job-combination approach wherein the firing workers perform molding job along with firing job thereby reducing their exposure to high temperature zone while maintaining their earning to a satisfactory level. Similar to firing workers, molding workers go for firing work partially in a job-combination approach. Firing, and molding jobs require special skills, therefore it is found essential to train a set of workers of a BM unit to perform these two jobs with reasonable skills.

Firing workers pour the coal inside the kiln by opening the holes at the top partially as and when required to ensure the proper baking of bricks. Their exposure to high temperature zone during this process causes risk of heat stress. Heat stress is a well known OHH in many industries including BM units. Current guidelines define working environment that cause an increase above $38°C$ as potentially hazardous (ACGIH-2004). It is essential to assess the thermal environment of a workplace with good reliability (d'Ambrosio Alfano et al., 2011). Hot conditions give rise to physiological heat strain (Candi et al., 2008), and cognitive decrements (Ftaiti et al., 2010). In general, heat stress decreases workers' performance significantly (Hancock et al., 2007).

We assess the risk of OHH in terms of composite discomfort score (CDS) of workers for a given job-combination. CDS is a direct indicative of perceived discomfort level of workers. We make the following observations using interview method: (1) workers maximize their earnings by subjecting themselves to extreme work conditions, and are exposed to greater risk of OHH; (2) three factors are identified which influence CDS of a given job, viz. number of working hours (WH), rest break time in minutes (RB), and number of rest breaks (NRB). Many studies have shown the impact of long working hours on workers' performance (Dorrian et al., 2011). Helander and Quanc (1990) have investigated the influence of rest breaks in reducing the amount of spinal shrinkage while establishing a relationship between duration and frequency of rest intervals (referred to as RB and NRB respectively in this paper) with spinal shrinkage. Faucett et al. (2007) have shown the importance of frequent, brief rest breaks in improving symptoms for workers engaged in strenuous work tasks.

CDS-earning trade-off (CET) belongs to a class of multi-objective optimization (MOO) problem. There is no single optimum solution in MOO rather there exists a number of solutions which are all optimal — Pareto-optimal solutions — optimal

CET solutions in occupational health literature. The curve joining nondominated CET solution points is termed as CET profile. There are six factors in a given job-combination which assume discrete values in real-life situations. Evolutionary algorithms (EAs) are meta-heuristics that are able to search large regions of the solution's space without being trapped in local optima (Dimopoulos and Zalzala, 2000). EMO techniques encompass different versions of multi-objective GAs including HMH.

2   METHODOLOGY

Twenty male workers (mean age: 26 years, average height: 163 cm, average weight: 52 Kg, average job experience: 4 years) are taken for the study. None of these workers report a history of chronic health problems. Specifically, these workers are trained to perform firing job along with molding with predefined working hours (WH) distribution. The risk of OHH is evaluated based on the CDS using factor rating (FR) method (Anand et al., 2010).

1.   Upper and lower bounds of WH, RB and NRB, along with values in between for the job under consideration are illustrated below.

$$WH_1 \in \{2, 3, 4, 5, 6, 7, 8, 9, 10\}$$
$$WH_2 \in \{12 - WH_1\}$$

The range of RB and NRB are same for each job under consideration.

$$RB_1 \in \{5, 10, 15, 20, 25, 30, 35, 40\}$$
$$NRB_2 \in \{1, 2, 3, 4, 5, 6\}$$

2.   For firing-molding job-combination we assign an average weight to each factor (Table 1).

Table 1 Weight assigned to each factors for Firing-molding job-combination

Factor	WH_1	RB_1	NRB_1	WH_2	RB_2	NRB_2
Weight (w)	0.40	0.06	0.08	0.21	0.11	0.14

An interview method is adopted to translate perception and opinion of workers and supervisors into the numeric value of weights. Typically weights sum to 1.0. These weights further verify the perceived severity of each job with firing being most severe job and molding being least severe job.

3.   The value of each factor is normalized in computing CDS by dividing it by its maximum value, yielding ratios ranging between 0.0 and 1.0.

4.   The normalized value of each factor is multiplied with its weight, weighted values are summed up together algebraically and the sum is multiplied by 100 to suitably scale the values. We term the result so obtained as CDS. General expression for CDS is shown below.

$$CDS_{1-2} = \left[\left\{\left(\frac{WH_1}{WH_{max}}\right) \times w(WH_1) - \left(\frac{RB_1}{RB_{max}}\right) \times w(RB_1) - \left(\frac{NRB_1}{NRB_{max}}\right)\right.\right.$$
$$\left. \times w(NRB_1)\right\}$$
$$+\left\{\left(\frac{WH_2}{WH_{max}}\right) \times w(WH_2) - \left(\frac{RB_2}{RB_{max}}\right) \times w(RB_2) - \left(\frac{NRB_2}{NRB_{max}}\right)\right.$$
$$\left.\left. \times w(NRB_2)\right\}\right] \times 100$$

where, $w(WH_1), w(RB_1), w(NRB_1), w(WH_2), w(RB_2)$ and $w(NRB_2)$ refer to average weights assigned to the respective factors mentioned in the bracket.

WH_1 and WH_2 contribute positively to CDS, whereas higher values of RB_1, RB_2, NRB_1, and NRB_2 would cause a decrease in CDS. Higher weights to job#1 in comparison to job#2 is attributed to the fact that job#1 (i.e. firing work) is more severe than job#2 (i.e. molding). We now illustrate an extreme case of worker's discomfort using CDS.

We categorize range of CDS values into one of the linguistic values of PDL (Table 2).

Table 2 Relation between CDS and PDL

CDS (Range a-b)*	PDL	Abbreviation
1-8	Negligible	N
8-16	Very Low	VL
16-25	Low	L
25-34	Moderate	M
34-43	High	H
43-51	Very High	VH
51 and above	Beyond Tolerance	BT

*Range a-b indicates $a \leq CDS < b$

Earnings of firing worker and molding worker are INR 23.15/hour and INR 12.50/hour respectively. We find it lucid to deduct an amount equivalent to worker's total rest break time from his earnings/day. ER_{1-2}/day is illustrated below for firing-molding job-combination

$$ER_{1-2}/\text{Day} = \left\{WH_1 - \frac{(RB_1 \times NRB_1)}{60}\right\} \times ER_1 + \left\{WH_2 - \frac{(RB_2 \times NRB_2)}{60}\right\}$$
$$\times ER_2$$

Now we formally define the CET problem for the firing-molding job-combination below.

Min CDS_{i-j}

Max ER_{i-j}/day

Subject to

$WH_i \in \{2, 3, .., 10\}$ $(i = 1)$

$$WH_j \in 12 - WH_i \qquad (j = 2)$$
$$RB_i \in \{5, 10, 15, 20, 25, 30, 35, 40\} \quad (i = 1, 2)$$
$$NRB_i \in \{1, 2, 3, 4, 5, 6\} \qquad\qquad (i = 1, 2)$$
$$NRB_i \leq \begin{cases} \delta_1, & if \ 7 \leq WH_i \leq 10 \\ \delta_2, & if \ 3 \leq WH_i < 7 \\ 1, & if \ WH_i < 3 \end{cases}$$
$$where \ \delta_1 = \min \{6, \ (WH_i - 4)\} \ and \ \ \delta_2 = \{3, \ (WH_i - 2)\}$$

3   HYBRID META-HEURISTIC FOR CET

We employ a hybrid meta-heuristic (HMH) technique for solving multi-objective discrete CET problem. HMH hybridizes a multi-objective GA with simulated annealing. Though HMH is unconventional in terms of its working (fitness function evaluation etc.) in comparison to existing multi-objective evolutionary algorithms (MOEAs) comprehended in (Deb, 2001), yet it suits well to our problem of searching the optimal CET profile. HMH embeds simulated annealing in GA to deciding the number of children to be generated from the parents of next generation. The preliminaries & definitions to understand HMH are explained below:

3.1   Structure of a solution and initial population

A solution here is a string which represents an instance $\theta = \{g_t\}_{t = 1, \dots, 5}$ of the work schedule; each element g_t of a 5-tuple string, θ, can assume any value from the sets $WH_1, RB_1, NRB_1, RB_2, NRB_2$ respectively for $i = 1, 2, \dots 5$. The associated CDS and ER/day of an instance θ (denoted CDS_θ and ER_θ respectively) are determined in the manner as described earlier in section 2. The initial population is generated by randomly selecting n_p individual strings from the feasible search space, i.e., each g_t of a string is chosen randomly from the sets mentioned in section 2.

3.2   CET profile, convex hull and distance measurement

Let θ_1 and θ_2 be two strings in a population F, θ_1 dominates θ_2 if $CDS_{\theta_1} < CDS_{\theta_2}$ and $ER_{\theta_1} > ER_{\theta_2}$. Let D be a binary relation defined on the set F by $D = \{(\theta_1, \theta_2) / \theta_1, \theta_2 \in F \wedge \theta_1 \text{ dominates } \theta_2\}$, then the non-dominating set NDS is given by $NDS = \{\theta_i \in F / (\theta_j, \theta_i) \notin D \ \forall j, \ j \neq i\}$ i.e. it represents the strings (solutions) of F which are not dominated by any other string of F. The pairs (CDS_θ, ER_θ) are represented on a Cartesian plane where CDS_θ and ER_θ are taken on X and Y axis respectively. We refer to the pair (CDS_θ, ER_θ) as the objective point corresponding to the solution θ. The objective points (CDS_θ, ER_θ) are joined by line segments. The curve formed by joining these solutions is referred to as CET profile and the solutions as the trade-off points in the context of occupational health literature. We define a convex boundary that encloses all members of a population from above. This boundary is in the form of straight line segments. The purpose of drawing a convex boundary for each population is to evaluate the fitness of each individual in the population (Feng et al., 1997). Then distance d_w of an individual objective point in a population is determined (Srivastava et al., 2010 b).

3.3   Crossover and Mutation

We design the crossover operator for the problem as follows: Let θ_1 and θ_2 are two parents selected for crossover randomly and $C\theta_1$ and $C\theta_2$ denote the children which will be produced by the crossover where $C\theta_1 = \{g_{1i}\}_{i=1, \ldots, 5}$ and $C\theta_2 = \{g_{2i}\}_{i=1, \ldots, 5}$.First copy θ_1 to $C\theta_1$ and θ_2 to $C\theta_2$. Now interchange the first elements i.e. $C\theta_1$ gets g_{11} and $C\theta_2$ gets g_{21}. If all the remaining elements satisfy the constraints then no other updation is done in $C\theta_1$ and $C\theta_2$. If any of the elements violate the constraints then new values are randomly selected from the sets described in section II. This process is repeated for both $C\theta_1$ and $C\theta_2$.

Mutation is performed on randomly selected $r_m \times n_c$ number of solutions of the child population. Let θ be a solution selected for mutation; g_1 is replaced by $(12-g_1)$ and corresponding to this value of g_1 if the rest of the elements violate the constraints, they are regenerated else they are retained. Srivastava et al. (2010 b) have given a detailed description of HMH in solving time-cost trade-off problem in project scheduling.

3.4   Pseudocode of HMH

A step-wise pseudocode of HMH follows.

Step 1: Set initial temperature $temp = temp_o$, no_improve_iter.
    Set n_p, n_c, cooling ratio α
    Set $Gen = 1$
Step 2: Select n_p parents;
Step 3: For $u = 1$ to n_p do $child\_num(u) = n_c/n_p$, and $par\_num(u)=1$
Step 4: Generate n_c children from the parents, with $parent(u)$ producing $child\_num(u)$ children. This creates n_p families consisting of parents and their corresponding children
Step 5: Determine the CET profile and convex boundary of the existing generation i.e. $\sum_{u=1}^{n_p} child\_num(u) + par\_num(u)$ strings constitute a generation
For each $family(u)$, $u = 1, \ldots, n_p$, do steps 6 to 8
Step 6: Find the number of members appearing in NDS, i.e., $par\_num(u)$. These members become the parents for the next generation.
Step 7: For each member(w), $w = 1, \ldots, par\_num(u)+child\_num(u)$, d_w is computed as defined by the distance function $d_w^u = min_{\forall v}(d_{wv}^u)$, where d_{wv} is the distance of wth member from the vth line segment of the convex boundary, u indicates the family to which a member belongs.
Step 8: Determine the number of children $child\_num(u)$ that will be generated by the family in the next generation, as detailed out in Procedure $find\_num$.

116

Step 9: Parents (those mentioned in step 6) produce *child_num(u)* children by crossing over randomly with the others members of NDS. If no member of a family appears in NDS then the new parent for this family is decided as follows: The member having least distance (d_w^u) among all the members of the family becomes the parent for the next generation.

Step 10: Apply mutation.

Step 11: *Gen = Gen* + 1;

Step 12: *temp = temp* × *a*

Step 13: Repeat steps 6 to 12 until *Gen ≥ Max_iter* or the CET profile remains identical for improve_iter number of generations.

Procedure *find_num*( )

Step 1: sum = 0;

Step 2: for *u* = 1 to n_p do *accept(u)* = 0;
 Repeat step 3 to 6 for each *family(u)*.

Step 3: Repeat step 4 for each member of the family.

Step 4: If the member does not belong to NDS,
 then if $(\exp(-d_w^u / temp) > \rho)$
 accept(u) = *accept(u)* + 1;
 end

Step 5: sum = sum + *accept(u)* + *par_num(u)*;

Step 6: for *u* = 1 to n_p do *child_num(u)* = $(n_c \times accept(u))$/sum

4 RESULTS AND DISCUSSION

HMH is coded in MATLAB 7.0 and run on Pentium (R)-based HP Intel (R) computer with 1.73 GHz Processor and 512 MB of RAM. The crossover rate and the mutation rate are kept as 1.0 and 0.05 respectively. The population size is chosen as 20. Computational experiments are performed to decide these parameters on the basis of faster convergence criteria. The search is set to terminate when nondominated CET profile remains unchanged for three consecutive iterations — a number is suitably decided based on extensive experiments.

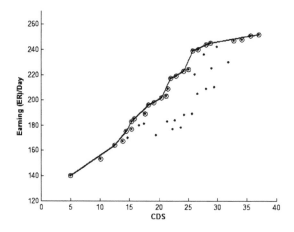

Figure 1 Firing-molding job-combination: CET profile

It takes on an average ten iterations for HMH to search for the best possible CET profile. Figure 1 depicts the results for firing-molding job-combination. Firing-molding is an interesting and useful job-combination wherein a severe job is combined with a less severe job.

Table 3 CET solution points: firing-molding job-combination

Solution Points	WH_1	RB_1	NRB_1	WH_2	RB_2	NRB_2	ER/Day	CDS	PDL
1	3	40	1	9	25	5	140	5.03	N
2	3	10	1	9	40	3	153	10.07	VL
3	3	25	1	9	40	1	164	12.48	VL
4	4	25	1	8	25	3	167	13.84	VL
5	4	25	1	8	40	1	175	14.38	VL
6	4	10	2	8	40	1	177	15.30	VL
7	6	40	1	6	25	3	183	15.39	VL
8	6	10	1	6	40	3	185	15.77	VL
9	6	25	1	6	25	3	189	17.64	L
10	6	25	1	6	40	1	196	18.18	L
11	6	10	2	6	40	1	198	19.10	L
12	6	10	1	6	40	1	202	20.43	L
13	6	5	2	6	35	1	203	21.23	L
14	8	40	1	4	25	2	209	21.53	L
15	8	25	1	4	40	1	217	21.98	L
16	8	10	2	4	40	1	219	22.90	L
17	8	10	1	4	40	1	223	24.23	L
18	8	5	2	4	35	1	224	25.03	M
19	10	25	1	2	40	1	239	25.78	M
20	10	10	2	2	40	1	240	26.70	M
21	10	10	1	2	40	1	244	28.03	M
22	10	5	2	2	35	1	245	28.83	M
23	10	15	1	2	20	1	247	32.78	M
24	10	15	1	2	15	1	248	34.16	H
25	10	5	1	2	15	1	251	35.66	H
26	10	5	1	2	10	1	252	37.03	H

Workers are in safer zone till *PDL* is *moderate*. Beyond this they fall into higher risk zones as *PDL* assumes values *high, very high,* and *extremely high*. Table 3 shows the solution points for firing-molding job-combination. *PDL* is moderate till 23rd solution point and workers fall into the unsafe zone from 24th solution point onwards. CET solutions provide a huge flexibility to workers and supervisors in terms of choosing the working hours and rest breaks. We present the comparison of two solution points (19th and 20th both having *moderate PDL* and minor differences in terms of *CDS* and *ER*/Day) below to illustrate the flexibility with respect to *RB* and *NRB* of firing job.

SP	WH_1	RB_1	NRB_i	WH_3	RB_3	NRB_3	ER	CDS
19th	10	25	1	2	40	1	239	25.78
20th	10	10	2	2	40	1	240	26.75

Workers preferring more *RB* over *NRB* for firing job have a choice to opt for 19th solution point, which provides a single rest break of 25 minutes for 10 working hours. Whereas those favoring *NRB* over *RB* can go for 20th solution point, which offers two rest breaks each of 10 minutes. Further the safer zone limits the earnings to INR 247/day (refer to 23rd solution point). If at all a worker is interested to earn more, he or she will have to move to the risky zones. Captivatingly, our proposed system provides the best possible earnings to a worker even in the risk zones. For example say in an extreme case, a worker may choose the last solution point i.e. 26th solution point — this would correspond to the highest possible earnings of INR 252/day against a *CDS* of 37.03 corresponding to a high *PDL*.

5   CONCLUSIONS

The proposed system acts as an advisor to a worker to choose a job-combination and the corresponding values of *WH, RB,* & *NRB* to decide his/her occupational risks and earnings suitably. Job-combination approach ensures that workers' earnings are not compromised to a greater extent. HMH searches for the optimal *CDS*-earning trade-off profile, and it does not place any restriction on the form of inputs to evaluate *CDS* and earnings for firing-molding job-combination. The unifying system amalgamating *CDS* with factor rating and HMH with job-combination approach in a unique way turns out to be a powerful and efficient scheme. Top level management in a BM unit faces the problems of monopoly of firing workers. The system presented in this work alleviates this problem intelligently — job-combination approach make other workers getting trained for firing work. In fact the system will help in 'work generalization' to take over 'work specialization'. Therefore, it is beneficial to both the parties — managers as well as workers. The proposed solutions provide a wider flexibility to a manager to arrange to complete the jobs in the BM units. The proposed system is general enough to be applied to any labor intensive industrial unit, therefore, as part of future work, we intend to employ it to solve similar problems of other industrial units of nearby region.

ACKNOWLEDGMENTS

This work is supported by UGC, New Delhi (India) under Grant F. No. 36-65/2008 (SR), dated 24/03/2009.

REFERENCES

American Conference of Governmental Industrial Hygienists (ACGIH), 2004. Threshold limit values for chemical substances and physical agent and biological exposure indices. *American Conference of Governmental Industrial Hygienists*, Cincinnati, OH.

Anand, Y. K., S. Srivastava, and K. Srivastava. 2010. Optimizing the Risk of Occupational Health Hazard in a Multiobjective Decision Environment using NSGA-II. In: *Lecture Notes in Computer Science*, eds. K. Deb et al. Springer-Verlag Berlin Heidelberg, pp. 476-484.

Candi, D. A., L. L. Christina, and S. S. Skai, et al. 2008. Heat strain at the critical WBGT and the effects of gender, clothing and metabolic rate. *International Journal of Industrial Ergonomics* 38: 640-644.

d'Ambrosio Alfano F. R., B. I. Palella, and G. Riccio. 2011. Thermal Environment Assessment Reliability Using Temperature-Humidity Indices. *Industrial Health* 49: 95–106.

Deb, K. 2001. *Multiobjective optimization using evolutionary algorithms*. Chichester, Wiley, UK.

Dimopoulos, C., and M. S. Zalzala. 2000. Recent developments in evolutionary computation for manufacturing optimization: problems, solutions and comparisons. *IEEE Transaction on Evolutionary Computing* 4: 93-113.

Dorrian, J., S. D. Baulk, and D. Dawson. 2011. Work hours, workload, sleep and fatigue in Australian Rail Industry employees. *Applied Ergonomics* 42(2): 202-209.

Faucett, J., J. Meyers, and J. Miles, et al. 2007. Rest break interventions in stoop labor tasks. *Applied Ergonomics* 38(2): 219-226.

Feng, C. W., L. Liu, and A. Burns. 1997. Using genetic algorithms to solve construction time-cost trade-off problems. *Journal of Computer in Civil Engineering* 11: 184-189.

Ftaiti, F., A. Kacem, and N. Jaidane, et al. 2010. Changes in EEG activity before and after exhaustive exercise in sedentary women in neutral and hot environments. *Applied Ergonomics* 41(6): 806-811.

Hancock, P. A., J. M. Ross, and J. L Szalma. 2007. A Meta-Analysis of Performance Response Under Thermal Stressors. *Human Factors* 49: 851-877.

Helander, M. G., and L. A. Quanc. 1990. Effect of work—rest schedules on spinal shrinkage in the sedentary worker. *Applied Ergonomics* 21: 279-284.

Srivastava, S., Y. K Anand, and V. Soamidas. 2010 a. Reducing the Risk of Heat Stress Using Artificial Neural Networks Based Job-Combination Approach, in: *Proceeding of IEEE International Conference on Industrial Engineering and Engineering Management* (IEEM 2010), Macau, pp. 542-546.

Srivastava S, B. Pathak, and K. Srivastava. 2010 b. Project Scheduling: Time-cost tradeoff problems. In: Computational Intelligence in Optimization-Applications and

Implementations, eds. Y. Tenne, and C. K. Goh. Springer-Verlag Berlin Heidelberg, pp. 325-357.

CHAPTER 13

Adaptive Assembly Planning for a Nondeterministic Domain

Daniel Ewert*, Marcel Mayer[†], Daniel Schilberg*, and Sabina Jeschke*

*Institute of Information Management in Mechanical Engineering
[†]Institute of Industrial Engineering and Ergonomics
RWTH Aachen University, Germany
e-mail1: jeschke.office@ima.rwth-aachen.de

ABSTRACT

In this paper, we present a hybrid approach to automatic assembly planning, where all computational intensive tasks are executed once prior to the actual assembly by an Offline Planner component. The result serves as basis of decision-making for the Online Planner component, which adapts planning to the actual situation and unforeseen events. Due to the separation into offline and online planner, this approach allows for detailed planning as well as fast computation during the assembly, therefore enabling appropriate assembly duration even in nondeterministic environments. We present simulation results of the planner and detail the resulting planner's behavior.

Keywords: Self optimization, Assembly planning, Cognitive production systems

1   INTRODUCTION

1.1   Motivation

The industry of high-wage countries is confronted with the shifting of production to low-wage countries. To slow down this development, and to answer the trend towards shortening product life-cycles and changing customer demands regarding individualized and variant-rich products, new concepts for the production in high-wage countries have to be created. This challenge is addressed by the

122

Cluster of Excellence "Integrative production technology for high-wage countries" at the RWTH Aachen University. It researches on sustainable technologies and strategies on the basis of the so-called polylemma of production (Brecher, C. et al. 2007). This polylemma is spread between two dichotomies: First between scale (mass production with limited product range) and scope (small series production of a large variety of products), and second between value and planning orientation. The ICD) "Self-optimizing Production Systems" focusses on the reduction of the latter dichotomy. It`s approach for the reduction of this polylemma is to automate the planning processes that precede the actual production. This results in a reduction of planning costs and ramp-up time and secondly it allows to switch between the production of different products or variants of a product , hence enabling more adaptive production strategies compared to current production. Automatic replanning also allows to react to unforeseen changes within the production system, e.g. malfunction of machines, lack of materials or similar, and to adapt the production in time. In this paper we present the planning components of a cognitive control unit (CCU) which is capable to autonomously plan and execute a product assembly by relying entirely on a CAD description of the desired product.

Use case description

The CCU is developed along a use case scenario for an assembly task in a nondeterministic production environment (Kempf, T., W. Herfs, and C. Brecher. 2008). This scenario is based on the robot cell depicted in Figure 1.

Of the two robots of the robot cell, only Robot2 is controlled by the CCU. Robot1 independently delivers parts in unpredictable sequence to the circulating conveyor belt. The parts are then transported into the grasp range of Robot2 who then can decide to pick them up, to immediately install them or to park them in the buffer area.

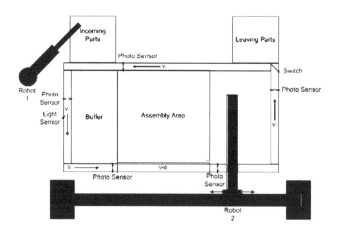

Figure 1 Schematic of the robot cell

The scenario also incorporates human-machine cooperation. In case of failure, or if the robot cannot execute a certain assembly action, the CCU is able to ask a human operator for assistance. To improve the cooperation between the operator and the machine, the operator must be able to understand the behavior and the intentions of the robot (Mayer, M. et al. 2009). Therefore, machine transparency is a further major aspect in our concept.

The only sources of information to guide the decision making of the CCU are a CAD description of the desired product, the number and types of single parts currently in the buffer and on the conveyor belt and the current state of the assembly within the Assembly Area. The planner is evaluated with the figures described in Figure 2. The pyramid construct (a) serves here as a benchmark for the computational complexity of our planning approach and has been used in different sizes (base areas of 2x2, 3x3, and 4x4 blocks). Construct b) and c) are used to demonstrate the planner's behavior.

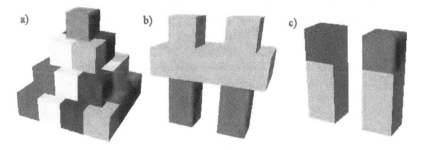

Figure 2 Toy model products for planner evaluation

2   RELATED WORK

In the field of artificial intelligence planning is of great interest. There exist many different approaches to planning suitable for different applications. Hoffmann developed the FF planner, which is suitable to derive action sequences for given problems in deterministic domains (Hoffmann, J. 2001). Other planners are capable to deal with uncertainty (Hoffmann, J., and R. Brafman. 2005),(Castellini, C., E. Giunchiglia, A. Tacchella, and O. Tacchella. 2001) . However, all these planners rely on a symbolic representation based on logic. The corresponding representations of geometric relations between objects and their transformations, which are needed for assembly planning, become very complex even for small tasks. As a result, these generic planners fail to compute any solution within acceptable time.

Other planners have been designed especially for assembly planning and work directly on geometric data to derive action sequences. A widely used approach is the Archimedes system by Kaufman et al (Kaufman, S.G. et al. 1996) that uses And/Or-Graphs and an "Assembly by Disassembly" strategy to find optimal plans. U. Thomas (Thomas, U. 2009) follows this strategy, too, but where the Archimedes system relies on additional operator-provided data to find feasible subassemblies, Thomas uses only the geometric information about the final product as input. However, both approaches are not capable of dealing with uncertainty.

Other products for assembly planning focus on assisting product engineers set up assembly processes. One example is the tool Tecnomatix from Siemens (Tecnomatix. 2011), which assists in simulating assembly steps, validates the feasibility of assembly actions etc. All of the mentioned works do not cover online adaption of assembly plans to react on changes in the environment. One exception is the system realized by Zaeh et al (Zaeh, M., and M. Wiesbeck. 2008), which is used to guide workers through an assembly process. Dependent on the actions executed by the worker, the system adapts its internal planning an suggests new actions to be carried out by the worker. The CCU uses the same technique for plan adaption.

3   AUTONOMOUS ASSEMBLY PLANNING

3.1   Hybrid Assembly Planning

The overall task of the CCU is to realize the autonomous assembly in a nondeterministic environment: Parts are delivered to the robot cell in random sequence and the successful outcome of an invoked assembly action cannot be guaranteed. While assembly planning is already hard even for deterministic environments where all parts for the assembly are available or arrive in a given sequence (Thomas, U. 2009), the situation becomes worse for this unpredictable situation. One approach to solve the nondeterministic planning problem would be to plan ahead for all situations: Prior to the assembly all plans for all possible arrival sequences are computed. However, this strategy soon becomes unfeasible: A product consisting of n parts allows for n! different arrival sequences, so a product consisting of 10 parts would already result in the need to compute more than 3.6 million plans. Another approach would be to replan during the assembly every time an unexpected change occurs in the environment. This strategy, however, leads to unacceptable delays within the production process.

Therefore, our approach follows a hybrid strategy. All computational intensive tasks are executed once before the actual assembly. This is done by an Offline Planner component. The results of this step serve as basis of decision-making for the Online Planner component, which adapts planning to the actual situation and unforeseen events. Due to this separation, our approach (see Figure 3) allows for detailed planning as well as fast computation during the assembly, therefore enabling appropriate assembly duration even in nondeterministic environments. The Offline Planner contains a CAD Parser which derives the geometric properties. The currently supported format is STEP (Röhrdanz et al. 1996). This data is then processed by the graph generator. The details of this process are explained in section 3.2. The Online Planner consists of the components Graph Analyzer, Parallelizer and Cognitive Control, which are detailed in section 3.3.

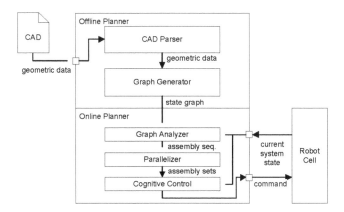

Figure 3 System overview of the hybrid approach

3.2   Offline: Graph Generation

The Offline Planner receives a CAD description of the desired final product. From this input it derives the relations between the single parts of the product via geometrical analysis as described in 0. The results are stored in a connection graph. Assembly sequences are now derived using an assembly-by-disassembly strategy: Based on the connection graph, all possible separations of the product into two parts are computed. The feasibility of those separations is then verified using collision detection techniques. Unfeasible separations are discarded. The remaining separations can then be evaluated regarding certain criteria as stability, accordance to assembly strategies of human operators or similar. The result of this evaluation is stored as a score for each separation. This separation is recursively continued until only single parts remain. The separation steps are stored in an and/or graph (Homem de Mello, L. S., and A. C. Sanderson. 1986) , which is then converted into a state graph as displayed in Figure 4 using the method described in Ewert D., D. Schilberg, and S. Jeschke. 2011. Here nodes represent subassemblies of the assembly. Edges connecting two such nodes represent the corresponding assembly action which transforms one state into the other. Each action has associated costs, which depend on the type of action, duration, etc. Also, each edge optionally stores information about single additional parts that are needed to transform the outgoing state into the incoming state.

The graph generation process has huge computational requirements for time as well for space. Table 1 shows the properties of resulting state graphs for different products. The results show the extreme growth of the graph regarding the number of parts necessary for the given product. However, as can be seen when comparing the state graphs of both constructs with 14 parts, the shape of a product affects the graph, too: The more possible independent parts are from each other, the more different assembly sequences are feasible. Therefore the graph of the construct a) with 14 parts has almost twice the size of the state graph resulting from construct b).

126

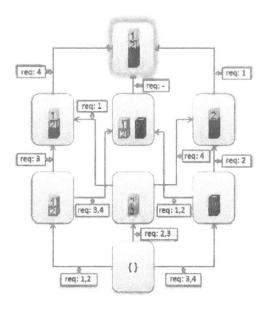

Figure 4 State graph representation of the assembly of a four blocks tower

Table 1 State graph properties for different products

Product	#Parts	#Nodes of Graph	#Edges of Graph
construct a) (size 2x2)	5	17	33
construct c)	6	16	24
construct b)	14	361	1330
construct a) (size 3x3)	14	690	2921
construct a) (size 4x4)	30	141,120	1,038,301

3.3   Online: Graph Analysis

The state graph generated by the Offline Planner is then used by the Online Planner to derive decisions which assembly actions are to be executed given the current situation of an assembly. The Online Planner therefore executes the following process iteratively until the desired product has been assembled: The Graph Analyzer perceives the current situation of the assembly and identifies the corresponding node of the state graph. In earlier publications (Ewert D., D. Schilberg, and S. Jeschke. 2011) we suggested an update phase as next step. In this phase all costs of the graphs edges reachable from that node were updated due to the realizability of the respective action. The realizability depends on the availability of the parts to be mounted. Unrealizable actions receive penalty costs which vary depending on how close in the future they would have to be executed. This cost assignment makes the planning algorithm avoid currently unrealizable assemblies. Additionally, due to the weaker penalties for more distanced edges, the algorithm prefers assembly sequences that rely on unavailable parts in the distant future to assemblies that immediately need those parts. Preferring the latter assembly results in reduced waiting periods during the assembly since missing parts have more time to be delivered until they are ultimately needed. Using the A* algorithm (Hart, P.E., N. J. Nilsson, and B. Raphael. 1968) the Online Planner now derives the cheapest path connecting the node matching the actual state with a goal node, which presents one variant of the finished product. This path represents the at that time optimal assembly plan for the desired product. The Parellelizer component now identifies in parallel or arbitrary sequence executable plan steps and hands the result to the Cognitive Control for execution. Here the decision which action is actually to be executed is made. The process of parallelization is detailed in (Ewert D., D. Schilberg, and S. Jeschke. 2011).

However, updating all edge cost reachable from the node representing the current state is a computational intensive task. To overcome this problem, the edge cost update can be combined with the A* algorithm, so that only edges which are traversed by A* are updated. This extremely reduces the computational time, since only a fraction of the graphs node is examined. So even for large graphs, the Online Planner is able to derive a decision in well under 100ms in worst case. Figure 5 shows the nodes reachable and examined by the Online Planner during the assembly of a 4x4 construct. Plateaus in the graph depict waiting phases where the assembly cannot continue because crucial parts are not delivered.

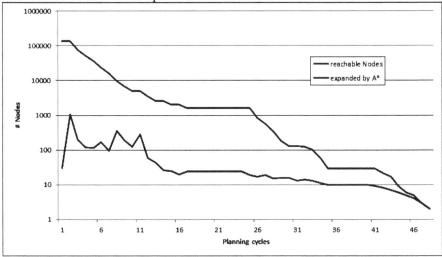

Figure 5 Number of nodes reachable from the node representing the given situation and number of nodes that are examined by the A*-algorithm. Number of nodes is shown using a logarithmic scale.

128

3.4   Planner Behaviour

Figure 6 shows the course of the assembly for the construct c). Newly arrived parts are shown in the third column. They can either be used for direct assembly (first column) or otherwise are stored in a buffer shown in column 2. The right column depicts the plan that is calculated based on the parts located. Here the number of a given block denotes the position where that block is to be placed. In step 0, no parts have been delivered. The planner therefore has no additional information and produces an arbitrary but feasible plan. In step 1 a new green block is delivered, which matches the first plan step. The related assembly action is therefore executed and the new block is directly put to the desired position. In step 2 a new red block is delivered. Given the current state of the assembly and the new red cube, the planner calculates an improved plan which allows to assemble this red block earlier than originally planned: Now it is more feasible to first mount two green blocks on top of each other (positions 1 and 3), because then the red block can be assembled, too (position 5). This plan step is executed in step 3 when a second green block becomes available. Now, in step 4, it is possible to mount a red block. From that step on only one feasible assembly sequence is possible, which is then executed.

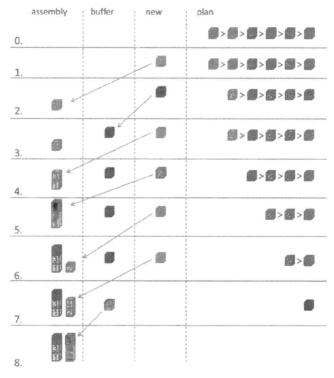

Figure 6 Exemplary assembly flow for construct c)

The described behaviour results in more rapid assemblies compared to simpler planning approaches: A purely reactive planner which would follow a bottom up strategy, would have placed the first two green blocks next to each other (positions 1 and 2). Thus, in step 5 no assembly action would have been possible and the assembly would have to stop until a further green cube would be delivered.

4   SUMMARY

In this paper we presented our hybrid approach for an assembly planner for nondeterministic domains. We described the workflow of the offline planner, which analyses CAD data describing the desired product. The outcome of the offline planner is a state graph which holds alle possible (and feasible) assembly sequences. This graph is generated by following an assembly by disassembly strategy: Recursively all possible separations of the final product are computed until only single parts remain. During the actual assembly, this state graph is updated to mirror the current situation of the assembly, specially the availability of newly delivered parts. Using the A* algorithm, the at that time optimal assembly sequence is derived and handed over to the cognitive control unit, which then decides which assembly step gets to be executed. This step is then executed and the outcome of that step is reported back to the planning system. This process is iterated until the product is completed.

5   OUTLOOK

Future work must optimize the described Planners. Using techniques of parallel programming and by incorporating specialized databases which can cope efficiently with large graphs, the planning duration can be improved. Subsequently, the planner will be extended to be able to deal with industrial applications as well as plan and control the production process of a complete production network.

ACKNOWLEDGEMENTS

The authors would like to thank the German Research Foundation DFG for supporting the research on human-robot cooperation within the Cluster of Excellence "Integrative Production Technology for High-Wage Countries".

REFERENCES

Anderl, R, Tripper, D., STEP Standard for the Exchange of Product Model Data, B. G. Teubner, Stuttgart/Leipzig, 2000.

Brecher, C. et al. 2007. Excellence in Production, Apprimus Verlag, Aachen, Germany,.

Castellini, C., E. Giunchiglia, A. Tacchella, and O. Tacchella.2001. Improvements to sat-based conformant planning. In: In Proc. of 6th European Conference on Planning.

Ewert D., D. Schilberg, and S. Jeschke. 2011. Selfoptimization in adaptive assembly planning. In: Proceedings of the 26th International Conference on CAD/CAM, Robotics and Factories of the Future.

Hart, P.E., N. J. Nilsson, and B. Raphael. 1968. A Formal Basis for the Heuristic Determination of Minimum Cost Paths, IEEE Transactions on Systems Science and Cybernetics SSC4 (2), pp. 100–107.

Hoffmann, J. 2001. FF: The Fast-Forward Planning System. In: *The AI Magazine,*2001.

Hoffmann, J., and R. Brafman. 2005. Contingent planning via heuristic forward search with implicit belief states. In: In Proceedings of ICAPS'05. 71–80.

Homem de Mello, L. S., and A. C. Sanderson. 1986. And/Or Graph Representation of Assembly Plans. Proceedings of 1986 AAAI National Conference on Artificial Intelligence, p. 1113–1119.

Kaufman, S.G. et al. 1996. LDRD final report: Automated planning and programming of assembly of fully 3d mechanisms. Technical Report SAND96-0433, Sandia National Laboratories , 1996.

Kempf, T., W. Herfs, and C. Brecher. 2008. Cognitive Control Technology for a Self-Optimizing Robot Based Assembly Cell., In: Proceedings of the ASME 2008 International Design Engineering Technical Conferences & Computers and Information in Engineering Conference, America Society of Mechanical Engineers, U.S.

Mayer, M. et al. 2009. Simulation of Human Cognition in Self-Optimizing Assembly Systems, In: Proceedings of 17th World Congress on Ergonomics IEA 2009. Beijing, China.

Röhrdanz, F., H. Mosemann, and F.M. Wahl. 1996. HighLAP: a high level system for generating, representing, and evaluating assembly sequences. In: 1996 IEEE International Joint Symposia on Intelligence and Systems, Seiten 134–141, 1996

Tecnomatix. 2011.
http://www.plm.automation.siemens.com/en_us/products/tecnomatix/index.shtml

Thomas, U. 2009. Automatisierte Programmierung von Robotern für Montage-aufgaben.Volume 13 of Fortschritte in der Robotik. Shaker Verlag, Aachen.

Zaeh, M., and M. Wiesbeck. 2008. A Model for Adaptively Generating Assembly Instructions Using State-based Graphs. In: Manufacturing Systems and Technologies for the New Frontier, Springer, London.

CHAPTER 14

Planning-modules for Manual Assembly Using Virtual Reality Techniques

Leif Goldhahn, Katharina Müller-Eppendorfer

University of Applied Sciences, Mittweida,
Faculty of Mechanical Engineering
Mittweida, Germany
leif.goldhahn@HS-Mittweida.de
katharina.mueller.1@HS-Mittweida.de

ABSTRACT

Efficient and competitive manufacturing demands a systematic assembly planning procedure. The paper focuses on using and improving available processes for assembly planning that can be customised to suit the needs of small and medium-sized companies (SMEs).

Special attention should be given to the following:
- Great flexibility and variability
- Cost effectiveness
- Change from parts manufacturer to component supplier
- Consideration of an integrated planning process
- Consideration of all preceding and subsequent processes and general conditions
- Integration of planning and implementation of assembly systems into the business targets
- Opening up the potential of virtual reality techniques.

The new planning algorithm for assembly integrates system planning for manual assembly. The reason for this expansion is that many companies have to rearrange their shop floor areas for assembly due to product changes or because they are establishing an assembly area for the first time. Moreover, in assembly planning, the assembly planner should follow a systematic paradigm.

132

In addition to the integrated approach, new techniques should also support planning. Virtual Reality technology is an appropriate tool for efficient planning of both the system and the process. Virtual session members can see the assembled object itself as well as the components and their functionalities by means of Virtual Reality. Sectional views of the object can be shown and different components can be displayed transparently.

An important aspect of the virtual reality technique is the virtual model of a human being. This model makes it possible to check the accessibility of parts within the assembly system as well as the visible areas. Physical stress on the human worker and the worker's posture can also be monitored within the assembly system.

As a result, it is not only possible to discover planning errors earlier, but also to avoid the expense of changes and follow-up costs.

To address these issues, planning-modules are developed for manual assembly. The planning-modules are part of the assembly planning procedure; they can be used repeatedly and are independent of the product.

The planning-modules include methods, tools and auxiliaries, such as the virtual representation of the product structure, graphs elucidating the sequence of assembly activities, known as precedence graphs and assembly plans. The planning engineers use the planning sequence in assembly in conjunction with their own know-how.

Data input into the planning-modules is information that is already available. Thus, the planning modules may be used and the planning results obtained may be documented. To be efficiently supported by Virtual Reality techniques, the planning modules and Virtual Reality techniques are used both in parallel and sequentially. For example, one may view the object in the virtual world in order to create the product structure.

Thus the virtual planning modules support the planning engineer and make it possible to avoid errors in the assembly object and the assembly system that do not exist in reality.

Keywords: Manual assembly, Virtual Reality, Ergonomics, Process planning

1 FUNDAMENTALS

Application of the latest technologies, such as Virtual Reality (VR), in industry offers a firm a competitive advantage on the market.

This is especially true when VR is used to support the planning of the manual assembly systems as well in the development and storage of research knowledge in small and medium-sized enterprises (SMEs).

Given the increasing number of requests for small batches, as well as more demanding customer requirements and an increased number of product variants combined with shorter product life cycles, it is sensible to assemble parts manually. Assembly areas have to be kept flexible and even smaller series of subassemblies and components have to be assembled profitably.

To achieve this goal, a holistic planning concept has to be developed. When supported by Virtual Reality techniques, the planning process can be made even more efficient. Planning-modules are created based on the planning process. These planning-modules make it possible to use experience and knowledge and integrate it into the assembly process that is being planned as well as into the training of employees.

Virtual Reality (VR) is defined "a medium composed of interactive simulations that sense the participant's position and actions and replace or augment the feedback to one or more sense, giving the feeling of being mentally immersed or present in the simulation." (Sherman, William R.; Craig, Alan B, 2003). VR technology is characterized by immersion, interaction, imagination, intuition and integration. (Goldhahn, Leif, 2003), (IC.IDO, 2010)

Figure 1 demonstrates the Virtual Reality laboratory established at the University of Applied Sciences Mittweida. The components for the VR technology were made by the firms IC.IDO, ART, imsys and plavis. The system's main elements are high-performance computers, two projectors, the powerwall as visualization technology, six infrared tracking cameras, interaction tools, the software Visual Decision Platform (VDP) and the visTABLE touch display (for various applications).

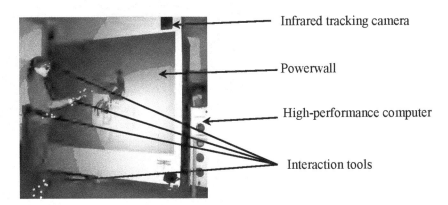

Infrared tracking camera

Powerwall

High-performance computer

Interaction tools

Figure 1: VR technology implemented at the University of Applied Sciences Mittweida

Each projector transmits an image in real time from the back to the powerwall (back projection on special plastic disc). Both images, which are offset perspectively and additionally amplified by mirror projection, are displayed on the projection screen. The images are assigned to the left and right eyes by means of polarization and, consequently, appear as a stereoscopic image for the viewer.

Infrared radiation emitted by the infrared tracking cameras and reflected on the interaction tools by the tracking marker makes it possible to record the current position and spatial location and to interact with the user.

The virtual human can be used to monitor physical stress and the posture of the worker or to observe the parts' accessibility with regard to assembly (cf. Goldhahn, Raupach, 2012). This monitoring is carried out by means of standard interaction. Body tracking offers another possibility for innovative interaction. Interaction with the virtual human can be implemented by means of body tracking using special markers located on the hands, abdomen and feet.

2   PLANNING PROCESS FOR MANUAL ASSEMBLY

Since assembly is a very complex process, it makes sense to make available to the planning engineer a systematic assembly planning strategy. The planning algorithm includes system planning for manual assembly that takes into account the ergonomics of workplace design and meets the challenge of carrying out modifications during assembly.

The new planning process consists of a conceptual model, sequence planning, assembly system design, detailed design, implementation and use, see Figure 2 (Lotter, Schilling, 1994), (Lotter, 2006) and (Bullinger, 1986).

The planning algorithm runs systematically in a top down manner. First, sequential planning is performed using design data, as well as input data like tolerances, fits, material data and required volumes. As a result of sequence planning, we obtain the assembly plan that then provides data for the assembly system design. Above all, it is necessary during the conceptual design and detailed engineering of the assembly system to comply with ergonomic guidelines. Human workers are very important in manual assembly.

Once all of the planning steps have been performed, the planning process is complete. If an error is found or a modification in the previous planning steps is observed, or in the event that a modification must be made, the planner then restarts the algorithm at the appropriate planning step to optimise the overall system.

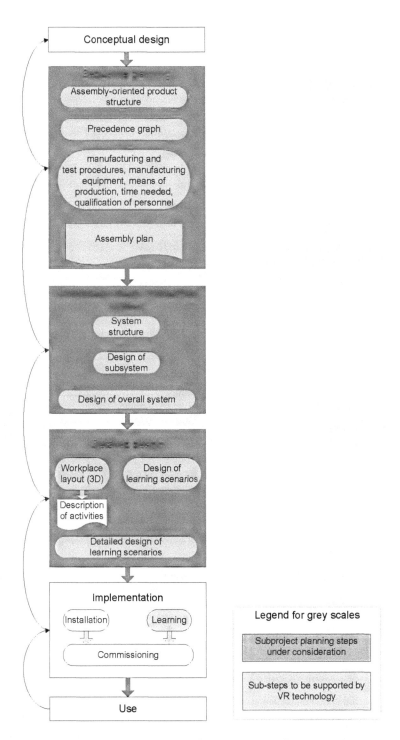

Figure 2: Holistic planning process for manual assembly using VR

3 INTRODUCTION AND USE OF PLANNING-MODULES FOR MANUAL ASSEMBLY

3.1 Overview of planning-modules

Planning-modules are defined elements of the assembly planning process. They are independent of the product and can be used repeatedly. Figure 3 elucidates input and output data for the planning modules.

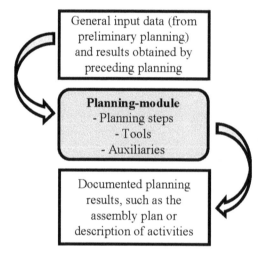

Figure 3: Structure of planning modules

Based on the structure of planning-modules and the planning process, eight modules have been engineered that are shown in Table 1.

They rely on methods, auxiliaries and tools from process planning, such as product structure, assembly precedence graph and workplace layout. This project particularly focuses on the use of VR technology to support the planning-modules.

Table 1: Overview of planning modules

Phase	Name of the module	Elements (activities)
Sequence planning	**Sequence 1** Create product structure for assembly	Product analysis, Design scheme, Structural overview, Providing data for assembly plan
	Sequence 2 Create precedence graph (define sequence)	Define operations, Display dependencies, Provide data for assembly plan
	Sequence 3 Create data for sequence structure	Define manufacturing techniques, test methods, product-specific manufacturing equipment / means of production, determination of time (estimate), Assessment of work, Create assembly plan, Provide data for detailed design
Conceptual design of assembly system	**System 1** Define system structure	Define constraints, Forms of organisation, Differentiate subsystems
	System 2 Design the subsystem	Criteria check list for requirements to be fulfilled by layout, Subsystem variants Performance value analysis
	System 3 Design the overall system	Linking of workplaces, Configuration of subsystems, Bringing together subsystems, Taking into account marginal systems, Providing data for detailed design
Detailed design	**Detailed design 1** Detailed design of assembly sequence and assembly system	Combine design data from sequence planning and conceptual design from the assembly system, data detailing, generate 3D representations
	Detailed design 2 Documentation	File technologies, manufacturing equipment / means of production, time values, conditions for execution, Required qualifications, Describe activities

3.2 Use of planning-modules for selected examples

A multi-part assembly subject is used to plan and explore manual assembly for the first time and to use the planning modules. A bevel gear (see Figure 4) is introduced as a demonstrator to be used in the project.

138

Figure 4: Bevel gear (project demonstrator)

Several assembly operations, such as screwing, joining and inserting, can be shown on this demonstrator. The bevel gear consists of three main components: drive train, drive end and casing. The drive train and drive end, in turn, consist of components made of bevel wheels, shafts, bearings and fits.

Sequence planning stage:

VR technology is applied to illustrate both the assembly object and the geometry and configuration of the parts, as well as to understand the function and the design of the assembly object. VR technology makes it possible to represent the assembly objectively and more transparently and to show every individual part (Figure 5). In this way, one obtains information about the assembly object that only exists as a 3D drawing or virtual model.

After loading the assembly object, the object has to be prepared in the VR software in order to mount it later on. For example, features may be assigned to the individual parts, such as Collidable (to indicate collisions) and Constraints (to limit the number of degrees of freedom for better handling). Figure 6 illustrates the prepared bevel gear with a collision arrow. This arrow indicates that it is impossible to remove the shaft from the drive train subassembly.

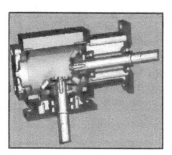

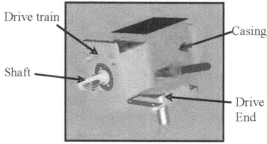

Drive train

Casing

Shaft

Drive End

Figure 5: Sectional view of the bevel drive

Figure 6: Bevel drive with collision arrow

Based on this assembly information, the product structure and precedence graph can be generated. This step is necessary to create the assembly plan. The assembly plan, in turn, is a prerequisite to the creation of the assembly system for the object itself. Conceptual design of the assembly process is the next stage of the planning process.

Conceptual design of assembly system:

Based on the assembly plan, the assembly workplace can be drafted. This conceptual design includes all required tools and resources. Workplace, manufacturing equipment and parts should be configured so as to allow short gripping and handling distances. Visualization using VR technology makes it possible to avoid rough design errors.

For a detailed workplace design, in particular to address ergonomic requirements, 3D models of the workplace and the equipment are loaded in the VR software.

The use of the virtual human makes it possible to represent the accessibility and visible areas of the worker's body in conjunction with the assembly object and the workplace (Figure 7 and Figure 8). The presented body tracking are intended to display sequences and to monitor them by means of VR. This makes it generally unnecessary to create complex animations.

Figure 7: Reach area of the virtual human

Figure 8: Vision area of the virtual human

The software can also be used to indicate physical stress and the posture of the human worker. **Fehler! Verweisquelle konnte nicht gefunden werden.** and Figure 9 illustrate the elbow positions (shown in yellow and red). However, the posture shown is not optimal for workers on a daily basis.

140

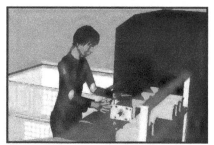
Figure 9: Posture of a female worker in the workspace

Figure 9: Posture of a male worker in the workplace

4 Conclusion

A systematic planning algorithm and planning-modules to be used by the planning engineer are a necessary prerequisite for economical and efficient performance of the assembly process in production. VR technology is applied to support this planning algorithm and these planning-modules. A holistic planning paradigm, product- independent planning-modules and VR technology guarantee a productive planning process and simplify earlier detection and avoidance of errors.

In addition to providing an assembly feasibility study for the product, VR technology also uses a virtual human to make it possible to evaluate the designed assembly workplace according to ergonomic guidelines.

REFERENCES

Bullinger, Hans-Jörg. 1986. Systematische Montageplanung - Handbuch für die Praxis. München. Wien: Hanser.

Goldhahn, Leif. 2003. Montageplanung und -ausführung virtuell und real. In. Scientific Reports - Journal of the University of Applied Sciences Mittweida, magazine 2: 46-50.

Goldhahn, Leif and Raupach, Annett. 2012. Methode zur digitalen und virtuellen Modellierung, Bewertung und Verbesserung von Arbeitssystemen. In. GfA - Gesellschaft für Arbeitswissenschaft e.V., (Ed.): Gestaltung nachhaltiger Arbeitssysteme. Dortmund, GfA: 507-511.

"IC.IDO - Startseite, Beschreibung Virtual Reality." Accessed February 11, 2010, URL: http://www.icido.de/de/index.html.

Lotter, Bruno. 2006. Planung und Bewertung von Montagesystemen. In. Lotter, Bruno and Wiendahl, Hans-Peter. Montage in der industriellen Produktion. Berlin, Heidelberg, New York: Springer.

Lotter, Bruno and Schilling, Werner. 1994. Manuelle Montage, Planung, Rationalisierung, Wirtschaftlichkeit. Düsseldorf: VDI.

Sherman, William R. and Craig, Alan B. 2003. Understanding Virtual Reality, Interface, Application and Design. San Francisco: Morgan Kaufmann.

CHAPTER 15

Learning and Forgetting in Production-inventory Systems with Perishable Seasonal Items

Ibraheem Abdul[1], Atsuo Murata[2]

[1]Department Mechanical Engineering,
Yaba College of Technology, Lagos, Nigeria
E-mail: dotun.abdul@gmail.com
[2]Department of Intelligent Mechanical Systems,
Okayama University, Okayama, Japan
E-mail: murata@iims.sys.okayama-u.ac.jp

ABSTRACT:

The consideration of learning and forgetting effects on production–inventory system is usually based on the assumption that item produced by the system have a constant or unidirectional demand pattern. The demand pattern for perishable seasonal product, however, is often a mixture of increasing, steady and decreasing functions of time. This study develops an EPQ model for perishable seasonal products having three-phase ramp-type demand pattern. We further investigate the effect of learning and forgetting in set-up on the optimal schedules and costs of the production-inventory system with this category of items. The study shows that learning-based reduction in set-up costs can be used to achieve some vital aspects of the just-in-time (JIT) philosophy.

Keywords: production, inventory, learning, forgetting, seasonal items

1   INTRODUCTION

It is a common experience in real life that a worker engaged in repetitive operations improves with time due to learning effects. The learning phenomenon implies that the performance of a system improves with time because the firms and employees perform the same task repeatedly and consequently learn how to provide a standard and improved level of performance. In manufacturing environment, factors that contribute to this improvement may include the more effective use of tools and machines, increased familiarity with operational tasks and work environment, and enhanced management efficiency. The set-up cost, one of the most important cost components in production operations, is often assumed constant in most classical production-inventory models. However, experience has shown that the set-up cost can actually be reduced due to learning effects.

The simplest and most commonly used learning theory in lot-sizing models is the one introduced by Wright (1936) which links the performance of a specific task to the number of times that task is repeated. To incorporate the effect of learning in set-up into lot-sizing models, most researchers applied the Wright's power function model to set-up costs with some modifications. Chand (1989) used a log-linear learning function to generate set-up costs in his computational study that investigated the effects of learning in set-ups on set-up frequency. Cheng (1994) assumed that set-up and unit variable manufacturing costs decrease as a result of learning over time and that some percentage of learning is lost between consecutive orders. Das et al. (2010) assumed a set-up cost in a cycle is partly constant and partly decreasing in the cycle due to learning effects of the employees. Most researchers considering the effect of learning in set-up agree to the existence of a minimum value of the set-up cost that cannot be affected by learning. This is a reasonable assumption when we consider the fact that there are some aspects of the set-up cost that may not be subject to the reduction due to learning. Some researchers have equally modelled the effect of loss of learning (forgetting) in the set-up cost (e.g. Chiu et al., 2003, Jaber et al., 2010) but unlike the learning process a full understanding of the forgetting process is yet to be developed (Jaber and Bonney, 2003).

Seasonal items like fruits, fish, winter cosmetics, fashion apparel, etc. generally exhibits different demand patterns at various times during the season. The demand usually begins with increasing trend, attains a peak and becomes steady at the middle of the season. Various time-dependent functions have been used to depict this demand pattern. Hill (1995) developed the first model for products whose demand variation is a combination of two different types of demand in two successive periods over the entire time horizon and termed it as a ramp-type demand pattern. Subsequently, many researchers have adopted this pattern for seasonal products whose demand is a mixture of non-decreasing, constant and non-increasing functions of time. Panda et al. (2008) developed an inventory model for deteriorating seasonal products using ramp-type demand pattern with a three-phase variation in demand. The demand pattern in this case is assumed to increase exponentially with respect to time up to a certain point. Then it becomes steady and finally decreases exponentially and becomes asymptotic. Another form of this pattern was used by Cheng and Wang (2009), in developing an economic order quantity model for deteriorating items. Production-inventory models for seasonal products using a ramp-type demand pattern in recent time includes Manna and

Chaudhuri (2006), Panda et al. (2009), Manna and Chiang (2010). The production-inventory models are, however, single-period models and did not allow for the reduction in set-up costs. Also none of these models for seasonal products consider the effects of learning or forgetting on set-up.

This paper investigates the effect of learning and forgetting in set-up on the optimal schedules and costs of a production-inventory system for perishable seasonal products. A new model is developed to solve the multi-period production lot-sizing problem that involves perishable seasonal products with varying demand under set-up learning and forgetting.

2   MODEL FORMULATION

2.1 Assumptions and Notations

The set-up cost used in the model is assumed to reduce due to learning effect and the cost of n-th set-up is computed using learn-forget curve model (LFCM) with plateau effect proposed in Jaber and Bonney (2003).

$$A_n = \begin{cases} A_1 (m_i + 1)^{-c} & \text{if } m_i < n_s \\ A_{min} & \text{if } m_i \geq n_s \end{cases}$$

A_1 is the cost of the first set-up, A_n is the cost of nth set-up, A_{min} is the minimum set-up that is obtainable when $n = n_s$, c is the slope of the learning curve.

$$m_i = \sum_{j=0}^{i-1} de^{-\alpha(t_{r_j} - t_{e_j})} - 1 \quad \text{where } 1 \leq i \leq n.$$

d_i is the amount of knowledge acquired in repetition i; and α is the forgetting exponent. $d = 1$; and n is the number of set-ups. The time at which the first set-up occurs, and the retrieval time of knowledge gained in i set-ups are represented as:

$$t_{e_0} = 0; t_{r_i}.$$

Demand rate $f(t)$ is a general time dependent ramp-type functions (see Fig 1), and is of the form:

$$f(t) = \begin{cases} g(t), & 0 \leq t \leq \mu, \\ g(\mu), & \mu \leq t \leq \gamma, \\ h(t), & t \geq \gamma. \end{cases}$$

$$g(t) \geq 0, \; h(t) \geq 0, \; 0 \leq \mu \leq \gamma, \; g(\mu) = h(\gamma).$$

The function $g(t)$ can be a continuous and non-decreasing function of time, while $h(t)$ is a continuous and non-increasing function of time in the given interval. Parameters 'μ' and 'γ' represent the trend of the ramp-type demand function.

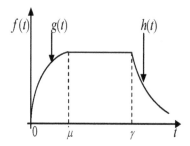

Figure 1. The three-phase ramp-type demand pattern

Other major assumptions and notations of the model are as follows:

- A single item multi-period production-inventory system is considered.
- The production rate of the item is a known function of demand rate.
- Demand rate is a general time-dependent three-phase ramp-type function.
- Deterioration rate of the item is represented by a two-parameter Weibull-distribution function.
- No repair or replacement of deteriorated items during the period.
- Set-up cost decreases due to the learning in set-ups while set-up time is negligible.
- Production cost per unit and inventory holding cost per unit are known and constant.
- The following notations are used in formulating the models:
- t_{i-1} is the time when production for the i-th cycle begins
- s_i is the time when production for the i-th cycle stops
- t_i is the end of the i-th cycle
- $T_i = t_i - t_{i-1}$ is the length of a cycle
- q_{i-1} is the inventory level at the beginning of the i-th cycle
- q_i is the inventory level at the end of the i-th cycle
- $K(t) = \alpha f(t)$ is the production rate $(\alpha > 1)$
- C_P is the production cost per unit
- C_H is the inventory holding cost per unit

2.2   Model Analysis

The system consists of several production-inventory cycles. Each cycle begins with the production at time t_{i-1} and ends with the consumption due to demand and deterioration at time t_i. The production stops at time s_i within the interval while demand and deterioration of products occur throughout the interval. The variation of inventory level with time for a typical cycle is shown in Fig. 2.

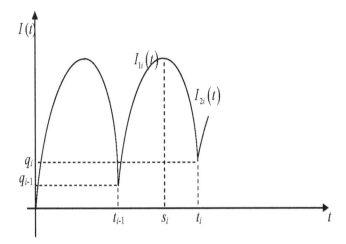

Figure 2. Variation of inventory level with time for a typical cycle

The objective is to determine the optimal values of the production schedules, total production quantity, and total relevant cost for the first and subsequent cycles using different values of learning and forgetting rate. The rate of change of inventory level with time is as follows:

$$\frac{dI_{1i}(t)}{dt} = K(t) - f(t) - \theta(t)I_{1i}(t) \quad t_{i-1} \le t \le s_i; I_{1i}(t_{i-1}) = q_{i-1}, \tag{1}$$

$$\frac{dI_{2i}(t)}{dt} = -f(t) - \theta(t)I_{2i}(t), \qquad s_i \le t \le t_i; I_{2i}(t_i) = q_i.$$

Here we consider all production and consumption cycles that do not involve a change in demand pattern. The demand pattern may be any of the patterns given in f (t). Inventory level at any time during the production stage of the i-th cycle is represented by $I_{1i}(t)$ while $I_{2i}(t)$ is the inventory level at any time during the no-production stage of the i-th cycle. The solutions to Eq. (1) above are as follows:

$$\tag{2}$$

$$I_{1i}(t) = q_{i-1}e^{a(t_{i-1}^b - t^b)} + e^{-at^b}\int_{t_{i-1}}^{t} e^{ax^b}(K(x) - f(x))dx, \quad t_{i-1} \le t \le s_i,$$

$$I_{2i}(t) = q_i e^{a(t_i^b - t^b)} + e^{-at^b}\int_{t}^{t_i} e^{ax^b} f(x)dx \qquad s_i \le t \le t_i.$$

The inventory holding cost (HC_i), the production cost (PRC_i), and the set-up cost (SUC_i) for one cycle is as follows:

$$HC_i = C_H\left(\int_{t_{i-1}}^{s_i} I_{1i}(t)dt + \int_{s_i}^{t_i} I_{2i}(t)dt\right),$$

$$PRC_i = C_P\int_{t_{i-1}}^{s_i} K(t)dt, \quad SUC_i = A_n.$$

The total relevant cost per unit time for one cycle is as follows:

$$TCT_{1i}(s_i, t_i) = \frac{1}{(t_i - t_{i-1})}(HC_i + PRC_i + SUC_i), \tag{3}$$

$$= \frac{1}{(t_i - t_{i-1})}\left(C_H\left(\int_{t_{i-1}}^{s_i} I_{1i}(t)\,dt + \int_{s_i}^{t_i} I_{2i}(t)\,dt\right) + C_P\int_{t_{i-1}}^{s_i} K(t)\,dt + A_n\right).$$

3   SOLUTION PROCEDURE

The total relevant cost per unit time for one cycle is derived by using Eq. (3). The optimal production schedules and the cost for the first and the subsequent cycles can be obtained using different values of learning and forgetting rate by minimizing the total relevant cost for the cycle subject to the constraints below.

I. $I_{1i}(s_i) = I_{2i}(s_i)$,

II. $0 < t_{i-1} < s_i < t_i$.

The algorithm of the single demand pattern policy (SDP) developed in Abdul and Murata (2011) is used in determining the optimal values.

4   NUMERICAL EXAMPLE

A set of numerical experiment was conducted to illustrate the application of the model and analyze its performance. Consider the production-inventory system for a seasonal product with a three-phase ramp-type demand pattern as shown below.

$$f(t) = \begin{cases} 100 + 5t, & 0 \le t \le 4, \\ 120, & 4 \le t \le 8, \\ 120e^{-0.2(t-8)} & 8 \le t \le 13. \end{cases}$$

The system undergoes a learning-based continuous improvement process which results in reduction in set-up cost. The initial set-up cost prior to any learning is $200 per production cycle while the minimum set-up cost that is not subject to further reduction is $50 per cycle. The production rate is 1.5 times the demand rate at any point in time; the unit production cost and inventory holding cost per unit of the item are $10 and $1.5, respectively. The deterioration of the item follows a Weibull distribution ($\emptyset = abt^{b-1}$) with parameters $a = 0.005$, $b = 2$. The inventory level at the beginning and end of the season is zero while at other times during the season a minimum inventory level of 50 units is maintained.

The above system is used to illustrate the model by obtaining optimal schedules and costs under different learning and forgetting rates. The result of the experiment is shown in Table 1.

Table 5.2: Optimal results using the SDP policy with various values of learning and forgetting index

	n	t_{i-1}^*	s_i^*	t_i^*	TCT_i^*	T_i^*	$TCT_i^*\,T_i^*$
	1	0	2.62	3.36	1417.43	3.36	4762.0
	2	3.36	3.80	4.00	1612.60	0.64	1032.7
	3	4.00	5.53	6.21	1474.50	2.21	3263.4
$c=0$	4	6.21	7.46	8.00	1492.70	1.79	2667.2
$a=0$	5	8.00	10.41	13.00	829.08	5.00	4145.4
Total relevant cost for the season							**15870.7**
	1	0	2.62	3.36	1417.43	3.36	4762.0
	2	3.36	3.80	4.00	1573.2	0.64	1007.5
	3	4	5.36	5.96	1454.00	1.96	2856.8
$c=0.25$	4	5.97	7.39	8	1463.2	2.04	2977.9
$a=0.1$	5	8	10.41	13	818.19	5	4091
Total relevant cost for the season							**15695.2**
	1	0	2.62	3.36	1417.43	3.36	4762
	2	3.36	3.80	4	1538.8	0.64	985.5
	3	4	5.20	5.74	1436	1.74	2505.4
	4	5.74	6. 90	7.40	1440.2	1.65	2381.8
$c=0.50$	5	7.40	7. 82	8	1512.8	0.60	910
$a=0.1$	6	8	10.41	13	807.96	5	4039.8
Total relevant cost for the season							**15584.5**
	1	0	2.62	3.36	1417.4	3.36	4762
	2	3.36	3.80	4	1582	0.64	1013.1
	3	4	5.38	5.99	1456.3	1.99	2901.8
$c=0.25$	4	5.99	7.40	8	1467.2	2.01	2945.3
$a=0.2$	5	8	10.41	13	820.23	5	4101.2
Total relevant cost for the season							**15723.4**
	1	0	2.62	3.36	1417.4	3.35	4762.
	2	3.36	3.80	4	1554.4	0.64	995.4
	3	4	5.23	5.79	1440	1.79	2583.8
	4	5.79	7.00	7.53	1447.2	1.74	2514.4
$c=0.50$	5	7.53	7. 86	8	1588.2	0.47	743.8
$a=0.2$	6	8	10.41	13	810.16	5	4050.8
Total relevant cost for the season							**15650.2**

148

5   DISCUSSION OF RESULTS

The result shows that the lowest value of total cost for the season occurs at the highest value of the learning index which corresponds to the highest rate of reduction in set-up cost due to learning. The effect of learning is also reflected in the form of the reduction in cycle length (T_i) as the number of cycles (n) increases for a given demand pattern. This resulted in the increase in the number of lots as the learning index increases and it corroborates the fact that reduction in set-up cost often leads to higher production frequency and shorter production runs which forms an aspect of the just-in-time (JIT) philosophy. However, the percentage reduction in total cost and cycle length reduces with consideration of forgetting effect.

For a given learning rate, the result shows that an increase in forgetting rate lead to increase in the total relevant cost of the system and slight changes in the replenishment schedules. This indicates that the forgetting phenomenon has some effect on the production-inventory system. However, unlike the learning phenomenon, the effect may be insignificant at times.

6   CONCLUSIONS

In this study we have considered the multi-period production lot-sizing problem that involves deteriorating seasonal products with a three-phase ramp-type demand pattern. The study also included the effects of learning and forgetting in set-up on the production and inventory schedules and costs. Through numerical analyses, we showed that the total relevant cost of a production-inventory system is the highest when there is no improvement and no reduction in set-up cost due to learning. It was equally shown that the learning-based reduction in set-up costs led to higher production frequency and shorter production runs which are vital aspects of the just-in-time (JIT) philosophy. The example considered also showed that the effect of the forgetting phenomenon may sometimes be insignificant.

In some real-life situations, the effect of learning and forgetting on product quality, production rate and set-up at the same time may be significant. This study can be extended to consider this effect on perishable seasonal products.

REFERENCES

Abdul, I. and Murata, A. 2011. A fast-response production-inventory model for deteriorating seasonal products with learning in set-ups. *International Journal of Industrial Engineering Computations*, 2: 715-736.

Chand, S. 1989. Lot sizes and set-up frequency with learning and process quality. *European Journal of Operational Research*, 42: 190-202.

Cheng, M. and Wang, G. 2009. A note on the inventory model for deteriorating items with trapezoidal type demand rate. *Computers & Industrial Engineering*, 56: 1296-1300.

Cheng, T. C. E. 1994. An economic manufacturing quantity model with learning effects. *International Journal of Production Economics*, 33: 257-264.

Chiu, H. N. Chen, H. M. and Weng, L. C. 2003. Deterministic time-varying demand lot-sizing models with learning and forgetting in set-ups and production. *Production and Operations Management*, 12: 120-127.

Jaber, M. Y. and Bonney, M. 2003. Lot-sizing with learning and forgetting in set-ups and in product quality. *International Journal of Production Economics*, 83: 95-111

Das, D. Roy, A. and Kar, S. 2010. A Production-Inventory Model for a Deteriorating Item Incorporating Learning Effect Using Genetic Algorithm. *Advances in Operations Research Volume 2010*, Article ID 146042, 26 pages, doi:10.1155/2010/146042

Hill, R. M. 1995. Inventory model for increasing demand followed by level demand. *Journal of the Operational Research Society*, 46: 1250-1259.

Jaber, M. Y. and Bonney, M. 2003. Lot-sizing with learning and forgetting in set-ups and in product quality. *International Journal of Production Economics*, 83: 95-111.

Jaber, M. Y. Bonney, M. and Guiffrida, A. 2010. Coordinating a three-level supply chain with learning-based continuous improvement. *International Journal of Production Economics*, 127: 27-38.

Manna, S. K. and Chaudhuri, K. S. 2006. An EOQ model with ramp-type demand rate, time-dependent deterioration rate, unit production cost and shortages. *European Journal of Operational Research*, 171: 557-566.

Manna, S. K. and Chiang, C. 2010. Economic production quantity models for deteriorating items with ramp-type demand, *International Journal of Operational Research*, 7: 429-444.

Panda, S. Senapati, S. and Basu, M. 2008. Optimal replenishment policy for perishable seasonal products in a season with ramp-type time-dependent demand. *Computers and Industrial Engineering*, 54: 301-314.

Panda, S. Saha, S. and Basu, M. 2009. Optimal production stopping time for perishable products with ramp-type quadratic demand dependent production and setup cost, *Central European Journal of Operational Research*, 17: 381-396.

Wright, T. 1936. Factors affecting the cost of airplanes, *Journal of Aeronautical Science*, 3: 122-128.

Section II

Work Study — Improving the Skills, Quality and Effectiveness

Influence at Years of Experience on Operation Concerning Kyoto Style Earthen Wall

Akihiko GOTO*, Hiroyuki SATO**, Atsushi Endo**, Chieko NARITA**,

Yuka TAKAI*, Hiroyuki HAMADA**

*Osaka Sangyo University
Osaka, JAPAN
gotoh@ise.osaka-sandai.ac.jp
**Kyoto Institute of Technology

ABSTRACT

The earthen wall is consisted of the clay taken near Kyoto prefecture. It takes long time to paint up the earthen wall for having the higher mechanical properties. The earthen wall is laminated the clay for many times. The process of painting is complex. The earthen wall has much kind of functions, for example there are the adjustment of the temperature and humidity, deodorize effect, fireproof and aromatherapy effect. The Kyoto style earthen wall has small hole. The moisture of the atmosphere enters into the small hole.

In this study, difference of both method and frequency of the painting between the expert and non-expert was analyzed. Correlation of the amount of used soil according to career of the painting was compared.

Keywords: earthen wall, experience, proficiency, motion

1. INTRODUCTION

The earthen wall expert actually improves the technique by painting up working on the site. However, it takes much time to master the higher skill. In this study, the difference at years of experience on painting up work of Kyoto style earthen wall was examined. The video cameras to take a picture of the painting operation were employed. The painting method and the paint frequency of the wall were analyzed by the video. It paid attention to the change in operation and the usage of the tool, and the proficiency of the operation was examined. It aimed to clarify wisdom of technique inside the experts for Kyoto style earthen wall.

2. EXPERIMENTAL PROCEDURE

2. 1 THE SUBJECT

Three persons of the expert and twentythree persons of the non-expert as the subject were employed. In the case of the experts, two persons were teachers of the plasterer school and one person was craftsman of the painting wall. Their years of experience are 40 years or more. The non-experts were students of the plasterer school. Their years of experience are ranges from about 2 months to about 8 years. The size of the objective wall was 86cm x 165cm.

2. 2 OBJECTIVE WALL

The objective walls were used by the practice class of the plasterer school. The size of the wall was 86cm x 165cm. Figure 1 shows the objective wall and appearance of operation.

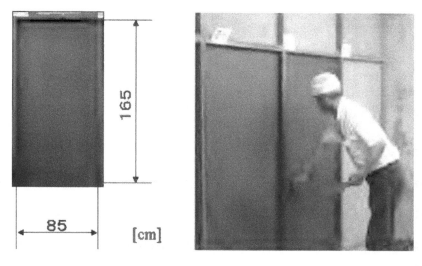

(a) Size of the wall.   (b) Appearance of paint work.
Figure 1 Objective wall.

2. 3 PAINTING OPERATION AND EVALUATION OF FINISH DEGREE

Each subject executed the painting operation. It took a picture of all processes from beginning of work to the end for each subject with the video camera. The amount of the soil used and the working hours were measured. In addition, the direction and the frequency of moving trowel were measured. After the end of operation, the expert evaluated the finish degree to each wall. The evaluation has mainly 5 kinds of grade. The best finished status of painting wall is A grade. On the other hand, the worst finished status of painting wall is E grade. The measurement was executed for two years every six months.

3. RESULTS AND DISCUSSION

3. 1 PAINTING OPERATION OF THE EXPERT

Figure 2 shows the illustration of direction of moving trowel in the case of the expert. First of all, the soil caught with the trowel is put above the left of the wall. This soil paints and expanded from the lower side to the upper side by using the trowel. After finishing painting operation from the left upper side to the left lower side, the operation was carried out from the right upper side to the right lower side. The trowel is occasionally moved to the vertical direction, and it is occasionally horizontally moved. It is thought that the way to move trowel is important.

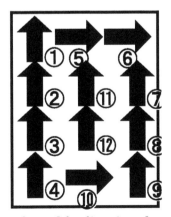

Figure 2 Procedure of the direction of moving trowel.

3. 2 AMOUNT OF SOIL USED AND YEARS OF EXPERIENCE

Figure 3 shows the relationship between the amount of the soil used for the wall painting operation and years of experience. The amount of the soil used showed the difference with the expert and non-expert. Averages of the amount of the soil that non-expert had used were about 16.0 kg. On the other hand, averages of the amount of the soil that the expert had used were about 12.7 kg. The amount of the expert showed the tendency to decrease by about 3 kg compared with the amount of non-expert. As for the earthen wall of the expert, it was clarified that the amount of the soil was little and light.

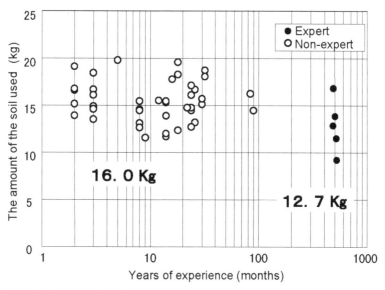

Figure 3 Relationship between the amount of the soil used for the wall painting operation and years of experience.

3. 3 WORKING HOURS AND YEARS OF EXPERIENCE

Figure 4 shows the relationship between working hours and years of experience. The working hours of non-expert had a big difference even though it was the same experience years. In the work of every day, non-expert is hardly doing work to use the soil or there is a possibility to use not the soil but other materials. It has been understood that working hours of the expert is almost a half of the working hours of the non-expert.

Figure 4 Relationship between working hours and years of experience.

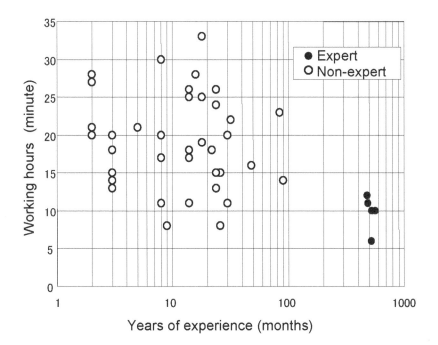

3. 4 MOVEMENT FREQUENCY OF TROWEL

Figure 5 shows frequency of painting operation in the vertical direction and horizontal direction. Moving trowel to the vertical direction showed a similar tendency to the expert and non-expert. It is necessary to move the trowel in the vertical direction to put the soil on the wall first. On the other hand, the movement in the horizontal direction of the trowel showed the tendency to increase gradually as years of experience increased. It is thought that the horizontal technique is acquired in about at least one year. Thus, it is thought that the technique for horizontally carrying the soil was able to be acquired as years of experience increase.

3. 5 EVALUATION OF FINISH DEGREE

Figure 6 shows evaluation of finish degree by the expert. When years of experience exceeded about six months, the evaluation of the finish degree showed the tendency that C or more increased, and non-expert who obtained B evaluation increased. However, it was clarified to require experiencing for about seven years or more so that non-expert might obtain A evaluation.

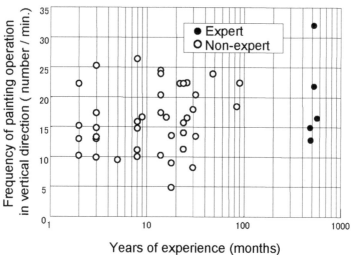

(a) Vertical direction.

(b) Horizontal direction.

Figure 5 Frequency of painting operation.

Figure 6 Evaluation of finish degree by the expert.

4. CONCLUSIONS

The summary was obtained as follows; it has been understood that non-expert acquires the operation carried to insufficient while painting the soil on the wall. This is a master of the operation that not only is moved trowel to the vertical direction for the wall but also is horizontally moved. In addition, the technique of paint and the management of the soil have improved rapidly by acquiring the posture maintenance to the wall from experience.

REFERENCE

A.Goto and H.Sato, et al. 2009. Movement analysis of master concerning Kyoto style earthen wall. *Proceedings of 11th Japan International SAMPE Symposium & Exhibition*, TC-1-3.

CHAPTER 17

Comparison of Painting Technique of Urushi Products between Expert and Non-expert

Akihiko GOTO*, Atsushi ENDO**, Chieko NARITA**,
Yuka TAKAI*, Yutaro SHIMODE**/***, Hiroyuki HAMADA***

*Osaka Sangyo University
Osaka, Japan
gotoh@ise.osaka-sandai.ac.jp
takai@ise.osaka-sandai.ac.jp
**Kyoto Institute of Technology
Kyoto, Japan
torima_taskurushimakie_14@yahoo.co.jp
soy155ap@mail.goo.ne.jp
hhamada@kit.ac.jp
***Shimode Maki-e Shisho
Kyoto, JAPAN
shimode-yutaro@xa2.so-net.ne.jp

ABSTRACT

The technique of urushi painting, which is one of Japanese traditional crafts, is too difficult and it is performed by hand, so that technology transfer from the expert to non-expert would be hard. This study aimed to clarify the influence on painting technology for the difference of skill and the brush to use. An expert and two non-experts were subjects and they painted urushi with their and other person's brush. Talk Eye Ⅱ was used as the measuring equipment to take images. The painting process and painting time were measured to analyze. As a result, the expert used the brush for 4 directions but both non-experts used brush only 2 directions. In addition, the expert changed painting way with the difference of brushes however both non-expert didn't change their painting ways with the difference of brushes

Keywords: Urushi painting, motion analysis, skill

1   INTRODUCTION

Urushi craft is one of Japanese representative traditional crafts and it has the name as "japan". In japan, Urushi has been used from the Jomon period. Urushi has a long history and is used widely. A lot of Japanese Urushi products were exported to Europe by East India Company around Momoyama period. At that time, privileges in those days have been captivated by its beauty charm, "Marry Terejia", "Marry Antoinette", etc.

A large number of "Kashu" and "Uretan" products, which has similar appearance as Urushi, have been sold in recent years. However traditional Urushi crafts are painted by craftspeople' hands still now. Urushi painting needs high level technique and technique acquisition for it needs long and strict training.

The painting technology of Urushi crafts, proficiency of people who paint Urushi and the difference of brushes were focused and analyzed scientifically in the study. This study was to clarify the effect of the level of skill and the difference of brushes on the painting techniques of Urushi. Painting procedure and time of both expert and non-experts were measured and compared.

2   METHODS

2.1   SUBJECTS

Regards subjects, the expert was an Urushi painting teacher who has 19 years painting experience and two non-experts were students. The student who has 0.5 year studying of Urushi painting defined as non-expert A and the student of 1.5 years was defined as non-expert B. The subject's data are shown in Table 1.

Table 1 Subject data

	Age	Height (cm)	Weight (kg)	Career (year)
Expert	35	152	51.5	19
Non-expert A	18	154	53.0	0.5
Non-expert B	30	162	49.0	1.5

2.2   MATERIALS AND TOOLS

Kuro-roiro-urushi (Shikata Kizou Urushiten) was used for painting in this experiment. The substructure was acrylic boards with a size of 300 mm × 300 mm and a thickness of 5mm. As prior processing, as well as a Usushi painting process the substructure was done "Kiji-gatame". "Kiji-gatame" is to rub the substructure with "Ki-urushi" after polishing the surface of substructure.

Subjects used their own wooden pallets respectively for carrying Urushi from vessel to acrylic board. The brush generally used for Urushi painting made of woman hair with 30mm in width was used. In particularly, two kinds of brushes were used in this experiment. One was subject's brush and the other was the one used by other. The materials and tools are shown in Fig.1.

Figure 1 The picture of Urushi, pallet and brush.

2.3   MEASUREMENT METHOD

Talk Eye II (Takei scientific Instrument Co, ltd) was used to measure painting methods and time with a sampling rate of 30Hz. All of the three subjects performed Urushi painting two times. In the first trial personal brush was used and secondly they used other person's brush. The paint process was distinguished by the sign which did subjects following action. When subject painted the acrylic board from top to bottom or when the subject rotate acrylic board more than 90 degrees.

3   RESULTS

3.1   PERSONAL BRUSH

Painting methods of the expert by using her personal brush which included 7 processes were shown in consequently in Fig.2. As regards to non-experts, the painting methods of A and B are shown in Fig.3 and Fig.4, respectively.

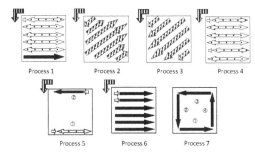

Figure 2 Schematic diagram of painting method in personal brush (expert).

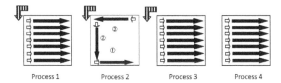

Figure 3 Schematic diagram of painting method in personal brush (non-expert A).

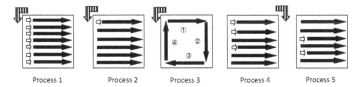

Figure 4 Schematic diagram of painting method in personal brush (non-expert B).

The arrows in the square indicate the moving direction of the brush on acrylic board and the left upper arrows indicate the rotation of acrylic board. Based on painting direction and the length where brush moved, the painting methods were divided to the following 4 ways.

a.   "long"; the brush was moved more than half of the acrylic board.
b.   "short"; the brush was moved less than half of the acrylic board.
c.   "repetition"; the brush was moved repeatedly on the same line.
d.   "bias"; The brush was moved diagonally on an acrylic board.

It is found that repetition paintings were carried out many times in the case of the expert as shown Fig.2. However this painting method cannot been found in the case of non-experts. In addition, bias direction painting has been used by expert while dose not by non-experts. Both non-experts used 2 painting methods "long" and "short", then their process were not different so much. On the other hand, for both "long" and "short" painting methods, they are found in both expert and non-experts. As a result, the expert used 4 painting methods while non-experts used only 2 methods. The total painting time and process are shown in Fig.5.

164

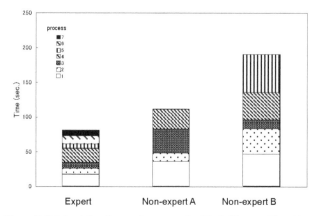

Figure 5 Total painting time and process of subjects (Personal brush).

The painting processes of the expert were more than non-experts however total painting time was shorter. In the case of using personal brush, the painting time of the expert was two thirds of that of non-expert A and half of the non-experts B. The relative time of each subject's painting methods are compared in Fig.6.

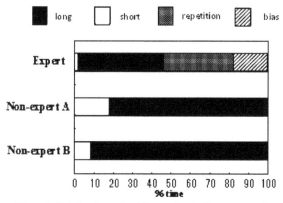

Figure 6 Relative time of painting methods (Personal brush)..

It is found that the expert used long, repetition mainly as well as bias while non-experts used long almost. Painting method and time between non-expert A and B was not obvious different.

3.2   OTHER PERSON'S BRUSH

Painting processes of the expert by using other person's brush are illustrated in Fig.7. In addition those painting processes of non-expert A and B are shown in Fig.8 and Fig.9, respectively.

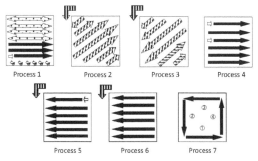

Figure 7 Schematic diagram of painting method in other person's brush (expert).

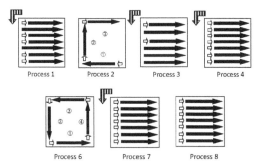

Figure 8 Schematic diagram of painting method in other person's brush (non-expert A).

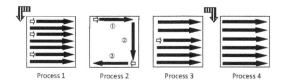

Figure 9 Schematic diagram of painting method in other person's brush (non-expert B).

In the case of the experts, the painting processes with other person's brush were found almost same as the case of own brush. Additionally the number of rotation of the board decreased and the timing of the rotation was different against the personal brush. In the case of non-expert A, her number of painting process when she used the other brush was twice as much as that when she used her own brush. In the case of non-expert B, her painting process when she used the other brush was almost equal to the case of her personal brush; however frequency was decreased once. The relationship of total painting time and process are shown in Fig.10.

The painting times of expert and non-expert A were almost twice as much as personal brush. The painting time of only non-expert B was decreased a bit. The person whose painting time was the shortest was also the expert when other brush was used. The relative time of each subject's painting method are shown in Fig.11.

166

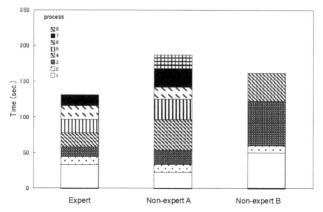

Figure 10 Relationship of total painting time and process (Other person's brush).

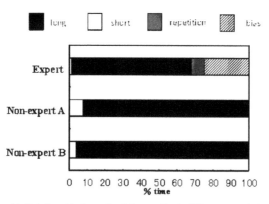

Figure 11 Relationship time of painting methods (Other person's brush).

When the expert used other person's brush, the ratio of long and bias method increased and the usage of repetition method was decreased. Non-expert A used long painting about 90% and non-expert B also used long painting about 95%. Compared to the case of personal brushes, the painting method of expert varied greatly however the painting method of both non-experts hardly varied in the case of using other person's brush.

4   DISCUSSION

It is found that expert used brush in 3 directions while non-experts only 2 directions. Therefore, it is considered that expert have more painting methods than non-experts. The expert changed painting method in the case of different brushes however both non-expert didn't change their painting methods with the difference of brushes. In addition, it is noticed that the expert stopped repetition painting on the way of process 1 when she used other brush in spite of she used repetition painting mainly when she used personal brush.

From above two points, the expert would take account of the characteristics both brushes and change her painting method with the difference of brushes by her long experience. The relative time of each subject's painting method in both case of personal and other person's brush are compared in Fig.12.

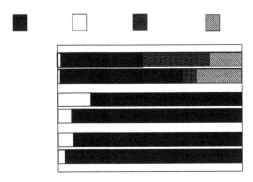

Figure 12 Relative time of painting methods (both brushes).

Only the expert changed her painting method when other person's brush was used and her percentage of painting time was changed greatly. Both non-experts didn't change their methods and percentage of their painting time was almost same no matter of different brushes. Therefore it was considered that expert flexibly coped with the difference of brushes by her way. On the other hand both non-experts used other brush just like their own brushes.

5   CONCLUSION

The painting direction of the expert was not only orthogonality but also diagonal and the usage of her brush were more varied than that of non-expert regardless of brushes. No matter of brushes, the painting time of expert was shorter than non-experts however her number of processes was more than that of non-experts.

In short, the expert changed painting method flexibly according to the difference of brushes based on her experience. However the working efficiency fell down in case when other person's brush was used even though the expert.

REFERENCES

Akinori, K. 2000, Development of Cognitive Model on Skilled Craftsperson, *Japan ergonomics society*, 36, 532-533.

Katsuyuki, S. and Emiko, U. 2009. Towards support for learn expertise skill and knowledge: a case study on top coating process of Japan ware, *Japan society for educational technology*, 09(1), 223-230.

Rika, I. Atsushi, H. Hidetoshi, N. Mariko, M. Tomohiro, T. Masaaki, M. and Michitaka, H. 2009, Extraction of Artisans' Implicit Knowlege for Skill Training, *The institute of electronics, information and communication engineers*, 109(75), 123-127.

CHAPTER 18

Subjective Evaluation of Kyo-Yuzen-dyed Fabrics with Different Material in Putting-past (Nori-oki) Process

FURUKAWA Takashi*, ENDO Atsushi*, NARITA Chieko*,
SASAKI Tomokazu**, TAKAI Yuka***, GOTO Akihiko***
HAMADA Hiroyuki*
*Kyoto Institute of Technology
Kyoto, Japan
**SASAKI CHEMICAL CO. Ltd
***Osaka Sangyo University
Osaka, Japan
t-furukawa@hishiken.co.jp

ABSTRACT

Yuzen-zome is a traditional method of dyeing in Japan. It is still one of the most popular techniques to dye. Especially, the method which a cloth is dyed in Kyoto city called "Kyo-Yuzen-zome".Kyo-Yuzen-zome has 4 kinds of dyeing techniques; Hand-write dyeing, Model-used dying, Screen dying and Mechanical dyeing. In this study, we focused on Hand-write dyeing. This method can divide into 11 Processes. Each process is handled by specialist by hand work. As you see, there are many processes to dye a fabric, which means there are many factors, which value the dyed-one. In particular, at putting-past (Nori-oki) process, expert must care which past they use, starch past or rubber-past. Starch-past gives more grace and makes it more "Hannari". Hannari is an adjective to eulogize the dyed fabric. However, nobody has defined what is hannari. Therefore, we are concerned with what is Hannari and what function makes fabrics hannari to dye fabrics better and better. In this study, effects between starch-past and rubber-past on Nori-oki process were evaluated.

Keywords: traditional, dyeing, starch, sharpness, elegance

1. Introduction

Yuzen-zome is a traditional method of dyeing in Japan. It is still one of the most popular techniques to dye. We, especially, call the method "Kyo-Yuzen-zome", when a cloth is dyed in Kyoto city. We have mainly 4 kinds of "Kyo-Yuzen-zome" , "Hand-write dyeing", "Model-used dying", "Screen dying", "Mechanical dyeing". Here, we talk mainly about hand-write dying. Hand-write dyeing, so-called "Tegaki-Yuzen" or "Itome-Yuzen" in Japanuse , is one of most popular way of dying . This method is divided into some processes each specialist works for. Those processes are all handicraft.

Mainly, hand-write dyeing needs 11 processes as we show the process chart as Figure1. We talk concretely about the process below.

1.1.   Shitae   Drafting design patterns

Fig. 1

As Fig.1 shows, a writer draw an out-line of a design pattern on plain fabric by water-soluble ink which is extracted a "blue flower."

1.2.   Nori-oki   Putting a past on the out-line

Fig. 2

As Fig.2 shows, another specialist put past on the line to prevent each dyestuff mixing over the line. At this process, you can choose which past you use, starch-past or rubber-past. Starch-past makes the fabric look softer and more tasteful.

1.3.   Nori-fuse Putting a paste to protect patterns from dyestuff

Fig.3

As Fig.3 shows, the fabric needs past to cover some patterns with to avoid dyeing particular parts as well as small pattern.

1.4. Hikizome dyeing

Fig.4

As Fig.4 shows, Hikizome is dyeing the fabric uniformly by a brush.

1.5. Mushi Steaming

Fig.5

As Fig.5 shows, the fabric is heated by 100°c steam to fix the dyestuffs on the cloth.

1.6. Mizumoto washing

Fig.6

As Fig.6 shows, at this part, a specialist washes out past, ink, and excess dyestuffs from the dyed fabric by water.

1.7. Sashi Yuzen dyeing by injection

Fig.7

As Fig.7 shows, this is another way of dyeing, injecting ink into the small part of patterns which you covered by past at Nori-fuse.

1.8. Mushi and Mizumoto

Fig.8

As Fig.8 shows, to fix the dyestuffs completely on the cloth, do Mushi and Mizumoto again.

1.9. Yunoshi ironing

Fig.9

As Fig.9 shows, the fabric needs to be smoothed by steam. This makes the fabric soft.

1.10. Kin-Kako Painting gold

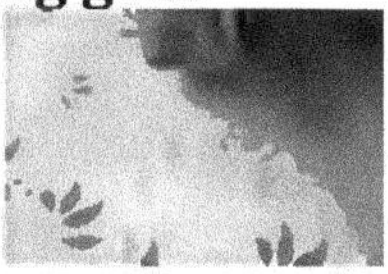

Fig.10

As Fig.10 shows, to make that cloth gorgeous, an expert ornaments that with gold and silver leaf, powder, and gold paint-stuff.

1.11. Shisyu embroidering

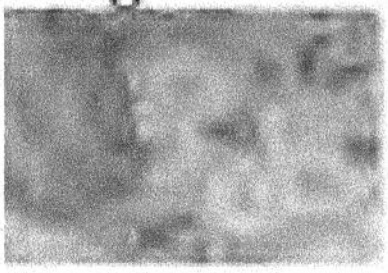

Fig.11

As Fig.11 shows, professionals embroider that cloth to make it look more luxurious and elegant.

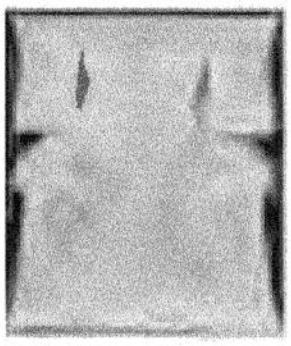

Fig.12

Fig.12 shows a finished product.

1.12. The object of this research

As you see, there are 11 processes to dye a fabric by Tegaki-Yuzen. Nori-oki is the second process. These 2 types paste, starch-paste or rubber-paste, make the different effect to fabrics dyed. The fabric to which starch-paste is used values one and a half times more than another fabric to which rubber-paste is used. To elucidate the reason, by using "How it is Hannari" as a keyword, we measure how much starch-paste and rubber-paste make fabrics "Hannari".

We clarify what kind of image people have about the word "Hannari" and what Hannari fabrics make people feel when they see them. Can we define what is Hannari?

2. Examination method

We used the following procedures in a questionnaire survey.

2.1. Samples

As Figure 13 shows, we provide two samples. On Sample A, starch-past is put and sample A is dyed and washed. Rubber-past is put on Sample B and this sample also dyed and washed by the same way of sample A. Other factors, the pattern, performer, the fabric and so on, is under same condition.

Fig. 13

Sample A   Sample B
Starch-paste   Rubber-paste
Material: pure silk Kyo-Uzen

2.2. The target of a questionnaire survey

Targets of a questionnaire survey are 50 persons, who work for fabrics, irrespective of age or experience.

1. To discuss the image of Hannari, we ask targets, for them, what the image of hannari is and ask them to answer by their own words.

2. We ask them how hannari each sample is by the five-grade evaluation system.

3. To research what factors gives hunnari to fabrics, we also ask them what they think each factors, sharpness, elegance, brightness, warmth, richness and deepness, about two samples. Targets answer by the five-grade evaluation system again.

4. We also ask targets what image these samples give them and ask to answer by description style.

3. Result

Fig.14 shows distribution and an average about Hannari.

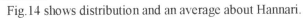
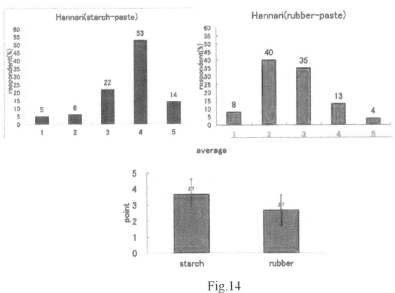

Fig.14

Using starch-paste gives fabrics more hannari than using rubber-paste does.

Fig.15 shows distribution and an average about sharpness.

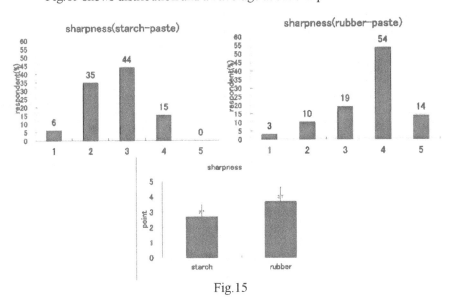

Fig.15

Using rubber-paste gives fabrics more sharpness than using starch-paste does.

Fig.16 shows distribution and an average about elegance.

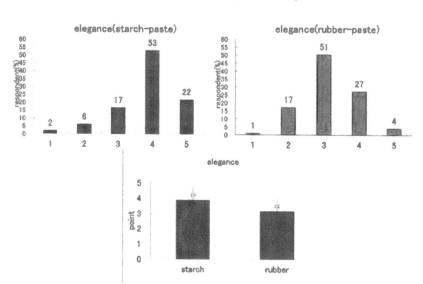

Fig.16

Using starch-paste gives fabrics more elegance than using rubber-paste does.

Fig.17 shows distribution and an average about brightness.

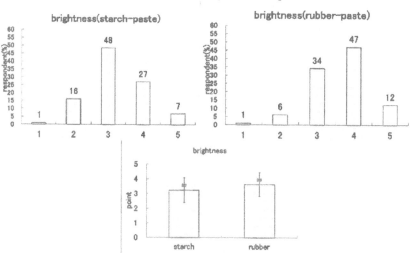

Fig.17

Using rubber-paste gives fabrics more brightness than using starch-paste does.

Fig.18 shows distribution and an average about warmth.

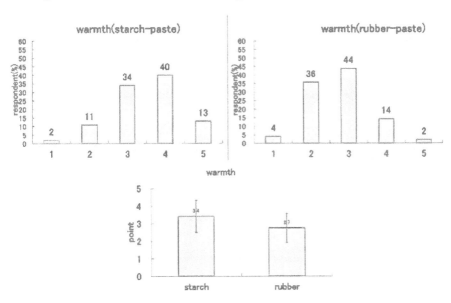

Fig.18

Using starch-paste gives fabrics more warmth than using rubber-paste does.

Fig.19 shows distribution and an average about richness.

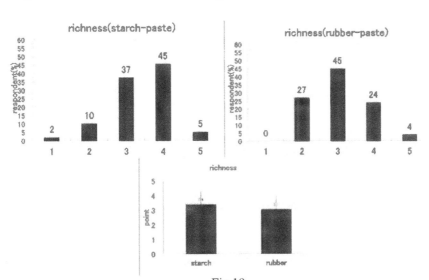

Fig.19

Using starch-paste gives fabrics more richness than using rubber-paste does.

Fig.20 shows distribution and an average about deepness.

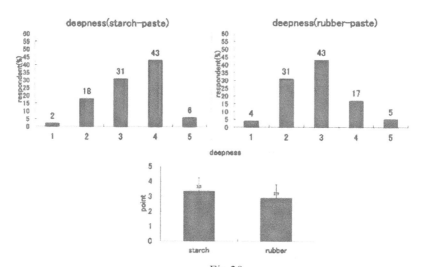

Fig.20

Using starch-paste gives fabrics more deepness than using rubber-paste does.

From these 6 distribution and an average of each factors,Sample A, using starch-paste, gets more score than Sample B, which means starch-paste can make proper Hannari of fabrics. Starch-paste wins at elegance, warmth, richness and deepness. Rubber-paste wins at Sharpness and brightness. We have many answers of what image have about "hannari." From 152 answers, we can find noticeable words, softness　(37 answers), gentleness(14 answers) and elegance(11 answers).

4. Conclusion

From this result of questionnaire survey, we can say Starch-paste gives more Hannari than rubber-paste does. However, we still have some problems to be solved.

1. Almost all targets of the survey are from Kyoto prefecture. Different culture and possibly has different image of Hannari. We must research and compare them.

2. We get a new keywords from answers of "what is the image of Hannari", softness and gentleness. We need another research by questionnaire survey with these new functions.

3. We must evaluate the condition of fabrics, not only surface but cross-section of them, to dye any fabrics "hunnarily".

4. We must analyze the result if a research if there is the gender gap to recognize "Hannari."

REFERENCE

Kyoto dyeing and weaving center (1984), Techniques of Hand-write dyeing (pp.84).

CHAPTER 19

Human Motion of Weaving "Kana-ami" Technique by Biomechanical Analysis

Kenichi TSUJI*, Yuka TAKAI**, Akihiko GOTO**,

Tomoko OTA***, Hiroyuki HAMADA***

*Kanaami Tsuji
Kyoto, Japan
tsujiken@kanaamitsuji.com

**Osaka Sangyo University
Osaka, Japan
takai@ise.osaka-sandai.ac.jp
gotoh@ise.osaka-sandai.ac.jp

***Kyoto Institute of Technology
Kyoto, Japan
promotl@gold.ocn.ne.jp
hhamada@kit.ac.jp

ABSTRACT

The concept of "MA" is emphasized in terms of time and space among lifestyle such as human relations and etiquette, sports, arts, and architecture; however, the concept of "MA" that is studied scientifically is not found. In this study, we defined "MA" as time span in aspect of acceleration, and defined two types of "MA": intermittent "MA" and continuous "MA". It was hypothesized that there exists "MA" during experts' working, and investigated it from scientific standpoint of motion analysis. This study conducted motion analysis especially of wire net knitting, and "MA" emerging during the work operation was analyzed. A wire net product, called "kana-ami" in Japanese, is a traditional craftsmanship in Kyoto all made by the hand work. "kana-ami" making process is divided into the main task and the secondary task. "MA" emerging in continuous motion and in intermittent motion was compared. As a result, it was clarified that intermittent

"MA" was proper to observe when "MA" emerged in the main task. On the other hand, continuous "MA" was suitable for comparing the main task and secondary task, and for observing "MA"-emerging time in whole work.

Keywords: wire net, 3D motion analysis

1   INTRODUCTION

"*Kana-ami*", a wire net fabrication known for a tofu scoop and a grid for toasting rice cake as shown in Fig. 1, is one kind of traditional craftsmanship in Kyoto, Japan. "Kana-ami" products are all made by exquisite hand work of superior crafts men (experts). It is considered that experts are required to control their finger motion for a long time at work because they have to make high-quality products. Finger control during producing process can be thought to become unstable by muscle fatigue, and it can be considered that experts' finger control must include a factor to decrease workload.

Watanabe et al. evaluated workload from body acceleration, and reported that higher the value of body acceleration is, the bigger workload is likely to be. Through this, it can be considered that there is a close relationship between the body acceleration and workload.

Tanaka et al. studied "MA" interval emerging during an expert knitting "kana-ami" from aspect of acceleration in hand and fingers, but the concept of "MA" has not been scientifically clarified yet.

In this study, in order to clarify the difference of "MA" of intermittent motion and continuous motion, "MA" from body acceleration during the expert's knitting "kana-ami" process was calculated.

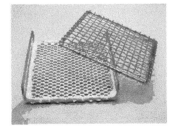

For scooping Boiled tofu   **For burning rice cake**

Figure 1 "kana-ami" products

2   METHOD

2.1   Subject

The subject was one male (63 years, 173.0cm, and 66.0 kg) who had experienced wire knitting craftsman for 44 years.

2.2   Measurement setting

The working environment was reproduced in the laboratory so that the subject might work in his usual working environment as much as possible. In addition, explanation of measurement condition was instructed to the subject and he was allowed to work at his own pace.

The three dimensional motion capture system (MAC 3D SYSTEM; Motion Analysis Co. Ltd.) was used. As shown in Fig. 2 and Fig. 3, infrared reflection markers were affixed at (1) plans of mental foramen, (2) left and right acromion, (3)left and right olecranon, (4)left and right radial styloid process, (5)left and right head of ulna, (6)left and right wrist joint, (7)left and right pollex MP joint, (8)left and right index finger MP joint, (9)left and right middle finger MP joint, (10)left and right MP joint, and (11) left and right index finger DIP joint (total 21 points).

Six cameras (hawk-I; Motion Analysis Co. Ltd.) captured the position of each marker in the X (transverse),Y (front-back) and Z (vertical) directions, and all markers position data were synchronized and taken in by a computer (sampling rate: 100Hz).

Figure 2 Photo of measurement setup

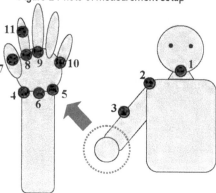

Figure 3 Position of infrared reflection markers

2.3 Measurement procedure

The subject was instructed to knit wire and make 3 lines of a wire net of a tofu scoop and other teaching was not conducted. Fig. 4 is a schematic image of making process of a hexagonal pattern of "kana-ami". The subject crossed wires twice to make one side of a hexagonal pattern, pinched other wires, and repeated them from top left to bottom right.

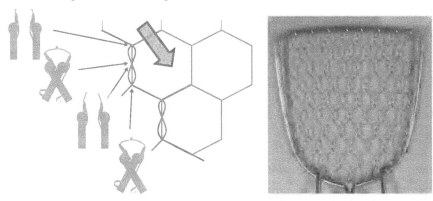

Figure 4 Schematic drawing of making process of a hexagonal pattern of "kana-ami"

2.4 Analysis procedure

In previous research, the expert's knitting movement was divided into two tasks:

Main task: from the moment that left and right fingers start knitting to the moment they totally cross
Secondary task: a phase other than the main task.

This study also used the same method. Fig. 5 illustrates schematic image of start position and end position at main task.

The analysis object was 1 line of 3 lines of tofu scoop net because the line was comprised of the largest number of the main tasks. Moreover, one task of making one side of a hexagonal pattern was chosen at random from the line. The task, which is "whole task", includes two main tasks. First main task is "main task 1" and second main task is "main task 2". Obtained data were filtered by 4th butterworth low-pas filter (6Hz).

In previous research, "MA" was calculated from velocity and acceleration of left and right DIP joint of index fingers, elbows, and shoulders. This study focused on acceleration in the same regions. Velocity v and acceleration a of each measurement site were calculated from position data obtained from the three dimensional motion analysis by using formula (1) and (2).

$$v_i = \sqrt{v_{xi}^2 + v_{yi}^2 + v_{zi}^2} , \quad v_{xi} = \frac{x_{i+1} - x_{i-1}}{2 \times 0.01} , \quad v_{yi}, v_{zi} \quad (1)$$

$$a_i = \frac{v_{i\ 1} - v_{i\ 1}}{2 \times 0.01} \tag{2}$$

"MA" was calculated in the way hereinafter prescribed. Max acceleration in each region during the task was calculated and acceleration was normalized. "MA" is defined as time when acceleration is below 10%, 20%, ..., 90% of max acceleration. Whole operating time was normalized so that the task started at 0% time and finished at 100% time. Derived items were left and right index fingers, elbows, and shoulders. In addition, common time when "MA" emerged in all regions was derived as "MA" of the upper limb. This study analyzed two types of "MA":

intermittent "MA": "MA" that appears in the main task
continuous "MA": "MA" that appears in the task of making one side of a hexagonal pattern

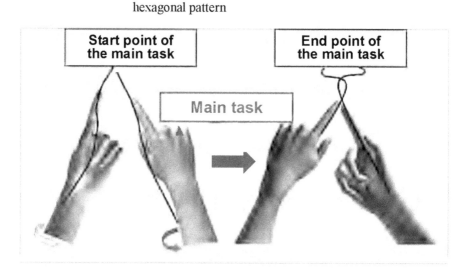

Figure 5 Schematic image of start position and end position at main task

3   RESULTS

3.1   Time of main tasks

Time of main task 1 was from 0% time to 16% time, and time of main task 2 was from 42% time to 52% time of whole task time.

3.2   Time of main tasks

Continuous "MA" in the right index finger, in the left index finger, in the right elbow, in the left elbow, in the right shoulder, the left shoulder, and in the upper limb are shown in Fig. 6-(a) to Fig. 6-(g).

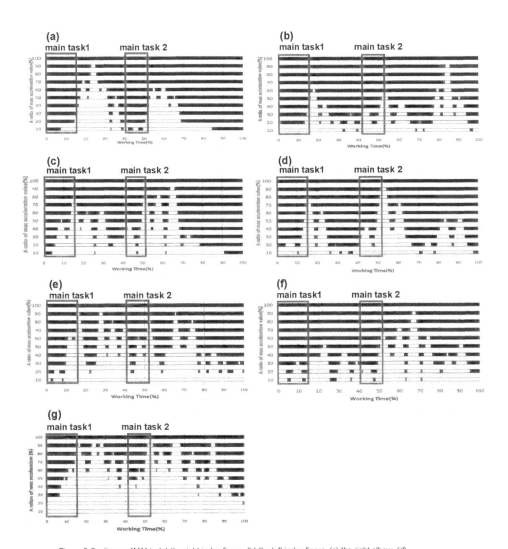

Figure 6 Continuous "MA" in (a) the right index finger, (b) the left index finger, (c) the right elbow, (d) the left elbow, (e) the right shoulder, (f) the left shoulder and (g) the upper limb

Fig. 7 shows "MA"-emerging time on 40% of threshold value, for the lowest threshold value was 40% in main task 2 in the upper limb. "MA" tended to appear around at 0%time, 40% time, and 100% time.

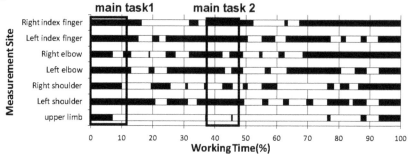

Figure 7 Continuous "MA" of each measurement site on 40% of threshold value

3.3   Intermittent "MA" calculated from max acceleration

Intermittent "MA" in the right index finger, in the left index finger, in the right elbow, in the left elbow, in the right shoulder, in the left shoulder, and in the upper limb are shown in Fig. 8-(a) to Fig. 8-(g). Where, (I) shows "MA" in main task 1, and (II) shows "MA" in main task 2. In comparison between main task 1 and main task 2, "MA"-emerging lowest threshold values around at 0% time of main task 1 was lower than that of main task 2, with the time exception of "MA" in the left shoulder and the upper limb. In all regions, there exists time when "MA" tended to emerge, and "MA"-emerging lowest threshold values around at 100% resulted in highest.

4   DISCUSSIONS

4.1   Comparison of intermittent "MA" between main task 1 and main task 2

"MA"-emerging lowest threshold values around at 0% time of main task 1 resulted in lower than that of main task 2, with the exception of "MA" in the left shoulder. The data of the left shoulder could omit, for the shoulders are relatively small motion compared to index fingers. In this case, intermittent "MA" in the upper limb showed the same trend as all regions but the left shoulder. This result can be considered that the expert decided the wire crossing position during main task 1, so that "MA" was likely to emerge, whereas main task 2 was limited just to cross wire, so that "MA" was thought to be difficult to emerge.

During main task 1 and main task 2 of all regions, "MA"-emerging time exists, so that the expert was thought to diminish muscle fatigue. Also, around at 100% time, "MA"-emerging lowest threshold values become commonly high. As the reason for it, in order to cross wire firmly, wire was presumably required to be crossed at high speed.

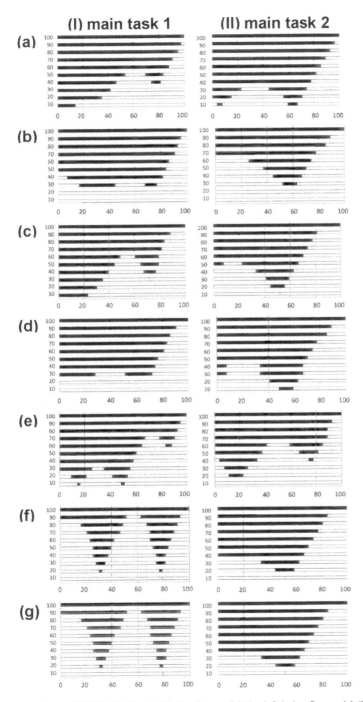

Figure 8 Intermittent "MA" in (a) the right index finger, (b) the left index finger, (c) the right elbow, (d) the left elbow, (e) the right shoulder, (f) the left shoulder and (g) the upper limb

4.2   Comparison of continuous "MA" between the main task and the secondary task

As shown in Fig. 7, "MA" emerged from immediately before to the first half of the main task. The main task 1 was an important motion for making beautiful hexagonal patterns, so that the expert moved particularly. Secondary task was, however, insignificant because arranging and pinching wires motion did not directly influence on a "kana-ami" showing. As a result, time of emerging "MA" was thought to decrease.

5   CONCLUSIONS

Intermittent "MA" was proper to observe when "MA" emerges in the main task. On the other hand, continuous "MA" was suitable for comparing the main task with the secondary task, and for observing "MA"-emerging time in whole work. Furthermore, it was clarified from this study that "MA" tended to appear in immediately before to first half of the main task, and "MA" was more likely to emerge in main task 1 than in main task 2.

REFERENCES

Tatsunori TANAKA, et al. 2007. Human motion of weaving "Kana-ami" technique by biomechanical analysis. *Proceedings of JISEE-10 Traditional craft (CD-ROM)*

Tatsunori TANAKA, et al. 2007. Motion analysis of the technique used to knit Kanaami. *symposium on sports engineering: symposium on human dynamics*: 258-261

Keiko WATANABE, et al. 2004. Workload Evaluation System Using Acceleration of Human Body and its application. *The 45th Conference of Japan Ergonomics Society*: 194-197

Tatsunori TANAKA, et al. 2008. Study of "MA" in the expert's knitting motion. *The 61th Conference of The Textile Machinery Society Japan*: 244-245

Tatsunori TANAKA, et al. 2008. "Interval" of Weaving "Kanaami". *Dynamics and Design Conference*: "326-1"-"326-5"

CHAPTER 20

Subjective Evaluation for Beauty of Texture on Metal Surface with Chasing Operation

Masaharu NISHINA*, Gen SASAKI**, Yuka TAKAI***,

Akihiko GOTO***, Hiroyuki HAMADA*

*Kyoto Institute of Technology
Kyoto, Japan
masaharu@nishina.com
hhamada@kit.ac.jp

**Hiroshima University
Hiroshima, Japan
gen@hiroshima-u.ac.jp

***Osaka Sangyo University
Osaka, Japan
takai@ise.osaka-sandai.ac.jp
gotoh@ise.osaka-sandai.ac.jp

ABSTRACT

Ornament attached on the top of a flag, and called "Hatakanagu" in Japanese, has been made by hand in Japan, traditionally. When the ornament has been made by hands, the skill of metal fitting essential for workers. Learning this skill is not easy for workers. Moreover, in the metal fitting handwork, it takes considerable amount of time. However, the skill and the process of the metal fitting have not been investigated.

From this study, it was found that the skilled worker used manufacturing tools by a bodily movement and a physical force degree. This movement might be advanced skill for workers.

Keywords: metal-fittings-of-flag, chasing, subjective evaluation

1   INTRODUCTION

Metal-fittings-of-flag which is one of the Japanese traditional craft is considered. Many industrial-arts technique elements are included in them. In a flag, they are dyeing and weaving, embroidery, and a cord & tassels. In a flagpole, they are woodwork and a lacquer-coated. The metal-fittings-of-flag is the peculiar specification which cannot be seen other than Japan. In Japan, the craftsmen who have these techniques have only several.

The metal fitting of flag is divided into the metal fittings of flagpole and the flag ornament attached in the pole tip as shown in Fig. 1. For the symbol forming by chasing is often used. That is not making figure by using hitting hammer to make dent. In this forming the figure could be swell out by using the strain of metal during chasing as demonstrated in Fig. 2. The technique is very special one in Japan, so that it is difficult to transfer the technology to the next generation. Evaluation of the metal fitting of flag products is an important skill.

In this paper, beginners and the expert were fabricated chasing specimens. The chasing specimens were evaluated by five metal fitting of flag craftsmen. Effects of years of experience of craftsman on evaluation of specimens were clarified.

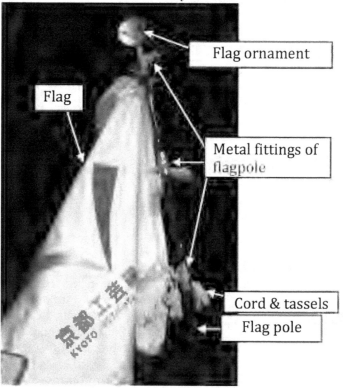

Figure 1 The flag system

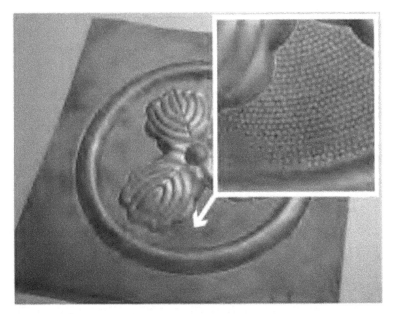

Figure 2 Example of product

2 EXPERIMENTAL

2.1 Evaluator

Table 1 shows information of evaluators. Craftsmen of the metal fitting of flag e cooperated as evaluator. The evaluator A, B and C are doing chasing work as usual. In contrast, the evaluator D and E are not doing chasing work as usual. Work experience means years of experience of the metal fitting of flag work. Working hours per day means work hours of just chasing.

Table 1 Information of evaluators

Evaluator	Gender	Age	Height (cm)	Weight (kg)	Dominant hand	Work experience (year)	Working hours per day
A	Male	66	170	71.5	Right	52	4
B	Male	42	163	65.0	Right	24	0.1
C	Male	41	164	74.0	Right	22	1
D	Male	32	173	56.0	Right	12	0
E	Male	29	173	71.0	Right	4	0

2.2 Specimen

The chasing range on specimen was illustrated in Fig. 3. Brass sheets (brass 1-class C2600P Cu70%-Zn30%) and copper sheet with a thickness of 0.6mm and 0.8mm were used as specimens. Length and width of specimen were 120 mm and 120mm, respectively. Subjects chased at one side of specimen with 50 mm length. Subjects were 22 people beginners and one expert. The beginners were university students. Table 2 shows number of specimens of each subject. Specimen No.1 to No.118 were chased by beginners. Specimen No.119 to No.178 were chased by expert.

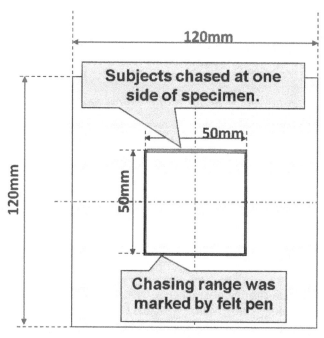

Figure 3 Chasing range on specimen

Table 2 Number of specimens of each subject

Subject	Brass sheets 0.6mm	Brass sheets 0.8mm	Copper sheets 0.6mm	Copper sheets 0.8mm	Total
Beginners	70	13	22	13	118
Expert	15	15	15	15	60
Total	85	28	37	28	178

2.3 Evaluation Criteria

Evaluators were evaluated specimens by tree-grade evaluation by only visual judgment. Evaluators had no information of subjects. Grade 1 means that it cannot use for sale, chasing does not fit into the chasing range, chasing was disappear and chasing was broken. Grade 2 means that it can use for sale with repair by expert, chasing fit into the chasing range, chasing was uniformly-distributed and chasing was a little broken. Grade 3 means that it can use for sale or evaluator wants to buy it.

3   RESULTS AND DISCUSSIONS

Fig. 4-12 show results of subjective evaluation and average grade of each specimen. Specimens chased by beginners were almost all evaluated as grade 1.However evaluator A evaluated 8 specimens as grade 2 and 2specimens as grade 3.The specimens of expert which all the evaluators evaluated as grade 3 did not exist. Even if it was the expert's specimen, which is evaluated as grade 1 existed.

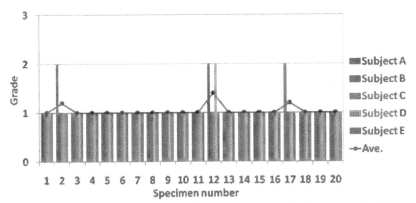

Figure 4 Results of subjective evaluation and average grade of specimen No.1-20

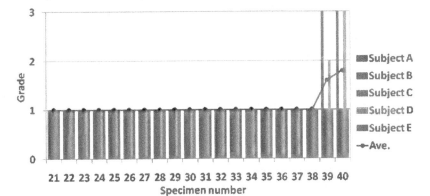

Figure 5 Results of subjective evaluation and average grade of specimen No.21-40

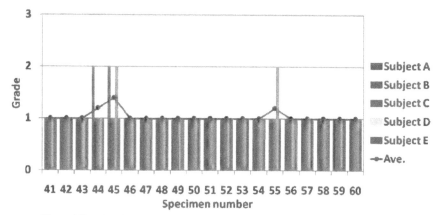

Figure 6 Results of subjective evaluation and average grade of specimen No.41-60

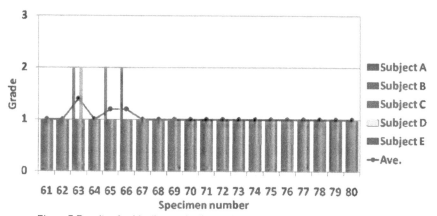

Figure 7 Results of subjective evaluation and average grade of specimen No.61-80

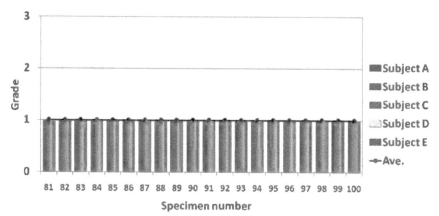

Figure 8 Results of subjective evaluation and average grade of specimen No.81-100

193

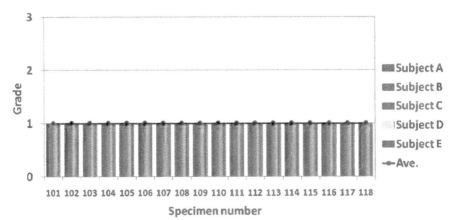

Figure 9 Results of subjective evaluation and average grade of specimen No.101-118

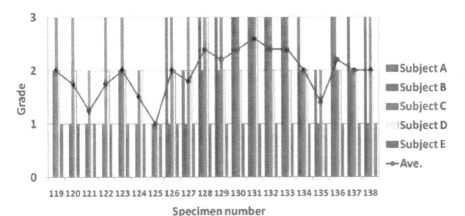

Figure 10 Results of subjective evaluation and average grade of specimen No.119-138

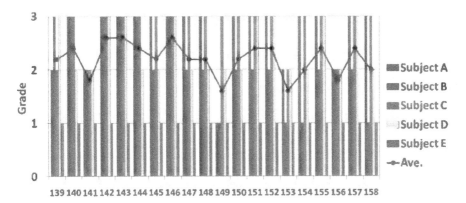

Figure 11 Results of subjective evaluation and average grade of specimen No.134-158

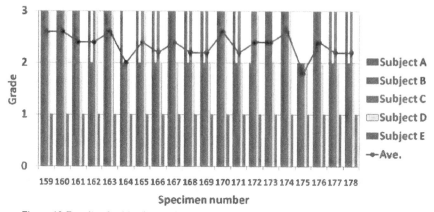

Figure 12 Results of subjective evaluation and average grade of specimen No.159-178

Average evaluation grade of beginners and expert is shown in Fig. 13. Average evaluation grade of beginners specimens were not up to 2. On the other hand, average evaluation grade of expert specimens most exceeded 2. The evaluator E evaluated all the specimens as grade 1. The evaluator E has short experience of metal fitting of flag work, it is surmised that his criteria of judgment differs from other four evaluators greatly.

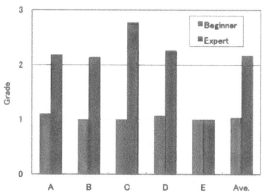

Figure 13 Average evaluation grade of beginners and expert

4   CONCLUSIONS

In this paper, the chasing specimens fabricated beginners and expert were evaluated by five metal fitting of flag craftsmen. The evaluator of longest years of experience of metal fitting of flag evaluated beginner specimen which can use for sale with repair by expert with the highest rate. The evaluator of shortest years of experience of metal fitting of flag evaluated all the specimens which cannot use for sale.

CHAPTER 21

Biomechanical Analysis of "kyo-Gashi" Techniques and Skills for Japanese Sweets Experts

Akihiko GOTO, Yuka TAKAI, Hiroyuki HAMADA

Osaka Sangyo University
Osaka, JAPAN
gotoh@ise.osaka-sandai.ac.jp

ABSTRACT

Japanese sweets called "Kyo-gashi" are made in Kyoto city, and it has long history. It is important to preserve the skill for making the sweets. In this paper we describe the finger motion during the making process, and particularly the stability of the finger motion of expert. Quanta time analysis was conducted in this paper, and the non-expert can learn the expert skill.

1   INTRODUCTION

"Kyo-gashi"(1-10) is one of the traditional sweets industries in Japan. As Japanese sweets experts decline in numbers over time, it is important to hand down the traditional techniques and skills for the next generation of sweets experts and to preserve the traditions of Kyo-gashi for Japanese society. Therefore at current study, the techniques and skills of making sweets for experts was examined to analyze the fingers and hands motions by biomechanical approach. An example of Kyo-gashi is shown in fig.1.

Figure 1 Illustration of Kyo-gashi.

"Hou-An" is one of basic techniques making "Kyo-gashi" by hand and is made by using this technique wrapping An. Zyou Namagashi and Zyouyou Manzyu which are used in tea ceremony are made by using this technique. The illustration of Zyouyou Manzyu are shown in fig. 2.

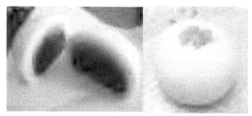

Figure 2 Illustration of Zyouyou Manzyu.

In this paper, an analysis for making process of "Kyo-gashi" sweets was conducted. Particularly, biomechanical analysis of the wrapping "An" (Hou-An) motion of fingers and hands was moved done. Also stability of the finger movement was mainly discussed.Fig.3 shows the final shape of "Wrapping An" in this paper. The height of this sweets was approximately 30 mm and the projected diameter was around 50 mm. This shape called Koshidaka that rather high than globe not globe. It is said that Japanese people feel some beauty to this shape rather than globe.

Figure 3 Final shape of "Wrapping An".

2   BACKGROUND

According to our presence experiment(11,12) the wrapping process was divided into 3 processes—Phase 1, 2 and 3:
- Phase 1: Wrapping from the bottom to the middle position of sweets.
- Phase 2: Wrapping from the middle to the top position of sweets.
- Phase 3: The finalization of making sweets accurately.

Here, these 3 processes were explained in detail by using Fig.4. In the 1st phase Konashi was extended on the left palm approximately 5 cm in diameter, and An was put on it Konashi was the ingredient for "Kyo-gashi" and it was made from steaming and kneading the mixture of An (Azuki bean paste) and some flour.. The left hand was in clockwise, and was pushed and settles down by right hand fingers into Konashi. In the 2nd phase, Konashi and An was rotated in reverse clockwise to wrap the surface completely. Mainly both thumbs and index fingers were used. In the 3rd phase wrapped sweets was upside down and was rotated in both palms in order to make the final shape as shown in Fig.3. During this process sweets was pushed to the left hand to increase the height.

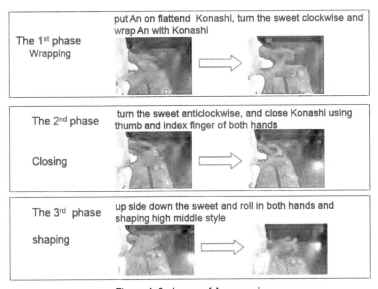

Figure 4. 3 phases of An-wrapping.

3 EXPERIMENT

A motion capture was used to measure the finger motion for the MAC3D system (version 5.0.4, Motion Analysis), a three-dimensional motion capture system with six cameras. The sampling rate of each camera was 100Hz. And data was stored in the MAC3D system (version 5.0.4, Motion Analysis) as shown in Fig. 5.

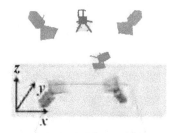

Figure 5 MAC3D System.

Regarding to the making process of "Kyo-gashi", the sweets experts put "Konashi". Next they put An on "Konashi", and then wrapped An with their fingers and hands. Therefore, the measurement of finger and hand motions was carried out, and the maker position was shown in Fig.6. A subject i.e. a male "Kyo-gashi" sweets expert with fourteen years of experience (Fig. 7).

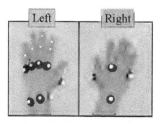

Figure 6 Markers set-up.

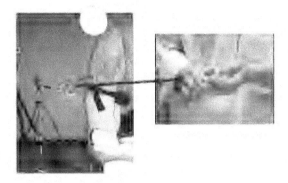

Figure 7 Movement of hand when experimenting.

The wrapping experiment was conducted 39 trials. The final shape such as weight, height and projected diameter was measured.

4   RESULTS

4.1   Shape of sweets

Table 1 shows the final shape of sweets. These values were mean value and standard deviation values in 39 trails. The measuring part was shown in Fig. 8. Surprisingly, SD. of measuring value was extremely low, and it indicated that the subject had special skill.

Table 1 Mean and S.D. values of weight of An and sweets.

Variable	(n = 39)	
	Mean	S.D.
Weight of An (g)	20.5	0.9
Weight of An &dough (g)	49.9	1.4
Hight(mm)	34.0	1.16
Diameter A (mm)	51.0	1.0
DiameterB (mm)	51.2	1.0

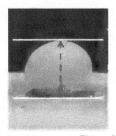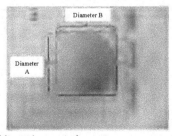

Figure 8 Measuring part of sweets.

4.2   Finger movement

The finger motion was measured by 3D system. In this paper PIP joint angle of left index finger of was pointed. Fig. 9 shows definition of angle of index finger. Extension condition was defined as 0 degree and in the finger 60 degree case was shown as example.

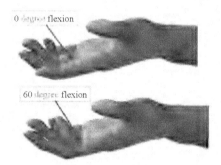

Figure 9 PIP joint angle of index finger.

Fig. 10 shows relation between flexion angle of left index finger in PIP joint and wrapping time. In the figure the classification of each phase (1,2 and 3) was also indicated. In this example 1st phase need 5.2 seconds, 3.0 seconds for the 2nd phase and 2.2 seconds for the 3rd phase.

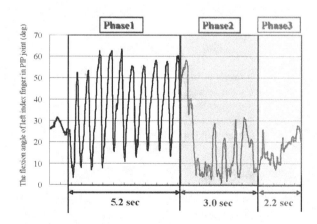

Figure 10 Relation between flexion angle of left index finger in PIP joint and wrapping time

In this case extension and flexion movement was 9 times which was obtained by counting the number of peak in the figure.

4.3 Stability of the finger movement

Fig. 11 shows PIP joint angle of index finger. The white circle was maximum extension angle, whereas the black circle was maximum flexion angle. Here also 9 times extension and flexion movement motion was performed.

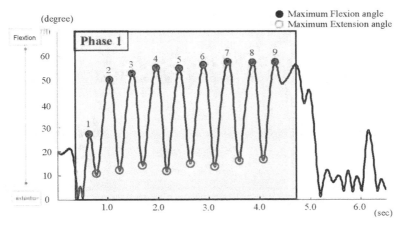

Figure 11 PIP joint angle of index finger.

Fig. 12 shows the maximum flexion angle and the maximum extension angle. The maximum flexion angle increased gradually with flexion time, and it reached 50 degree. Obviously, the scatter of the angle was small at the after 5 times. For the maximum extension angle, it can be said that the change of the angle value was small below 20 degree.

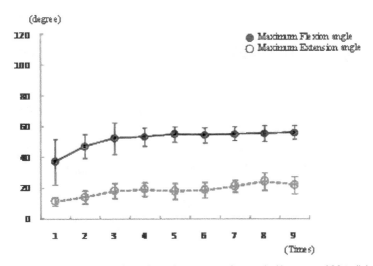

Figure 12 Maximum flexion angle and maximum extension angle (Average of 39 trails).

In order to discuss the stability of the finger movement wrapping time was measured. According to the figure (Fig. 13) two operation time were defined; the time for flexion movement and the time for extension movement.

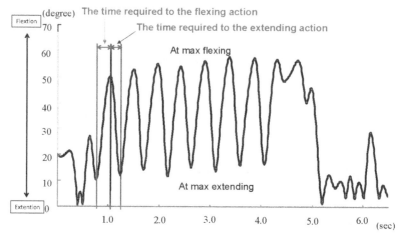

Figure 13 Phase division of flexing action and extending action.

Fig. 14 shows flexion time and extension time which were defined in Fig.13. The extension time increased slightly and reached at constant value; 0.2 seconds. Whereas the flexion time increased during 1st and 4th time movement from 0.2 seconds to 0.25 seconds. After that the value showed constant value, and from 7 times it decreased.

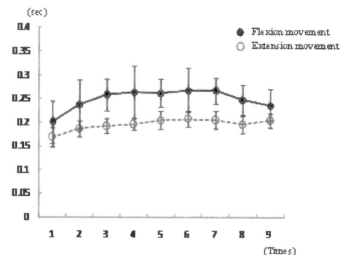

Figure 14 Coefficient of variation to the flexing and extending action (Average of 39 trails).

In Fig. 12, after 7 times the difference of the maximum flexion and extension angle was almost same. This indicated the range of movement. Therefore after 7 times, it can be said that flexion speed increased. The speed change was apparently observed and it seems that rhythm of task and comfortable tempo were created.

5   DISCUSSION

In this section, we observed the height of Konashi from side view photos as shown in Fig. 15. In the phase 1 An was wrapped in; that is the height of Konashi increased.

Figure 15 Side view photo of Konashi.

Fig. 16 shows the height of Konashi change. The height was drastically increased during 1st to 4th times, after 4th time the height was constant. The constant value was around 30 mm. According to Table 1 the final height of "Kyo-goshi" sweets was 34 mm. Therefore from the height of sweets and of Konashi until 4 times (4 times flexion and extension movement) making process was finished. That is before half of making process (4/9) task was finished. It seems that after 5 times some adjustment was performed; in the other word task was finished until 4 time.

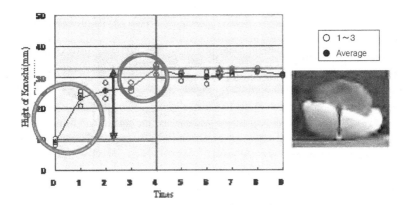

Figure 16 Change of Konashi heights.

6   CONCLUSION

In this paper "Kyo-gashi" sweets wrapping process was analyzed by using biomechanical approach. Particularly, left index finger flexion and extension movement was pointed out in the phase 1. The characteristic stability of movement was discussed. The following conclusions were obtained.

1. The angle of left index finger was changed from $20°$ to $50°$
2. The scatter of angle was small at the after 5 times
3. Making process was finished before half of the whole process

REFERENCES

1) Tatsuro Akai "Kyo-gashi". Culture magazine of Kashi. Kyoto, kawara shoten, 2005, p.207-232.
2) Keiko Nakayama. "Kashi of Genroku generation and Toraya's historical papers " Chado-gaku taikeishi kashhi and Kaiseki. Editer: Tsutsui Kouichi. Kyoto, Tankousha, 1999, p346-376
3) Japan Wagashi Association. "Method of forming process" Book of making method with sweets. Ver. Wagashi. Tokyo, Japan Kashi association center, 1979, p77-80.
4) Ryo Kato. Become healthy by soy. Tokyo. Kikurosu publisher, 2003, p.86.
5) Keiko Nakayama. Store of Wagashi. Tokyo. Shinjinbutsu-ouraisha, 1994.
6) Kouji Tani.Check Kashi in Chakai historical paper. Wagashi. 1999, 6, p.6-14
7) Nobuko Takadome. An industrial exhibition and formation of Wagashi The Japan Society of Home Economics Division of Food Culture. 2007, 3, p.1-12.
8) Kouya Nakamura. family tree of Wagashi. Kyoto, Tankousha,1967.
9) Shinichi Suzuki ,Nakano Matsumoto. Modern sweets recipe1, 2, Toyo-Bunko710,711. Tokyo Heibonsha, 2003.
10) Yasuko Nakamachi. Discussion of tools using for make sweets in Edo-Generation Wagashi. 2008, 15, p.35-60
11) Akihiro Onishi, Minayuki Shirato, Masashi Kume, Asami Nakai, Toru Ohta, Tetsuya Yoshida. Biomechanical Analysis on Technique for a Skilled Worker of Japanese-style Sweets. Dynamics & Design Conference 2007, "140-1"-"140-5", 2007-09-25
12) Akihiro Onishi, Akemi Hamada, Minayuki Shirato, Masashi Kume, Kenji Uemura, Toru Ota, Asami Nakai, Tetsuya Yoshida. Evaluation of the techniques and skills of making "Kyo-Gashi" sweets by analyzing finger motions and the weight and forms of sweets. The Institute for Science of Labour, 85(3), 108-119, 2009

CHAPTER 22

Analysis of Operation and Eye Movement Concerning Master of Wire Net

Akihiko GOTO*, Yuka TAKAI*, Tomoko OTA**,

Hiroyuki HAMADA***, Ken-chi TSUJI****

*Osaka Sangyo University
**Chuo business
***Kyoto Institute of Technology
***Kanaami Tsuji
Osaka, JAPAN
E-mail: gotoh@ise.osaka-sandai.ac.jp

ABSTRACT

Metal wire network has been used in kitchen and our daily dinner table in Japan for long time. These metal wires net are made by hand work. In Kyoto where had been capital of Japan for one thousand years, many handwork crafts were born and grow up and metal wire net is one of these crafts. During processing wire net not only finger motion but eye movement are important factors in order to keep and continue the technique and skill to the next generation. In this paper eye movement of expert in metal wire net was measured and a characteristic was discussed. It was shown that analyzed eye movement assist the skill of finger motion.

Keywords: Metal wire net, eye movement, skill

1 INTRODUCTION

In the products, conditions of high performance and multi-function have been required. However, grace is requested from the products today. A lot of products with a good grace exist in the handmade products. Therefore, it is very important to search for wisdom of the technology and the skill in the cultural properties and the tradition industries. In this paper, it paid attention to the wire net master who was one of the tradition industries, and the master's operation was measured quantitatively. Particularly, the eye movement of master in the making process of the products was measured, and the feature of the movement was examined.

2 WIRE NET(KANAAMI)

Fig.1 shows example of wire net called KANAAMI in Japanese According to Chinese character KANA is metal, and AMI is net. YuDoufu is famous winter food in Kyoto. For example left wire net has been used in YuDoufu food to scoop Toufu.

About 50 years ago, there were about 30 handmade wire net shops in Kyoto. However, it has decreased even slightly now. In this paper, it cooperated in Mr. Tsuji of "Kanaami Tsuji". The products of Mr. Tsuji are selected as excellent design one in Kyoto in 2003 and 2004. The products of kanaami which were made by Mr.Tsuji is shown in Fig.1.

Figure 1 The picture of kanaami (Tofu Server Squared, Kikumaru and Octagonal).

Copper and the stainless steel were used as the material of the wire. The mesh patterns are made by repeating operation that bends the wire and twists it. The extension degree of the mesh patterns changes according to the angle of the forefinger affixed to the wire. In addition, it changes variously according to the frequency of the twisted the wire and the angle of the bending it. Fig.2 shows how to fabricate the wire net work.

Figure 2   How to fabricate the wire net work.

3   EXPERIMENTS

Objective wire net products of this study were as follows.

(a) Tool to scoop the bean curd

Each object is shown in Fig.3. The materials used in this study were copper and stainless steels. Table 1 shows the diameter of materials.

(a) Tool to scoop the bean curb   (b) Gridiron

Figure 3 Tool to scoop the bean curb and gridiron which were used as object.

Table 1   Mean and S.D. values of weight of azuki been paste and sweets.

	(a)	(b)
Copper	0.5	1.0
Stainless steels	0.4	1.0

The measurement of the eye movement used Talk Eye II (Takei Scientific Instrument Co., Ltd.). The picture of the measurement is shown as Fig.4.

Figure 4   Measurement of eye movement when the tool to scoop the bean curd was made.

4   RESULTS AND DISCUSSIONS

From Fig.5 to 7 the eye movement and gaze point and shown. According to results, it has been understood that the eyeball moves the gaze point little by little. In addition, it has been understood that Mr. Tsuji has not been blinking for about 30 seconds. This period meant time to make mesh of one row. By analyzing the result of this eye movement in detail, it has been understood that the gaze point moves regularly. First of all, it gazes to the intersection part of the wire first. Next, it has been understood to gaze to Y character (in Fig.8) by centering on the intersection part while twisting the wire. It is guessed to confirm the overall balance of the mesh. It has been understood that time when the wire is intersected is very important. If, when the mesh in the intersection part cannot be corrected, the shape of the mesh changes for the worse gradually.

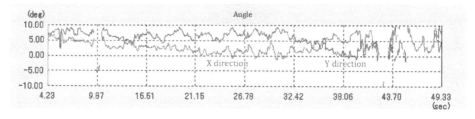

Figure 5   Eye movement part1

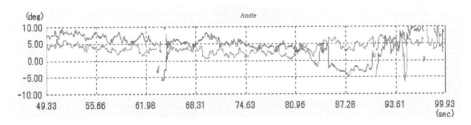

Figure 6 Eye movement and gazing point

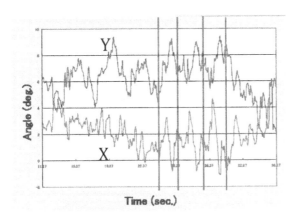

Figure 7 Eye movement of the regular fixation point was observed.

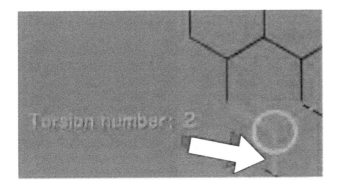

Figure 8 The movement of the fixation point.

Fig.9 shows processing time in each twisting. At the same number for exchange 1, left column was time for the first twisting, and the right was the second. In every case the first twisting time was longer. Fig.10 shows the detail of twisting movement. At the second twisting time movement speed was higher than the first twisting . By using eye after twisting point was confirmed at the first twisting the second twisting was fast. Totally the processing time decreased and also we can see rhythm of processing.

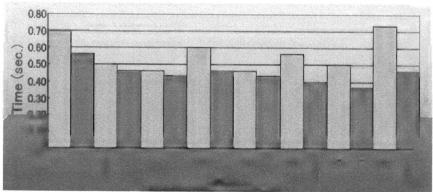

Figure 9 Time of making mesh

Figure 10 Making of one mesh

5   CONCLUSIONS

1. Master of wire net doesn't blink while being continuously twisting the wire.
2. When the wire is intersected, it gazes at the intersection momentarily. After wards, to confirm balance of the twist of the wire, the surrounding of the intersection is regularly seen.
3. Operation to which the wire is twisted by the forefinger of both hands is divided into three steps

There is the first step of the intersection, the second step of raising a left forefinger, the third step of lowering a left forefinger.

CHAPTER 23

Highly Cultured Brush Manufactured by Traditional Brush Mixing Technique "KEMOMI"

KAWABATA Shinichiro*, KAMADA Toshiyuki**, NASU Maki**, NAKAHARA Kenichi**, TSUKUDA Hiroshi **, GOTO Akihiko***, HAMADA Hiroyuki*

*Kyoto institute of technology
Kyoto Japan
hhamada@kit.ac.jp
** soliton corporation CO. LTD.
Kyoto Japan
arser3@gmail.com
***Osaka Sangyo University
Osaka Japan

ABSTRACT

To hand down the technique of the traditional handicrafts usually takes over a decade. Therefore, the shortage of successors is becoming a serious problem. Thereby there are needs of analyzing the technique scientifically to hand down the technique in less time. FUDE which is Japanese calligraphy brush have been designated as traditional products of Nara prefecture. It is impossible to manufacture high performance brush from a single raw material, however high performance brush can be obtained by blending many kinds of materials with different characteristics. This process of blending filaments with different characteristics is called "KEGUMI". The next manufacturing process of "KEGUMI" is "KEMOMI". "KEMOMI" is the work to even up the humidity mixing ratio in all the parts of the filaments bunch. Filaments are not to be able to demonstrate the performance only by mixing the filaments with different characteristics, synergy of high performance only begins to appear when filaments are uniformly distributed. In this study "KEMOMI" process was analyzed by the method of Image thresholding and motion analysis.

Keywords: brush, Traditional handicrafts, motion analysis, soliton corporation

1. INTRODUCTION

Traditional handicrafts, more precisely expressed as artisanic handicraft, is a type of work where useful and decorative devices are made completely by hand or by using only simple tools. It is a traditional main sector of craft. Usually the term is applied to traditional means of making goods. The individual artisanship of the items is a paramount criterion, such items often have cultural significance, also aimed to create new tradition of technology. The main idea of the work should be handmade using natural materials, and the fields of handicrafts are widely ranged, such as dyed and woven stuff textiles, metal work, lacquer ware, ceramic arts, calligraphy brush. As of April 2009, there are 211 traditional handicrafts in Japan, specified by the Ministry of Economy, Trade and Industry. Traditional handicraft is specified to satisfy five requirements described below. (1) What is used mainly in daily life. (2) The main portions of a manufacture process are handmade. (3) Manufactured by traditional technology and technique. (4) Traditional raw material is used. (5) The place of production is formed in the fixed region.

However, in recent years traditional handicraft industry is facing serious problems. As for traditional handicrafts main processes are handmade and since it is what is depended on advanced traditional technology, long years are needed for the acquisition to obtain the technique. Furthermore, with change of a lifestyle, the demand of traditional handicrafts articles has showed low transition, causing difficulties of training successors which has become a big subject of the whole industry.

It is not an exception in the traditional "FUDE" (Japanese calligraphy brush) industry of Nara prefecture in Japan. In recent years, an opportunity of taking the brush itself is diminishing, due to the change of a lifestyle such as using a pen instead of a brush, reduction of calligraphy lesson in school, spread of personal computers which can print brush characters easily at home and etc. Moreover, a cheap Chinese brush flows into a market in large quantities, consumer intention to low price products also helped to decrease both the amount of consumption and the quantity of production of a domestic brush.

The surrounding situation of traditional handicrafts industry is very severe as mentioned above. In order to overthrow this situation, such method as mechanizing a section of manufacturing process while leaving a traditional technique, and quantifying master craftsman's work behavior and technique as a data is highly required. Mechanizing the section of manufacturing process enables to lower the cost of manufactured goods, which will be able to oppose against cheap import products from overseas. Quantifying master craftsman's work behavior and technique has potentialities to achieve shortening successor's training period.

As described above, in recent years the demand of traditional handicrafts articles hangs low with change of a lifestyle, increase of the imported product at a low price, and the shortage of raw material, causing doubt of the continued existence. There are some companies which were obliged to discontinuance of business, and there are some cases which relocate a factory to overseas. However, there are companies corresponding to change of such a severe situation, one of them is soliton corporation CO. LTD, which is a brush manufacturing company in Japan. soliton corporation CO. LTD has succeeded in mechanizing a part of manufacturing process, also inheriting the traditional craftsmanship technique of the Nara Fude to manufacture the high quality brush which fulfills contemporary needs. Nevertheless, not all processes are mechanized, even now the important process is manufactured by the manual labor, aimed to apply the warmth of the person's hand and to receive the benefit of the wisdom from traditional craftsmanship as well as being highly efficient with high quality. In this study analysis of "KEMOMI" process, which is supposed as the most important process during brush manufacturing, was taken place using Image thresholding and motion analysis.

2. EXPERIMENTS

2-1 Analysis of "KEMOMI" progress degree

The scenery of "KEMOMI" process is shown in Fig.1 and Fig.2.

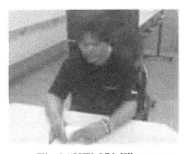 

Fig.1 "KEMOMI" scenery   Fig.2 "KEMOMI" scenery (close up)

2-1-1 Material

In this study three types of PBT (Polybutylene Terephthalate) filaments listed below was used for experiment. The taper is processed to the PBT filaments by the chemical mean (hydrolysis). As the result, row materials of PBT filaments are supplied with the rose in the angle and shape of the taper, even though in the same lot. Therefore, at the stage of "KEGUMI", measurement, selection and classification are taken place to minimize the rose of the filaments. Three kinds of PBT filaments after the selection and the classification process are mixed by 40g

214

each. Two kinds of filaments out of the three were painted red and blue on the opposite of the taper side, which allows checking the mixing degree by eyesight.

(1)520M-0.14-50 (TORAY MONOFILAMENT CO.,LTD.) Large diameter round shaped fiber
(2)SOW-W-0.10-50 (Suminoe Textile Co.,Ltd) Small diameter round shaped fiber
(3)521M-0.15-50 (TORAY MONOFILAMENT CO.,LTD.) Star shaped fiber

Three types of PBT filaments set as a bundle and cross section observation of the sample brush using SEM are shown in Fig.3 and Fig.4.

 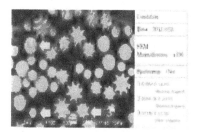

Fig.3 PBT fiber bundle   Fig.4 Cross section observation of three PBT fibers

2-1-3 Image thresholding

To analyze the "KEMOMI" process, Image thresholding analysis was taken place. For the first step, the experiment for deciding the optimal threshold value used for analysis was conducted.

The picture image of the base of the fiber bundle was taken every elapsed time during 7 minutes of "KEMOMI" processing time. In order to determine the optimal threshold value for analysis, 7 minute image, which is thought as the most progressed "KEMOMI" process was used for the first step of analysis. The image was cut to 320×240 pixel size for thresholding. To analyze the "KEMOMI" progress, distribution of the white fiber was focused and threshold value used for next analysis step was computed as the number of lumps of a white pixel becomes the maximum in the 7minutes image. The optimal threshold value presupposed as the same value from which the number of lumps of the white pixel became the maximum. The next step to analyze the share of a white pixel in bundle of fiber was carried out using this threshold.

The image of every elapsed time was partitioned to Upper, Left, Right parts for analysis and each image was analyzed based on a similar threshold computed in the first step, and the occupancy rate of the white pixel at every elapsed time was calculated. In this experiment the master craftsman showed the intention that the "KEMOMI" process was completed after 6 minutes, although extra 1 minute was added for measurement.

2-2 Motion analysis of "KEMOMI" process by master craftsman and unskilled operator

Motion analysis of "KEMOMI" process by master craftsman and unskilled operator was taken place due to achieve useful information for shortening successor's training period. Until now improvement in the technique had taken time without the ability of knowing tips of the technique. To see the difference of the motion compared with master craftsman by scientific data, it becomes possible to understand more concretely and more clearly.

Motion analysis was measured by using MAC 3D SYSTEM (made by the Motion Analysis Inc.) which is the optical real time motion capture system with 6 infrared cameras and a video camera(sampling rate:100Hz). Image of MAC 3D SYSTEM is shown in Fig.5. MAC 3D SYSTEM extracts the position of the infrared reflective marker stuck on the subject with three-dimensional coordinates. In this study 16 pieces of infrared markers were set. The "KEMOMI" process of master craftsman and an unskilled operator was recorded 3 times respectively to compare the difference.

Test subject
1. Master craftsman: 43years old male, 17 years of "KEMOMI" experience.
2. Unskilled operator: 51years old female, 8 month of "KEMOMI" experience.

2-3 Influence which a "KEMOMI" progress ratio has on finished product

150 brushes of finished product manufactured from 25%, 50%, and 100%"KEMOMI" progress degree were prepared for verification to see the influence of "KEMOMI" progress degree to finished products.

2-3-1 Brush fiber distribution observation

2 brushes were chosen as forbearance from each "KEMOMI" progress degree, in order to evaluate the mixture condition of three kinds of fiber items which forms a brush, the scanning electron microscope was used and the arbitrary section of the brush was observed to count a sum total of fiber and its items.

2-3-2 Rigidity evaluation

In order to evaluate the rose of the brush rigidity of the finished products, micrometer (MHD-50M Mitsutoyo Ltd.,) was set on the top of the scale, displaced down to 2mm and load was measured at each 0.1mm displacement. 20 brushes from each "KEMOMI" progress were chosen as forbearance.

2-3-3 Ink maintenance performance evaluation

20 brushes were chosen as forbearance to measure the amount of ink maintenance of each brush according to "KEMOMI" progress. The brush was first weigh in dry condition, subsequently was immersed in water for 3 minutes. After 3 minutes immersion in water, the brush was hung for 1minute with a condition of tip turned down. After that weight of wet condition was measured.

2-3-4 Incidence of defective rate evaluation

In order to evaluate the quality stability of a brush manufactured from different "KEMOMI" progress, defective rate of longer than reference value, shorter than reference value, slant, loop, twist, was counted on all 150 final products.

3. RESULT AND DISCUSSIONS

3-1 Analysis of "KEMOMI" progress degree

Fig.5 shows the optimal threshold value calculated from the image of 7 minutes "KEMOMI" progress. The number of lumps of the white pixel became the maximum when the threshold value was 49. The occupancy rate of the white pixel at every elapsed time was analyzed using this threshold. The analysis result of occupancy rate of the white pixel at every elapsed time is shown in Fig.6.

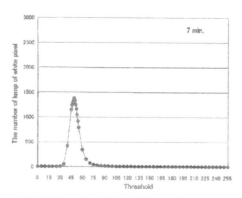

Fig.5. The number of lump of white pixel with increasing
threshold on the image at 7 min.

The share of a white pixel settled around in the passage of 3 minutes to approximately 20 percent. During 3minutes to 5minutes, change was seen only on left image which declined 10%, moreover during 5minutes to 6minutes, 10% decline was seen in right image. At the time of 6 minute progress, the white pixel share settled down to 9% and settled to 12.7% after 7minutes of progress. The share of a white pixel share showed the lowest at 6 minutes, this result shows that the experience of master craftsman matches with this analysis. From this result, "KEMOMI" progress can be divided to 3 stage as follows.

STEP 1 Start to 3minutes. The rapid mixing stage.
STEP 2 3minutes to 5 minutes. The overall adjustment stage.
STEP 3 5minutes to 6 minute. The finishing stage.

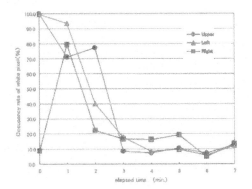

Fig.6 Change in white pixel count during "KEMOMI"

3-2 Motion analysis of "KEMOMI" process by master craftsman and unskilled operator

The results of motion analysis are described below.

1. The average number of time which master craftsman applies to a "KEMOMI" process is 6 minutes, and an unskilled operator takes 8 minutes.
2. The unskilled operator took an action of arranging the fiber bundles bottom for 35 times an average in the work for 8 minutes, which master craftsman took only 10 times in 6 minutes.
3. The unskilled operator`s right hand in earlier stage only moves 44% compare to master craftsman
4. The unskilled operator`s left hand only moves 59.2% during the process compare to master craftsman

The result shows that unskilled operator takes 2 minutes longer to complete "KEMOMI" process. Unskilled operator took time to arrange the fiber spreading

218

apart from the bundle, from there to understand that master craftsman has excellent technique to hold the bundle of fibers securely, which enables to complete the process with less arrangement.

Unskilled operator's right hand in earlier stage only moves 44% compare to master craftsman, this can be explained by the following reasons. Three fibers are solidified densely in the early stage of the process, therefore experience and technique are needed for mixing in this stage.

The reason of unskilled operator's left hand only moving 59.2% compare to master craftsman is that unskilled operator has less skills to hold the fiber bundle securely. She had an anxious of spreading the fibers causing her left hand motion becoming small.

For urging the improvement in "KEMOMI" technique, unskilled operator needs to holds below the central point of the fiber bundle, and reduce the number of times of the motion to arrange a fiber bundle, moreover, needing to be conscious of moving a left hand greatly and practicing the early stage of the process.

3-3 Brush fiber distribution observation

Table.1 shows the result of fiber distribution observation of each "KEMOMI" progress. In 25% progress brush, rose was seen on both the total of fiber and its items. If there is a rose in the number of total fibers, brush of softer than a standard or a stiffer than standard brush will be manufactured irregularly. Furthermore, in the state where three kinds of fibers are not blended equally, it affects the performance of writing feeling and operability of the brush.

Result of 50% progress brush, the sum total fiber number became close to a standard value, but three kinds of items greatly differed in the number of the large diameter size fiber and a star-shaped fiber. The number of sum total fibers and the items became the range of a standard value only after becoming a progress ratio 100%. As shown in the result, the stability of finished product stabilizes only when filaments are uniformly distributed.

Table.1 result of fiber distribution observation

25% KEMOMI Progress			
	Total fibers	Large Diameter Fiber	Small Diameter Fiber	Star shaped fibers
Sample A	649	142	339	169
Sample B	553	123	176	254
50% KEMOMI Progress			
	Total fibers	Large Diameter Fiber	Small Diameter Fiber	Star shaped fibers
Sample A	464	180	40	244
Sample B	479	281	36	162
100% KEMOMI Progress			
	Total fibers	Large Diameter Fiber	Small Diameter Fiber	Star shaped fibers
Sample A	484	237	45	202
Sample B	470	243	42	185

3-4 Rigidity evaluation

The results for rigidity evaluation are shown in Fig.7. When the "KEMOMI" progress stage is low, the rose can be seen in the load. This result dues because the three kinds of fiber materials has not become uniformed in the low stage of "KEMOMI" progress, the rigidity increases when the percentage of large diameter filaments increase and the rigidity decrease when the percentage of small diameter filaments increase. The rigidity is an important element that controls the performance of the brush. Therefore it is important for the brush fibers to be uniformly distributed to secure a stable performance.

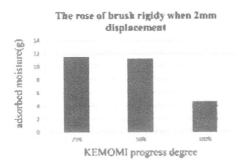

Fig.7 Result of brush rigidy evaluation

3-5 Ink maintenance performance evaluation

The result for coefficient of water absorption is shown in Fig.8. The coefficient of water absorption reduces in proportion to "KEMOMI" progress, this is to consider that distribution of the small diameter fiber has influenced. When the small diameter fiber is not distributed uniformly, large space will be made among the large diameter fibers, causing capillary phenomenon to occur which remains the ink inside a brush.

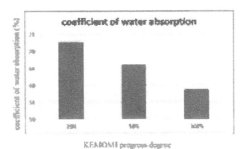

Fig.8 Result of coefficient of water absorption

220

3-6 Incidence of defective rate evaluation

Fig.9 displays the rejection rate of each brush manufactured by different "KEMOMI" progress. When the stage of KEMOMI progress proceeds, rejection rate of longer than reference value, shorter than reference value, slant, decreases sharply. This result have been caused because when the "KEMOMI" progress stage is low, fiber material of the brush are not uniformly distributed, and thereby fibers will not be inserted constantly in the picker which grips the amount of one brush. This will also give influence to next process of fiber sheath insertion and vibration, which causes irregular arrange of the fiber materials which leads to increase of rejection rate.

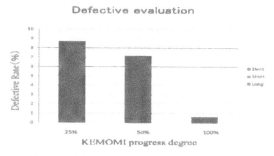

Fig.9 Incidence of defective rate

4   CONCLUSIONS

In this study "KEMOMI" process (process to even up the humidity mixing ratio in all the parts of the filaments bunch) during brush manufacturing was analyzed using Image thresholding and motion analysis. The result made it possible to achieve useful information for shortening successor's training period. Moreover the influence given by "KEMOMI" progress degree concerning the performance of the highly cultured brush manufactured by the machine was examined. As a result, "KEMOMI" progress degree gave extensive influence on brush performance such as fiber distribution, brush rigidity, ink maintenance and defective rate of the finished products. The results confirm that it is very important to distribute the material filaments uniformly, therefore improving one's ability of "KEMOMI" process in less time will be required to manufacture a high performance cultured brush.

REFERENCES

A. Yoshida and M. Nihei.et. (1996) Investigation of test method for stiffness of toothbrushes: (Part 2) Correlation among different test methods 316-317
M. Nihei and A. Yoshida.et. (1996) Investigation of test method for stiffness of toothbrushes

Virtual Agent Assistance for Maintenance Tasks in IPS² – First Results of a Study

Ulrike Schmuntzsch, Christine Sturm, Ralf Reichmuth, Matthias Roetting

Technische Universitaet Berlin
Berlin, Germany
{usc, cst, rre, mro}@mms.tu-berlin.de

ABSTRACT

In this paper we present the making of our animated assistant *Anastasia* and the interim findings of two associated experiments. In these participants, were instructed how to change a spindle on a micro milling machine. Results of the first explorative study with 10 participants were used to enhance and edit the instruction video. In the second experiment, 34 participants compared the modified video instruction with a commonly used instruction in written form (within-design). Firstly, the majority of participants preferred the video instruction. Secondly, the perceived workload was lower while being instructed by video. Thirdly, the items of the SAP questionnaire (Laugwitz, Schrepp & Held, 2006) such as *attractiveness* and *originality* were more ascribed to the instruction video. Fourthly, the perception of *Anastasia* regarding *facilitating learning* and *credible* (Ruy & Baylor, 2005) was in accordance with the image of an ideal virtual assistant.

Keywords: virtual assistant, animated assistant, pedagogical agent, maintenance task, industrial application, IPS²

1. INTRODUCTION

Industrial companies all over the world are nowadays confronted with growing global competition, which forces them to specialize and differentiate their products and services. In the resulting highly automated systems, such as Industrial Product-Service Systems (IPS²) (Meier & Völker, 2008), the human operator is in "supervisory control", which includes monitoring the machine and performing tasks that cannot be automated efficiently (Sheridan, 1997). Conceivable tasks are maintenance or overhaul. These oftentimes unusual, sudden and forced human interventions are crucial for system safety and economical factors (Reason, 2003).

With this in mind, our project focuses on the multimodal design of instructions in IPS². Due to heterogeneous user populations and contexts of usage, the given user support has to provide practical and easily comprehensible information on actions, technical processes and potential hazards. A prominent method to successfully support a multitude of users is to provide instructions by an animated character.

In this paper we present an approach on how an animated character can be used for providing instructions to human operators working in IPS². For demonstration purposes the overhaul scenario "changing a spindle" on a micro milling machine was used. To explore how our animated character is perceived by users two experiments were conducted. Thus, this paper contains on the one hand a description of how this animated character was created. On the other hand interim findings of both experiments and future considerations are depicted.

2. THEORETICAL BACKGROUND

2.1 Conceptual classification of animated characters

Animated characters can be subdivided into humanoid and non-humanoid types. For humanoid types exist numerous terminologies, such as *virtual human* (Rickel, 2001) or *digital human* (LoPiccolo, 2002), *digital actor* or *synthetic actor* (Thalmann, 1995), *animated pedagogical agent* or *virtual (interface) agent* (Rickel & Johnson, 1999). Though seeming similar, they have slightly different meanings.

For our purpose an *animated pedagogical agent* is appropriate. Animated pedagogical agents are used to support learning processes (see also chapter 3.1). They have sufficient knowledge about the content and can react properly towards user inputs (Lewis, 1999). A second interesting group are *virtual (interface) agents*, which can be split into *avatars*, *assistants* and *actors*. Relevant to us are *assistants* and *actors*. The first one supports users mastering their tasks, for instance by answering questions. The second type represents a computer-based human acting in a virtual setting. Its behavior can not be influenced by the user (Mase, 1997).

2.2 Fields of application and benefits of animated characters

Animated characters are widely used in different application fields, mainly in computer games and entertainment, e-commerce and learning environments. Functioning as instructor, advisor, motivator or companion, they not only support users by giving stepwise instructions, but also by evoking the feeling not to be left alone with a problem (Fröhlich & Plate, 2000). It is shown that humanoid characters can contribute to an increased user satisfaction and entertainment value (Lester, 1997). As to the human-like appearance Johnson, Rickel & Lester (2000) found out that a true-to-life representation of a virtual character leads to an increased user motivation and readiness to work. Its mere presence provokes a pleasant working atmosphere for users. Additionally, female characters are perceived to be friendlier and more helpful, whereas males are considered to be more authoritarian and dominant (Ziegeler & Zuehlke, 2005). Also, for general handling information or advice a warm and friendly tone is recommended, whereas important warnings should be presented in a steady voice, which rises according to the importance (Tanaka, Matsui & Kojima, 2011).

In learning environments of technical contexts, where factual but also practical and often complex knowledge is required, there are much less examples for user support (Ziegeler & Zuehlke, 2005). The animated pedagogical agent *Steve* is relatively well-known and functions as tutor teaching students how to use or overhaul complex machines (Rickel & Johnson, 1999). According to Rickel (2001) animated pedagogical agents like this unite many advantages, especially for industrial applications. One is the provision of *interactive demonstrations* to teach users how to perform physical tasks (e.g. the repair of equipment). That way, different sequences of events, which have to be carried out to fulfill a task, can be reproduced stepwise. As a result, a better understanding of the needed action knowledge can be acquired. Rickel (2001) also stated that the agent's *gaze and gestures* are very suitable as *attention guides,* because of their familiarity and human likeness. Other important advantages of animated pedagogical agents are their expressed *emotions and personality* as well as the associated *story and character*. These can be seen as key factors to trigger user's emotions and with that increasing the learning motivation as well as capturing and holding attention.

3. THE DESIGN OF THE ANIMATED ASSISTANT *ANASTASIA*

3.1 Implications from the theoretical background

Our animated character can be defined as a *virtual (interface) agent* who serves as an *actor,* implying that the user cannot converse with the actor while being instructed. Due to conceivable positive motivational effects on users, the decision was made in favor of a human-like character like the above-mentioned pedagogical agent *Steve.* Comparing *Steve* with our animated character some similarities are recognizable such as the focus on a technical context including a machine and associated technical procedures, the pedagogical approach and the demonstration of an overhaul task by providing instructions stepwise. We chose to create a female character, since they are perceived to be friendlier and more helpful instructors. Moreover, we decided to record a natural voice, because research points out its benefits in contrast to artificial sounding voices. Parallel to the character's action seen in the video sequences the recorded female voice-over explains in third person narration what to do. The verbal instructions are in imperative, e.g. *"Use the 5 Allen*

224

key for mounting the left screw.". Videos contain step-by-step instructions on how to change the spindle as well as safety instructions, like for instance *"Caution! The spindle is very heavy!"*. Voice intonation changes from friendly to steady to differentiate normal instructions from warnings.

We strove to take pedagogical advantages of our animated character by guiding user's attention through gaze and gestures and by utilizing "personality benefits". Our female character is named *Anastasia*. A short introduction sequence is displayed to familiarize users with *Anastasia*, whose name is an acronym for "<u>an</u>imated <u>as</u>sistant for <u>t</u>asks in <u>i</u>ndustrial <u>a</u>pplications". Besides being a catchy name, its acronym is very appropriate, since it indicates both *Anastasia*'s function and her field of application.

3.2 Modeling of *Anastasia* and animation process

The first step in the design process of the instruction video was to model the animated assistant *Anastasia* with the software tool *MakeHuman 1.0 alpha6.0* (Figure 1). The animated character then was imported to *blender* (Figure 2). To make *Anastasia* look more natural and familiar, the model was modified through several small adaptations, which are not explained in detail here.

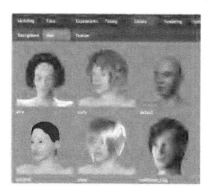 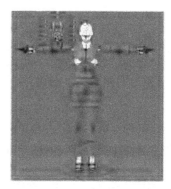

Figure 1. *MakeHuman* interface.   Figure 2. *MakeHuman* model imported in *blender*.

Since the user acts with several tools, models of the micro milling machine with spindle (Figure 3, center, right and Figure 4, right), torque handle, allen key and jaw spanner were created with *Autodesk Inventor 2012* and adapted in *blender*. These models were exported in low resolution as *CAD* model to *stl* file format. Micro milling machine and torque handle were divided into many *stl* files so the models could be colorized and animated in *blender*.

225

Figure 3. Real replica of a micro milling machine (left), replica of a micro milling machine in *Inventor* (center) and replica of a micro milling machine in *blender* (right).

Moreover, models of the power supply, operating computer, laptop, table and background were designed in *blender*. The animation process was divided into two parts. The first was the animation of *Anastasia* and the technical equipment (Figure 4). Rotation and position of the body parts were saved in specific key frames. The data of the other frames were calculated by *blender* with the interpolation mode "Bezier". By mirroring the key frames of the demounting video the mounting of the compressed air pipe, the air purge pipe and the electric cable of the micro milling machine were animated.

Figure 4. Introduction of *Anastasia* (left) and *Anastasia* using torque handle on the micro milling machine in *blender* (right).

The second part of the animation process was integrating the sounds. *Apple's GarageBand '11 version 6.0.4* was used to record a female voice in *mp3* file format. These *mp3* files were included in the "Video Editing" view in *blender* and put to the right frame position, so the sound was synchronous to *Anastasia's* actions. As a last step, the entire animation was exported as short *avi* video sequences of the individual work stages (codec *H.264*, audio codec: *mp3*, resolution: *1280x800*).

4. THE EXPERIMENTAL EVALUATION OF *ANASTASIA*

4.1 First experiment: Explorative pre-study

4.1.1. Procedure

The aim of this explorative pre-study was to work out what improvements can be made in the instruction video to enhance understandability and acceptance. The experiment was run in November of 2011 with 10 university students (9 male) aged 19 to 29 years and lasted about 45 minutes per participant. First, participants were supposed to change a spindle on a replica of micro milling machine with help of the instruction video. Second, after completing the spindle change task, participants were asked about their experiences with the instruction video.

4.1.2. Questionnaires

Two surveys and an interview were used to find out how *Anastasia* was perceived in general and how participants liked being instructed by an instruction video. In the first questionnaire participants rated the video with regard to its suitability as instruction. This contained questions on how helpful and understandable the video was. Ratings were assessed by a 5 point scale ranging from "very good" to "very bad". The second survey consisted of two semantic differential scales, where for one thing participants rated how they perceived the instruction video (e.g. *structured-unstructured, comprehensible-incomprehensible*) and also indicated what impression they gained from the virtual character *Anastasia* (e.g. *natural-artificial, competent-incompetent*). After finishing the surveys, participants were interviewed to find out the motivation behind their ratings.

4.1.3. Results

Results of the first explorative study were used to enhance and edit the instruction video and the virtual character *Anastasia*. Changes made in the video included fading out and fading in between video sequences to make transitions more smooth and natural and to give participants more time to carry out the task. A short introduction video (Figure 4, left) of *Anastasia* and her purpose was created and shown as the very first sequence (*"Hello and welcome! This is Anastasia, your virtual assistant. Anastasia now shows how to change a spindle. ... "*). Also, to finalize the instruction video a closing sequence was added at the end to say goodbye (*"Excellent, you've made it. We hope Anastasia was of help to you. Thank you and goodbye!"*). Another sequence was added to emphasize the significance of the most critical warning in this context. When the warning (*"Caution! The spindle weighs 3,5 kg. "*) is given *Anastasia* faces the viewer and raises her index finger.

4.2 Second experiment: Comparative study

4.2.1. Procedure and experimental design

Main purpose of this second experiment was to test the modified instruction video against a commonly used instruction in written form. The study was carried out with 34 participants (17 male) in January and February of 2012 and lasted about 90 minutes per individual. Participants who took part in experiment 1 were excluded from the participation in experiment 2. Among the participants were 25 students; the remaining were employed, unemployed or retired. Age range was 19 to 51 years (mean 27.8, SD = 7.3). All of the participants were right handed. Participants received a monetary reward for taking part.

The spindle change task was identical to the one in the first study. It consisted of the two parts: demounting (removing old spindle) and mounting (installing new spindle). Remember that in the first explorative study all instructions were exclusively given by the instruction video. However, in this second comparative study participants were instructed with both a video and a written instruction. The written instruction came in form of a *pdf*, displayed on a laptop. Participants got either the written instruction in part one and the video instruction in part two of the spindle change task or vice versa (see Table 1). This within-design allows direct comparisons between video and written instruction, since every participant gains experiences with both types of instruction.

Table 1 Experimental design of the second study

	Spindle change task	
	Part I: demounting	**Part II: mounting**
type of	written instruction (1st)	video instruction (2nd)
instruction	video instruction (1st)	written instruction (2nd)

The independent measure was the type of instruction which existed in two versions: a written instruction and a video instruction. Dependent measures can be extracted from *4.2.2 Technical equipment and questionnaires*. However, in this paper only very first results of the assessment of instruction types and the perception of *Anastasia* are presented. For this we selected the SAP semantic differential scale (Laugwitz, Schrepp & Held, 2006) and the *Agent Persona Instrument* (API) by Ryu and Baylor (2005).

4.2.2. Technical equipment and questionnaires

Participants performed the spindle change task on a replica of a micro milling machine as seen in Figure 3 (left). All required tools lay at hand in a box. To the right of the replica a touch screen laptop was positioned for presenting both the instruction video and the written instruction in form of a *pdf*.

In the beginning of the study, a questionnaire was handed out, which included several person specific variables (handicraft skill, technology affinity, experiences with virtual characters). Also an assessment on how the ideal virtual assistant looks like to participants was made by using the API questionnaire. The following spindle change task was divided into two parts: demounting and mounting. After every part several questionnaires were to be filled in by participants. These included the NASA Task Load Index (TLX) to measures the workload during the spindle change task. Furthermore, SAP semantic differential scale for general assessment of each instruction type was used. In the case instruction took place by means of the video, the API was additionally applied to find out how *Anastasia* was perceived.

After having been instructed with both video and written instruction, a final questionnaire was handed over. Participants were supposed to indicate which of the instruction types they would prefer supposed they had to carry out a similar maintenance task. Moreover, participants were asked to state which of the instruction types they thought was better than the other concerning comprehensibility, simplicity and efficiency. Finally, a closing interview was held based on the given answers in the questionnaires to find out the underlying reasons.

4.2.3. Results

When asked which instruction type would be chosen to perform a similar task, 29 of 34 participants (85%) preferred the video instruction to the written instruction. The difference between preferred instruction types was significant ($\chi^2(1) = 16.9$, p < .01). Participants were also asked which type of instruction they perceived as more stressful. 22 of all 34 participants (65%) stated that workload was higher during the written instruction, whereas only five (15%) participants found the video instruction to cause higher workload. Seven participants (20%) thought that both instruction types induced an equal level of workload. The differences among the instruction types are significant ($\chi^2(2) = 15.2$, p < .01). Moreover, workload was significantly related to the preferred instruction type ($\tau = -.35$, p < .05). That means workload was rated lower for the preferred video instruction compared to the less preferred written instruction where workload was rated higher.

To explore further aspects of both instruction types the SAP questionnaire was used. The categories were *attractiveness* (a), *originality* (o), *reliability* (r), *stimulation* (s), *efficiency* (e) and *foreseeability* (f). All six attributes are more ascribed to the video instruction. The four categories with significant (p < .05) differences are *attractiveness* ($\chi_a^2(2) = 10.7$), *originality* ($\chi_o^2(2) = 18.4$), *stimulation* ($\chi_s^2(2) = 6.1$) and *foreseeability* ($\chi_f^2(2) = 3.8$). The remaining categories *reliability* ($\chi_r^2(2) = 1.7$, p = .08) and *efficiency* ($\chi_e^2(2) = 0.9$, p = .26) are not significant.

To compare the perception of *Anastasia* with the image of an ideal virtual assistant answers of the API questionnaire were assessed (Figure 5).

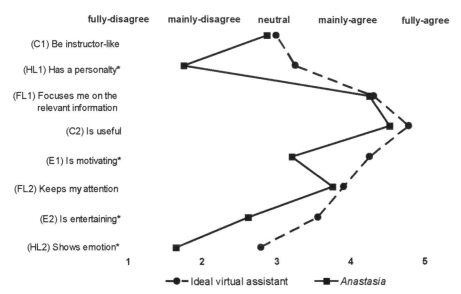

| | fully-disagree | mainly-disagree | neutral | mainly-agree | fully-agree |

Figure 5: *Anastasia* in comparison to ideal virtual assistant on selected API items (*p < .05)

API factors were *facilitating learning* (FL), *credible* (C), *human-like* (HL) and *engaging* (E). Exemplarily eight items were selected – two items of each factor. As seen in Figure 5 *Anastasia* came very close to an ideal virtual assistant in categories *facilitating learning* ($t_{FL1}(33) = .297$, $p = .77$; $t_{FL2}(33) = .59$, $p = .56$) and *credible* ($t_{C1}(33) = .51$, $p = .61$; $t_{C2}(33) = 1.66$, $p = .11$). However, as to the factors *human-likeness* ($t_{HL1}(33) = 5.9$, $p < .05$; $t_{HL2}(33) = 4.6$, $p < .05$) and *engaging* ($t_{E1}(33) = 4.5$, $p < .05$; $t_{E2}(33) = 3.3$, $p < .05$) the perception of *Anastasia* shows significant deviations from the image of an ideal virtual assistant.

5. IMPLICATIONS AND FUTURE CONSIDERATIONS

To sum up, results of the second experiment show that participants clearly prefer being instructed by *Anastasia*. One reason for this could be, as already described by Rickel (2001), that the help of an animated character is seen to be foreseeable, useful and less stressful. Like an ideal virtual assistant *Anastasia* is considered to be a credible instructor and to help focusing on relevant information, to keep their attention and to facilitate learning. Thus, similar to what Rickel (2001) found out, *Anastasia* supports users performing complex physical tasks through attention guidance. This matches to the positive effect of interactive demonstrations, which was mentioned by some participants in the closing interview. Participants emphasized the benefit of visual demonstration in combination with verbal explanations.

A second reason for the clear preference of *Anastasia* could be that she was perceived to be more original, attractive and stimulating compared to the written instruction. This is in line with findings of previous studies mentioning a higher entertainment value (Lester, 1997), user satisfaction and motivation (Johnson et al., 2000) when being instructed by animated characters with a human-like appearance.

Besides these encouraging results the experiments also revealed some aspects in need of improvement. Even though many aspects of *Anastasia* are rated better compared to the written instruction, some are rated worse compared to an ideal virtual assistant. As to Rickel's (2001) aspects *emotions and personality* the perception of *Anastasia* shows significant deviations from an ideal virtual assistant regarding motivation, entertainment and personality. Participants found that *Anastasia* has nearly no personality, does not show emotions and is not particularly entertaining. Possible explanations for this poor rating could be that users are already familiar with animated characters in movies and games, which are of course much more sophisticated than *Anastasia*. Nevertheless, these results have to be critically examined since these aspects are essential to trigger user's emotions and with that facilitating a deeper learning process and ensuring permanent user motivation (Rickel, 2001). Thus, future improvements should focus on the API factors *human-likeness* and *engaging*. However, some participants found the plain and functional presentation of the video instruction to be particularly suitable and not too distracting for a task in an industrial setting. Generally speaking, the positive effects of animated characters mentioned in literature are already apparent when using *Anastasia* as instructor.

ACKNOWLEDGMENTS

We would like to thank the German Research Foundation (DFG, Deutsche Forschungsgemeinschaft) for funding this research within the Transregional Collaborative Research Project TRR 29 on Industrial Product-Service Systems – dynamic interdependencies of products and services in production area.

REFERENCES

Fröhlich, B. and J. Plate. 2000. The cubic mouse: A new device for three-dimensional input. *Proceedings of the CHI* 2000 (pp. 526–531). New York, NY: ACM Press.

Johnson, W.L., Rickel, J.W. and Lester, J.C. (2000). Animated pedagogical agents: Face-to-face interaction in interactive learning environments. *International Journal of Artificial Intelligence in Education* 11: 47–78.

Laugwitz, B., M. Schrepp and T. Held. 2006. Konstruktion eines Fragebogens zur Messung der User Experience von Softwareprodukten. In *Mensch & Computer 2006* (pp. 125–134), eds. A.M. Heinecke and H. Paul. München: Oldenbourg Verlag.

Lester, J.C. 1997. The persona effect: Affective impact of animated pedagogical agents. *Proceedings of the Human Factors in Computing Systems CHI'97 Conference*, 59–366.

LoPiccolo, P. 2002. What are 'veople' for? *Computer Graphics World* 25(11): 4–5.

Mase, K. 1997. Aspects of Interface Agents: Avatar, Assistant and Actor. *IJCAI'97 Workshop on Animated Interface Agents*, 33–37.

Meier, H. and O. Völker. 2008. Industrial Product-Service-Systems – Typology of service supply chain for IPS² providing. *41st Conference on Manufacturing Systems*, 485–488.

NASA Task Load Index (TLX): Paper-and-Pencil Version. 1986. Moffett Field. CA: NASA-Ames Research Center, Aerospace Human Factors Research Division.

Reason, J. 2003. Human Error, 2nd edn. Cambridge: Cambridge University Press.

Rickel, J. 2001. Intelligent virtual agents for education and training: Opportunities and challenges. In *IVA 2001, LNAI 2190* (pp. 15–22), eds. A. de Antonio, R. Aylett, and D. Ballin. Berlin Heidelberg: Springer-Verlag.

Rickel, J. and W.L. Johnson. 1999. Animated agents for procedural training in virtual reality: Perception, cognition, and motor control. *Applied Artificial Intelligence* 13: 343–382.

Ryu, J. and A.L. Baylor. 2005. The psychometric structure of pedagogical agent persona. *Cognition and Learning* 2: 291–314.

Shaw, E., W.L. Johnson and R. Ganesha. 1999. Pedagogical agents on the web. *Autonomous Agents*, 283–290.

Sheridan, T. B. (1997). Supervisory control. In *Handbook of Human Factors* (pp. 1295–1327), ed. G. Salvendy. New York: Wiley.

Tanaka, K., T. Matsui and K. Kojima. 2011. Experimental study on appropriate reality of agents as a multi-modal interface for human-computer interaction. *Proceedings of the International Conference on Human-Computer Interaction* 2: 613–622.

Thalmann, D. 1995. Autonomy and task-level control for virtual actors. *Programming and Computer Software* 21(4): 202–211.

Ziegeler, D. and D. Zühlke. 2005. Emotional user interfaces and humanoid avatars in industrial environments. *Proceedings of the 16th IFAC World Congress*.

CHAPTER 25

Involving Users in the Design of Augmented Reality-Based Assistance in Industrial Assembly Tasks

Katharina Mura[1], Dominic Gorecky[1], Gerrit Meixner[2]

[1]SmartFactory[KL]
[2]German Research Center for Artificial Intelligence
Kaiserslautern, Germany
mura@smartfactory.de

ABSTRACT

The main goal of this paper is the evaluation of an assistance system providing a user with on-the-fly Augmented Reality (AR)-based instructions on a Head-Mounted-Display (HMD). In two user studies we address questions concerning the handling with the hardware as well as the interaction with the system prototype.

In a first dual-task study 22 participants executed a manual task while pictures were shown through a monocular HMD and a binocular HMD successively. In a second study 27 participants had to perform on two different assembly tasks either by AR-based assistance, paper manual or video instruction. By observations, questionnaires and structured interviews we assessed the following questions. How intuitive is the control of the HMD hardware, how does the user experience the combination of real task environment and superimposed information, how successful and satisfactory performs the AR-based support compared to traditional types of instruction, what advantages as well as which problems does the concept provide, and which tasks are especially suitable for AR-based instructions.

The results showed that participants preferred the monocular HMD compared to the binocular HMD because of better adjustment and higher comfort. The AR-based assistance during assembly tasks was effective, efficient, and satisfying compared to the other methods. It was experienced as helpful and motivating, but also as less controllable.

We discuss the usage of monocular versus binocular HMD in different industrial contexts, recommend real-time instructions for tasks with high complexity, and propose gestural user interfaces in order to enable navigation.

Keywords: Head-Mounted Display, Assistance System, Industrial Task, Instructions, Augmented Reality

1   INTRODUCTION

As part of the EU FP7 project COGNITO (Cognitive Workflow Capturing and Rendering with On-Body Sensor Networks) an assistance system for supporting manual tasks is developed. An expert user teaches the system how to perform a certain task in a certain environment and with certain tools. Afterwards the system can instruct an inexperienced user automatically in executing the same task in a similar environment. An on-body sensor network consisting of miniature inertial sensors and cameras captures and interprets the activity of the expert user. This data is transferred into a workflow model. When a new user executes the task wearing the on-body sensor network, the assessed data is compared to the existing workflow model and the system derives in which step the user currently is and which step should follow. The user receives online AR-based instructions for the next step (which operation, what tool, and what component part) on a HMD. The COGNITO system integrates the detection, recognition and representation of human actions in an overall system. Thus, it responds immediately and automatically to the activities of the user and leads him step by step through a complex, manual task.

2   RELATED WORK

The use of assistance systems in industrial environments can contribute to the support of complex tasks (Tang et al., 2003). Particularly in the course of demographic change, shorter product life cycles, and rapidly evolving technologies the factory of the future will be characterized by an increasing need for effective, efficient, and user accepted support (Gorecky et al., 2011).

AR has been tested in supporting industrial activities, e.g. in aircraft maintenance (Bowling, 2008) and pump assembly (Andersen et al., 2009). Its general potential compared to conventional support methods, such as manuals, instructions or training videos has been demonstrated (e.g. Nilsson & Johansson, 2008).

One important requirement of an assistance system is to show the right information at the exact moment in time when it is needed (Stork & Schubö, 2010). The better the system understands the environment and the workflow of the task, the better it is able to adjust its information provision to the user's current situation. This requires that the system collects and processes data on movements, gaze, and environmental conditions and integrates this information into a model of the underlying task (Worgan et al, 2011). The COGNITO project provides a first

Wizard of Oz simulated prototype with which the general concept can be evaluated in an early stage of development.

3   OBJECTIVES

It is to evaluate which factors are important for (1) choosing the right hardware and for (2) the user-friendly development of the interaction concept. We involve users to investigate the following questions: How intuitive is the handling of the hardware? How does the user learn the combination and exchange of real-world information and superimposed virtual content? How successful and satisfactory is the AR-based support compared to traditional types of instruction to accomplish manual tasks? What advantages does the concept have, and what problems and suggestions for improvement exist?

4   EXPERIMENTAL STUDIES

We conducted two studies involving potential users in order to meet the specified goals. In the first one we let participants work with different hardware in order to evaluate its suitability. In the second one, we focus on the overall concept of the COGNITO system, i.e. the automatic rendering of on-the-fly instructions.

4.1   Study "Hardware"

Subjects
In the first study 22 subjects participated. The sample consisted of mostly young, male students (M = 24.6 years, SD = 3.7, 4 women and 18 men) from technically oriented subjects which were expected to have a high technological affinity.

Manual Task
The manual task was to edit a jigsaw puzzle consisting of 24 pieces with a time limit of three minutes. The parallel task subjects had to perform was to detect pieces that were displayed subsequently in random order on the HMD (for 10 sec, 3 sec pause). Each subject executed this task twice with two different puzzles.

Head-Mounted Display
In figure 1 we displayed the two HMDs in question. They allow an optical see-through display with frameless optic where the real field of view may be augmented by virtual objects.

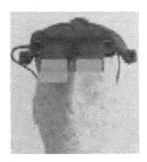

Figure 1 Hardware with information on weight (without headband): on the left side monocular (140g) and on the right side binocular HMD (320g). (Trivisio Prototyping GmbH, www.trivisio.com)

Procedure

After answering demographic questions, the participants had to set up and adjust one of the HMDs on their own until they could see a test pattern clearly. The time required and potential problems were recorded. Afterwards, they executed the puzzling task and the simultaneous detection task for three minutes. The remaining puzzle pieces after completion time have been counted. They served as a measure of performance in the manual task. In addition, the number of correctly identified images that were displayed on the HMD was determined. It provided an indicator of how well the subjects were able to pay attention to visual information on the HMD while simultaneously dealing with a manual task. A questionnaire assessed usability, visual quality and simulator sickness while using the HMD. Then, the subject conducted the same procedure again with the other HMD and with a new puzzle. Each participant gave a statement on his preference and opinion about the potential of HMDs in industrial context at the end. For successful completion of the study, which lasted approximately 40 minutes in a factory-like environment in the SmartFactory^{KL} subjects received a monetary allowance.

4.2   Study "AR-based support"

Subjects

A sample of 27 subjects participated in the second study. Similar to the first study it consisted of young students (M=24.2 years, SD=3.3) and the majority was male (6 women and 21 men).

Scenarios

We defined two ecologically valid scenarios each person had to perform. In *Nails and Screws* subjects were asked to attach two batons on a base plate with three nails and three screws in a certain order with a hammer or an electric screwdriver respectively (see left side of Figure 2). Execution time was about one minute. The second, more complex scenario *Ball Valve* lasted longer than three minutes and included the disassembling of a ball valve using two screwdrivers (right side of Fig. 2).

236

Figure 2 Overview of the scenarios' workplace; on the left side the easy task Nails and Screws and on the right side the more complex task Ball Valve.

Type of instruction

The participants were divided into three groups representing three instructional methods. The first group worked with the online Augmented Reality-based assistance whose behavior was simulated by means of the Wizard of Oz technique. The second group saw a video instruction beforehand and the third group received a guidance based on a paper manual.

Procedure

First, subjects were asked to perform the first scenario. Execution time and errors were assessed. Afterwards, participants gave subjective reports on the used method by help of a semantic differential. After the users were introduced to all three methods they completed a questionnaire on the acceptance of the HMD-based support and stated which method they would prefer for future tasks. Finally, subjects performed the same procedure for the second scenario.

In the last step, a structured interview was conducted in which subjects (of the Augmented Reality-based assistance group only) expressed their opinions on the hardware, the presentation of information, and the potential of the overall concept. Overall, the participation of each subject took approximately 60 minutes and was rewarded with a monetary allowance.

5 RESULTS

5.1 Hardware

The monocular HMD was significantly faster ready for use than the binocular HMD (1:04 vs. 1:54 minutes, t (21) =- 3.99, p <0.001). Accordingly, subjects evaluated the handling of the monocular HMD more positively than the binocular HMD (t (21) = 2.54, p <0.05). There was no difference between both HMDs regarding the quality of the visual display and simulator sickness. The latter did not seem to cause problems at all in these short scenarios. The puzzle task with simultaneous presentation of images on the HMD revealed no difference in performance between both HMDs (approximately equal number of remaining puzzle pieces). Yet, there was a tendency towards a more successful recognition of images using the binocular HMD compared to the monocular HMD (t (21) =- 1.777, p <0.10). In a direct comparison of the two HMDs the majority (N = 13) would select the monocular HMD. The reasons are summarized in Table 1. Furthermore, in structured interviews participants recommended to reduce the weight of the HMD and to optimize the fit. In order to work safely and effectively, the HMD should be wireless and easy adjustable to environmental conditions (e.g. changing exposure to light).

Table 1 Comparison of reported pros and cons concerning monocular and binocular HMDs.

	Monocular HMD	Binocular HMD
Advantages	– Better distribution of weight, lighter, more comfortable – Easier adjustment of display – Attentional focus on reality, information on display optionally ignorable	– Tendency to better visual image – More comfortable for the eyes – simultaneous attention on real task and virtual content, no active change required
disadvantages	– one-sided view distracting; active change between real task and virtual content required	– correct adjustment difficult

5.2 AR-based support

The type of instruction affects performance significantly in both scenarios (ANOVAs F (2) = 7823, p <.05 for *Nails and Screws* and F (2) = 11.519, p <.001 for *Ball Valve*). The group of video instruction had the fastest execution time, but without taking into account the duration of the video. The difference between video-based HMD support and AR based support was not significant by post-hoc test. Yet, it was shown clearly that there is an advantage of both methods compared to paper manual. The difference between paper manual and the AR-based instruction in favor of the latter was significant in the *Ball Valve* scenario (Fig. 3a and b).

238

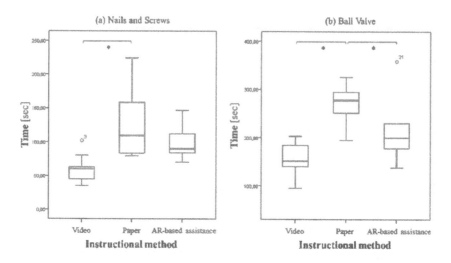

Figure 1 Execution time (y axis) dependent on the instructional method (x axis); for *Nails and Screws* on the left side and *Ball Valve* on the right side.

The analysis of the semantic differential revealed that the AR-based assistance was supportive, innovative, practical, and useful (in comparison to the other two methods it was often ranked equal or better). However, it was rated as less controllable. In addition, it turned out that the AR-based support is rated overall better when it comes to the more complex scenario *Ball Valve*. The majority of subjects preferred the AR-based approach, both retrospectively and concerning future tasks (N = 23 would prefer the COGNITO system). As advantages they noted, for example, that the hands are free to work, that probably fewer mistakes are made, that working memory is relieved, and that the method is easy to handle. In total, the participants expect that such an assistance system provides a faster and more reliable task execution. The more complex and unknown the task or the work environments are, the greater the benefits of AR based assistance systems. As a drawback it has been claimed that the motivation for independent thinking is lost. Some users felt monitored by the system, while others called for closer monitoring of the worker in order to be able to react to any kind of errors.

6   CONCLUSION

A HMD's comfort of wearing and its adjustability were crucial characteristics for the users leading to a preference for the lighter, more easily manageable, monocular HMD. The binocular HMD showed a higher immersiveness according to self-reports and the performance in the identification of puzzle pieces. The use of a binocular HMD is indicated for tasks in which a congruent AR visualization is desired (Reif, 2009) and technically possible. Virtual objects appear in this case at the exactly same place as the corresponding real objects. For instance, an arrow points at the relevant tool that is currently needed. By means of a monocular HMD such a fusion between real and virtual content is not realizable. Yet, it is still possible to show context-dependent information that is not matched with the real world (Reif, 2009). In contrast to the visualization with a binocular HMD the information displayed through a monocular HMD remains rather optional and less attracting attention. For instance, an image of a relevant tool that is currently needed appears on the display.

The AR-based assistance that the COGNITO project aims at performed favorably compared to other methods in terms of effectiveness and efficiency. Tang et al. (2003) explain that the time savings are due to the reduced demand on attention resources and working memory. The differences found between the two scenarios and the corresponding user reports suggest that the benefits of real-time instructions increase with increasing task demands. Another more complex scenario could demonstrate the suitability for the task with increasing difficulty even more clearly.

Support systems should not replace independent thinking and learning by artificial intelligences, but be conducive to learning and promote autonomous working. In this context, users criticized lacking control options when using the system. This goes along with the requirement that augmentation should not lead to automation (Stork & Schubö, 2010). Further development will focus on possibilities for user interaction. As movement tracking is already implemented in the system natural user interfaces by using gestures will be realized.

ACKNOWLEDGMENTS

This research was conducted within the EU FP7 project COGNITO (Cognitive Workflow Capturing and Rendering with On-Body Sensor Networks, Project Number: ICT - 248290, "Cognito").

REFERENCES

Andersen, M., Andersen, R., Larsen, C., Moeslund, T., and Madsen, O. 2009. Interactive Assembly Guide Using Augmented Reality. *Proceedings of the 5th International Symposium on Advances in Visual Computing: Part 1*: 999–1008.

Bowling, S.R., Khasawneh, M.T., Kaewkuekool, S., Jiang, X., and Gramopadhye, A.K. 2008. Evaluating the Effects of Virtual Training in an Aircraft Maintenance Task. *The International Journal of Aviation Psychology* 18: 104-116.

"Cognito". Accessed February 29, 2012, http://www.ict-cognito.org

Gorecky, D., Worgan, S. F., and Meixner, G. 2011. COGNITO - A Cognitive Assistance and Training System for Manual Tasks in Industry. In: Proceedings of the 29th European Conference on Cognitive Ergonomics. European Conference on Cognitive Ergonomics (ECCE-11), Designing Collaborative Activities, August 24-26, Rostock, Germany.

240

Nilsson, S. and Johansson, B. 2006. User experience and acceptance of a mixed reality system in a naturalistic setting: a case study. *International Symposium on Mixed and Augmented Reality*: 247-248.

Reif, R. 2009. Development and Evaluation of an Augmented Reality Order Picking System. TU München.

Stork, S. and Schubö, A. 2010. Human cognition in manual assembly: Theories and applications. *Advanced Engineering Informatics* 24: 320-328.

Tang, A., Owen, C., Biocca, F., and Mou, W. 2003. Comparative effectiveness of augmented reality in object assembly. *Proceedings of the SIGCHI conference on Human factors in computing systems*: 73-80.

Worgan, S., Behera, A., Cohn, A., and Hogg, D. 2011. Exploiting petri-net structure for activity classification and user instruction within an industrial setting. *International Conference on Multimodal Interaction*: 113-120.

CHAPTER 26

A New Method for Forecasting the Learning Time of Sensorimotor Tasks

Tim Jeske, Christopher M. Schlick

Institute of Industrial Engineering and Ergonomics
RWTH Aachen University
Aachen, Germany
{t.jeske, c.schlick}@iaw.rwth-aachen.de

ABSTRACT

A person introduced in a new manual assembly task has to acquire and to train the necessary skills for performing the task. These skills are usually sensorimotor skills and have to be trained at least until reaching a reference time within the task has to be accomplished. The period of time needed for this training is called learning time and could not be forecasted economically and validly until now. Due to this a new method for forecasting the learning time has been developed. It is based on empirical studies (138 participants) and reaches a high explanation of variance. The paper contains a detailed description of the new method and its development as well as a discussion and a further course of action. Additionally a new method for determining the complexity of sensorimotor tasks is presented.

Keywords: training, learning time, forecast

1. INTRODUCTION

When new production systems are launched or existing systems are restructured, employees are often assigned to tasks that are unfamiliar to them. In case of assembly areas, tasks are often performed manually and require sensory as well as motor skills. Thus, those tasks are called sensorimotor tasks and employees have to train their regarding skills. Due to that they cannot work from the very start as productively as after being trained (Rohmert et al., 1974). The period necessary to train a task respectively the requisite skills is called learning time. It starts by getting acknowledged to a new task and ends with reaching the capability to perform the task within a reference time. This reference time is usually defined by means of a target time and can be determined with the help of Predetermined Motion Time Systems (PMTS), such as Methods-Time Measurement (MTM), Work Factor (WF) or Maynard Operation Sequence Technique (MOST). In contrary to the target time, the learning time cannot be forecasted economically and reliably using MTM or any other known methods (Bokranz & Landau, 2006). This leads in new or restructured production systems to large scheduling uncertainties and for customers to a low adherence on delivery dates. Both cause disadvantages in time-to-market, customer satisfaction and market share.

To prevent this, a method for forecasting the learning time of sensorimotor tasks has been developed at the Institute of Industrial Engineering and Ergonomics of RWTH Aachen University.

2.   BACKGROUND

A forecast of learning time requires knowledge regarding an adequate mathematical description of learning time as well as regarding relevant factors influencing the learning time.

2.1.   DESCRIPTION OF LEARNING TIME

Even though learning effects have been already observed long time ago by Smith (1776) and Babbage (1832), a first theory including a mathematical description of learning time for industrial use – a so called learning curve – was developed not until 1936 when Wright investigated the production of airplanes (Laarmann, 2005). His model was based on a power function:

$$t_n = t_1 \, n^{-k} \tag{1}$$

The model describes the time t_n necessary for the nth execution of the work task depending on the time t_1 for the first execution, the number of repetitions n and the proportionality k which can be interpreted as learning velocity.

Due to the mathematical structure Wrights model predicts an infinite increasable performance. This led to criticism and to the development of two further mathematical models: De Jong (1960) used the same variables as Wright and added c as a limit towards the learning curve converges:

$$t_n = c + (t_1 - c) \, n^{-k} \tag{2}$$

Based on the same variables and the same idea of defining a limit Levy (1965) developed a learning curve using an exponential function (Hieber, 1991):

$$t_n = c + (t_1 - c) \, e^{-k(n-1)} \tag{3}$$

These models are suitable for describing the learning curve ex post. They have been parameterized and adopted for many different specific tasks (ex post); nevertheless, they cannot be applied for serious production planning.

2.2. INFLUENCES ON LEARNING TIME

The progression of a learning curve depends on numerous influencing factors. In previous studies several of these factors have been identified (Greiff, 2001). They can be separated into three areas (see figure 1): (1.) factors describing characteristics of the working person as motivation, individual predisposition or experiences, (2.) factors which are characteristic of the work task (including environmental circumstances), for example the degree of complexity, the number of movements according to a PMTS or the cycle time of a single execution, and (3.) factors that describe the learning method. The latter include the kind of work instruction as well as the method of exercise or incentives.

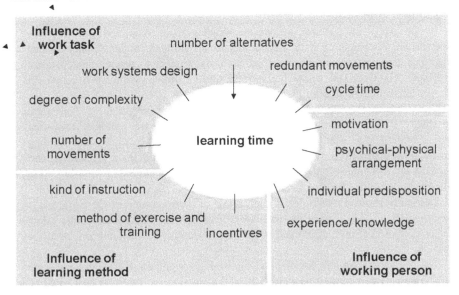

Figure 1   Areas of factors influencing the learning time. (Based on Greiff, 2001 and Jeske et al., 2009.)

3.   APPROACH

The development of the aimed forecasting method is based on an integrated analysis of three empirical studies, which focused on several influencing factors of the learning time. The analysis started with choosing a learning curve model for describing the studies' results adequately. After that relevant influencing factors of the learning time were identified out of the studies. The identified factors were used to build forecasting models for the parameters of the chosen learning curve model with the help of statistical methods. In this manner a forecast of learning curves was realized and can be applied to calculate the learning time.

3.1. EMPIRICAL STUDIES

Altogether three laboratory studies have been conducted. They focused on three different aspects of influences on the learning time: cultural influences (Jeske et al., 2010), influences of task descriptions (Jeske et al., 2011) and influences of task complexity (Jeske & Schlick, 2011; Jeske & Schlick, 2012). Thus all aforementioned areas of the learning times' influencing factors were covered by the studies.

In all studies subjects had to execute sensorimotor work tasks repetitively to ensure the appearance of learning effects. The level of significance for all statistical analyses in the studies was $\alpha=0.05$.

Cultural influences

In the study on cultural influences subjects from Asia and Europe were investigated. A sample of sixty undergraduate and graduate students was balanced regarding the cultural background as well as regarding gender (independent variables). The work task was to build up ten times a simple pyramidal structure out of a predetermined number of identical building blocks.

The time consumptions were measured for each execution as dependent variables. Before starting the main trail subjects were queried regarding demographic data as well as regarding their experience with assembly and building blocks. The instructions were supported by means of guidelines in German and respectively in Chinese. All subjects were told to have unlimited time but were asked to build up in a timely manner.

The study showed a significant main effect of the repetition of the work task's execution ($F_{(2.539,\ 147.249)}=51.034$; $p=0.000$). This learning effect had a large effect size of $\omega^2=0.352$ (Field, 2005) and could be proved also by a Bonferroni post-hoc test.

The subjects' cultural background showed no significant effect; equally no interaction between cultural background and the repetition of the work task could be found. The analysis showed no significant correlations between the subjects' experiences with assembly or building blocks and the time consumptions.

A more detailed description of the study and its results can be found in Jeske et al. (2010).

Influences of the task description

Three different task descriptions – one textual, one graphical and one animated – have been investigated regarding their influence on the learning time. A sample of sixty undergraduate and graduate students was balanced regarding gender and the educational background: Thirty subjects had an engineering education and thirty subjects were educated in other disciplines. The work task was assembling a carburetor (type: Stromberg 175 CD-2) by the help of one of the three task descriptions. The task description was assigned randomly and the task had to be executed ten times per subject. The educational background and the type of task description were designated as independent variables.

The time consumptions were measured for each execution as dependent variables. Before starting the main trail subjects were queried regarding demographic data as well as regarding their experience with technical drawings and assembly. Furthermore, their motor skills were measured according to Schoppe (1974) and Hamster (1980) and were consolidated to six factors according to Fleishman (Fleishman, 1962; Fleishman & Ellison, 1962). All subjects were told to have unlimited time but were asked to assemble in a timely manner.

The study showed a significant main effect of the repetition of the assembly task's execution ($F_{(2.197, 118.645)}=227.453$; p=0.000). This learning effect had a very large effect size of $\omega^2=0.787$ (Field, 2005) and could be proved also by a Bonferroni post-hoc test.

The type of task description showed no significant effect since it influences only the time consumption of the initial repetitions – in the long term the time consumption is independent from the type of task description. This is proved by an interaction effect between the type of task description and the repetition of the assembly task's execution ($F_{(4.394, 118.645)}=5.112$; p=0.001); it has a small effect size of $\omega^2=0.0345$ (Field, 2005).

The subjects' experiences with technical drawings were correlated significantly with six oft ten time consumptions – predominantly with the initial repetitions (from r=-0.255 up to r=-0.410). This goes along with the initial influence of the task description. The subjects' experiences with assembly were correlated significantly with every time consumption (from r=-0.265 up to r=-0.492).

A more detailed description of the study and its results can be found in Jeske et al. (2011).

Influences of task complexity

The third study was focused on the influence of task complexity and finding a measure to quantify it. For this purpose eighteen apprentices in industrial mechanics had to execute three work tasks of presumably different complexity. Due to a lack of female apprentices only male apprentices could be included in the study. The work tasks – a carburetor (the same as in the previous study), a hydraulic slider and a motor block – were designated as factor levels of the independent variable. Each subject had to execute each task for a period of two hours by the help of a graphical task description.

The time consumptions were measured for each execution of each task as dependent variables. Before starting the main trail subjects were queried regarding demographic data as well as regarding their experience with technical drawings and assembly. Furthermore, their motor skills were measured according to Schoppe (1974) and Hamster (1980) and were consolidated to six factors according to Fleishman (Fleishman, 1962; Fleishman & Ellison, 1962). The instructions were supported by means of guidelines. All subjects were asked to assemble in a timely manner.

The study showed significant main effects of the repetition of each work task's execution (carburetor: $F_{(2.625, 44.622)}=59.411$, p=0.000; hydraulic slider: $F_{(4.029, 68.493)}=28.033$, p=0.000; motor block: $F_{(3.490, 59.330)}=27.257$, p=0.000). Each learning effect had a very large effect size (carburetor: $\omega^2=0.738$; hydraulic slider: $\omega^2=0.554$; motor block: $\omega^2=0.357$; Field, 2005) and could be proved also by a Bonferroni post-hoc test. Significant correlations between the subjects' experiences with technical drawings or assembly and the time consumptions were not found.

A comparison of the task complexity was conducted on the basis of the time consumption of the first execution related to the reference time consumption in the long term according to MTM-UAS (Universelles Analysier-System). The resulting quotient was called λ, describes the reached performance related to the reference time and was significantly different between the three work tasks.

Based on this preparation the tasks characteristics were analyzed regarding correlations with λ. Analyzed task characteristics were inter alia, the cycle time, the number of parts to be handled, the number of elements of MTM-UAS necessary for describing the task as well as the first order entropies (Schlick et al., 2010) of the number of elements and of the elements according to MTM-UAS. It was found that both entropies were significantly correlated to λ. Since both entropies describe different dimensions of a task (the parts to be handled and the movements to be executed) they were summarized Euclidean. The summarized Euclidean entropy was also significantly correlated to λ (r=.737; p=0.000) and thus can be used to measure task complexity as well as its influence on learning time.

A more detailed description of the study and its results can be found in Jeske & Schlick (2011) and Jeske & Schlick (2012).

3.2.   MODEL FITTING

In order to identify a learning curve model to become the basis of the aimed forecasting model the acquired data had to be fitted to the initially described existing models.

Since a long term execution time can be predicted on the basis of PMTS this value was used and integrated as limit c in the aimed forecasting model. Thus the fitting of the existing models had to be analyzed taking the predetermined limit c into account and so was limited to the models of de Jong and Levy. For predetermining this limit MTM-UAS has been applied.

The fitting was calculated exemplarily for the assembly of the carburetor done by apprentices in industrial mechanics. The explanation of variance showed the model of de Jong ($R^2=0.7622$) as more adequate than the model of Levy ($R^2=0.4682$). This is due to the initially faster decrease of the exponential function in Levy's model compared to the power function in de Jong's model.

Based on this intermediate result a detailed analysis of the proportionality k which is interpreted as learning velocity was done. It could be found that the value of k decreases and converges against a limit. This decrease can be described mathematically with the help of the model of Levy while allocating different meanings to the variables:

$$k_n = (k_1 - k)\, e^{-r(n-1)} + k \tag{4}$$

The model describes the learning velocity k_n during the nth execution of the work task depending on the initial learning velocity k_1 during the first execution, the number of repetitions n and a proportionality r which can be interpreted as decrease of the learning velocity.

Integrating this relation into de Jong's model (formula 2) led to a multiplicative correction term:

$$t_n = c + (t_1 - c)n^{-k} \; n^{-(k_1-k)e^{-r(n-1)}} \qquad (5)$$

The explanation of variance due to this advanced model was R^2=0.7931. Further analyses showed the value of k_1 to be nearly twice the value of k and the value of r to be nearly half of the value of k:

$$t_n = c + (t_1 - c)n^{-k} \; n^{-(2k-k)e^{-\frac{k}{2}(n-1)}} \qquad (6)$$

This simplification led to an explanation of variance of R^2=0.7896. Based on these analyses the simplification of the advanced learning curve model of de Jong was chosen to be basis of the aimed forecasting model. The learning curve model and its main parameters are illustrated in figure 2.

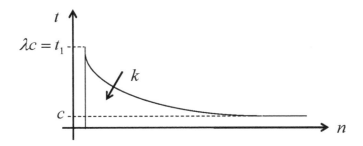

Figure 2 Parameters determining the learning curve. (From Jeske & Schlick, 2011.)

3.3.   DEVELOPMENT OF A FORECASTING MODEL

After conducting empirical studies and identifying an adequate learning curve model for describing the studies' results, it was analyzed which influencing factors of the studies are correlated to the learning curves' parameters. This was done with the objective to integrate the identified influencing factors into regression analyses which aim on developing forecasting models for the learning curves' parameters.

In this regard numerous significant correlations between the learning curves parameters and the single studies respectively the combined studies were found. The identified influencing factors were integrated into stepwise multiple regression analyses for the learning curve parameters λ and k.

For the parameter λ the analysis was done for about four steps and led to an explanation of variance of R^2=0.723. The regression model takes a constant and four further variables into account which cover all areas of influencing factors. These variables are the Euclidean entropy H, the experience with assembly E, the type of task description D and the gender G:

$$\lambda = 2{,}256 + 0{,}978\,H - 0{,}755\,E - 0{,}45\,D + 0{,}87\,G \qquad (7)$$

The stepwise multiple regression for the learning progression k was done about five steps and led to an explanation of variance of $R^2=0.351$. In this regression model were included three influencing factors of λ so that a second stepwise multiple regression was conducted which included λ as a potential influencing factor. This regression was done about six steps and led to an explanation of variance of $R^2=0.434$. The regression model includes λ as well as a constant, the first F_1 and the sixth Fleishman factor F_6 and the age A:

$$k = 0{,}141 + 0{,}073\ \lambda - 0{,}008\ F_1 + 0{,}006\ F_6 + 0{,}013\ A \tag{8}$$

A forecast of the learning time can be calculated on the basis of the forecasted parameters of the learning curve and the resulting learning curve. Since the reference time respectively the reference performance cannot be reached (due to the mathematical characteristics of the power function contained in the learning curve model) the learning time has to be calculated regarding a performance level, which is a little lower than the reference performance. This means calculating the number of repetitions necessary for reaching the chosen performance by the help of the parameterized learning curve. Afterwards, the time consumptions from the first execution up to the calculated number of repetitions have to be summarized. The resulting value is the learning time without any breaks.

4.　　SUMMARY AND OUTLOOK

The application of a statistical approach to forecast learning time led to three main results which will be summarized separately: (1.) a new measure for determining the complexity of sensorimotor tasks, (2.) a new mathematical description for learning curves and (3.) the aimed method for forecasting the learning time of sensorimotor tasks.

New Measure for Task Complexity

The application of the first order entropy for measuring a sensorimotor task's complexity regarding the time consumption for learning continues and supports first researches of Stier (1968) and Rohmert (1974). Since entropy can be interpreted as quantity of information the introduced measure represents the information which has to be learned by the working person. In this context the Euclidean sum of the entropies of parts to be handled and movements to be executed takes the interaction between both dimensions of information into account.

New Description of Learning Curves

The developed mathematical description of learning curves is based on two different kinds of mathematical functions. The exponential function affects mainly the first executions of the work task. Since subjects tend to not use task descriptions after a certain number of repetitions this exponential part of the function can be interpreted as acquirement of basic knowledge regarding the structure of the task. In contrast the power function has no specific focus. This can be interpreted to describe a continuously slowing down improvement of task specific motor skills which is determined by a constant learning velocity.

New Method for Forecasting Learning Time

For the first time a statistical approach was used for developing a model for forecasting the learning time of sensorimotor tasks. Characteristic for this approach is the prediction of the parameters of a learning curve model. This allows forecasting the time consumptions for each repetition of the task's execution as well as forecasting the number of repetitions necessary for reaching a certain performance level which is (and has to be) a little lower than the predetermined limit according to MTM-UAS. This constraint is less important since a 99% exact forecast is usually sufficient for production planning.

Since gender and motor skills are influencing factors of the forecasting model the application of the method may be restricted by legal or other constraints.

Overall the presented research shows the development of a new method to forecast the learning time based on a statistical approach. The forecasting method takes into account existing knowledge regarding influencing factors of the learning time and combines existing mathematical descriptions of learning curves to a new learning curve model. Furthermore, a new approach for determining task complexity has been developed and integrated into the forecasting method.

For completing the conducted research a validation of the described model is necessary and currently in progress. For this purpose a two-staged spur gear unit has been selected as experimental work task which has to be assembled ten times per subject.

ACKNOWLEDGEMENTS

The development of the forecasting method is embedded into the research project FlexPro which is publically funded by the German Federal Ministry of Education and Research (BMBF) and the European Social Fund (ESF); the grant no. is 01FH09019.

REFERENCES

Babbage, C. 1832. *On the economy of machinery and manufactures*. Philadelphia: Carey & Lea.

Bokranz, R., and K. Landau, 2006. *Produktivitätsmanagement von Arbeitssystemen*. Stuttgart: Schäffer-Poeschel.

De Jong, J.R. 1960: Die Auswirkung zunehmender Fertigkeit. *REFA Nachrichten* 13(1): 155-161.

Field, A. 2005. *Discovering Statistics Using SPSS*. London: Sage Publications.

Fleishman, E.A. 1962. The description and prediction of perceptualmotor skill learning. In: *Training research and education,* ed. R. Glaser. Pittsburgh: 137-175.

Fleishman, E.A. and G.D. Ellison, 1962. A factor analysis of fine manipulative performance. *Journal of Applied Psychology*, 46: 96-105.

Greiff, M. 2001. *Prognose von Lernkurven in der manuellen Montage unter besonderer Berücksichtigung der Lernkurven von Grundbewegungen*. Düsseldorf: VDI.

Hamster, W. 1980. *Die Motorische Leistungsserie - MLS*. Mödling: Schuhfried.

Hieber, W. L. 1991. *Lern- und Erfahrungskurveneffekte und ihre Bestimmung in der flexibel automatisierten Produktion*. München: Vahlen.

Jeske, T., S. Hinrichsen, S. Tackenberg, S. Duckwitz, and C.M. Schlick, 2009. Entwicklung einer Methode zur Prognose von Anlernzeiten. In: *Arbeit, Beschäftigungsfähigkeit und Produktivität im 21. Jahrhundert*, ed. Gesellschaft für Arbeitswissenschaft e.V. Dortmund: GfA-Press.

Jeske, T., M.Ph. Mayer, B. Odenthal, K. Hasenau, S. Tackenberg, and C.M. Schlick, 2010. Cultural Influence on Learning Sensorimotor Skills. In: *Advances in Human Factors, Ergonomics, and Safety in Manufacturing and Service Industries*, eds. W. Karwowski, and G. Salvendy. Boca Raton: Taylor & Francis.

Jeske, T., K. Hasenau, and C.M. Schlick, 2011. Einfluss von Arbeitsplänen auf das Anlernen sensumotorischer Fertigkeiten. In: *Mensch, Technik, Organisation - Vernetzung im Produktentstehungs- und -herstellungsprozess*, ed. Gesellschaft für Arbeitswissenschaft e.V. Dortmund: GfA-Press.

Jeske, T., K. Hasenau, and C.M. Schlick, 2011. Influence of task descriptions on learning sensorimotor tasks. In: *Human Factors in Organisational Design and Management - X, Volume I*, eds. M. Göbel, C. Christie, S. Zschernack, A. Todd, and M. Mattison. Santa Monica: IEA Press, p. I-343 - I-348.

Jeske, T. and C. Schlick, 2011. Influence of task complexity on learning times of sensorimotor tasks in assembly systems. In: *Innovation in Product and Production - Conference Proceedings, 21st International Conference on Production Research (ICPR 21)*, eds. D. Spath, R. Ilg, and T. Krause. Stuttgart: Fraunhofer. p. 1-6 (CD-ROM).

Jeske, T. and C. Schlick, 2012. Einfluss der Schwierigkeit sensumotorischer Arbeitsaufgaben auf ihre Anlernung. In: *Gestaltung nachhaltiger Arbeitssysteme - Wege zur gesunden, effizienten und sicheren Arbeit*, ed. Gesellschaft für Arbeitswissenschaft e.V. Dortmund: GfA-Press, p. 821-824.

Laarmann, A. 2005. *Lerneffekte in der Produktion*. Wiesbaden: Deutscher Universitäts-Verlag.

Levy, F.K. 1965. Adaptation in the Production Process. *Management Science* 11(6): B136-B154.

Rohmert, W., J. Rutenfranz, and E. Ulich, 1974. *Das Anlernen sensumotorischer Fertigkeiten*. Frankfurt a. M.: Europäische Verlagsanstalt.

Schlick, C.M., C. Winkelholz, F. Motz, S. Duckwitz, and M. Grandt, 2010. Complexity assessment of human-computer interaction. In: Theoretical Issues in Ergonomics Science, 11(3), p. 151-173.

Schoppe, K.J. 1974. Das MLS-Gerät: ein neuer Testapparat zur Messung feinmotorischer Leistungen. *Diagnostica*, 20: 43-47.

Smith, A. 1776. *An Inquiry into the Nature and Causes of the Wealth of Nations*. London: printed for W. Strahan and T. Cadell.

Stier, F. 1968. Informationsverarbeitung am Arbeitsplatz. *Werkstatt und Betrieb* 101: 473-478.

Wright, T.P. 1936. Factors Affecting the Cost of Airplanes. *Journal of the Aeronautical Sciences*, 3(4): 122–128.

Section III

Human Factors in Work Systems

CHAPTER 27

Industrial Robots – the New Friends of an Aging Workforce?

Dino Bortot, Benno Hawe, Silvia Schmidt and Klaus Bengler

Technische Universität München - Institute of Ergonomics
Boltzmannstraße 15, 85747 Garching, Germany
bortot@tum.de

ABSTRACT

Various parameters influence humans' wellbeing in, and thereby their acceptance of and performance within, human-robot interaction (HRI) applications. The experiment described here aimed to detect the most influential factors when subjects had to coexist with a medium-sized industrial robot in the same workspace. The distraction caused by the robot was detected using both objectively and subjectively measured variables. Robot velocity, the working area of the Tool Center Point (TCP) (which can be interpreted as equivalent to the distance between the two entities), and the hearing abilities of the participants were used as independent variables.

The results revealed lower distraction/better performance when the robot had a lower acoustic impact (i.e. when the hearing abilities of the subjects were constrained). Additionally, robot velocity has an impact on human performance, as a slow velocity characteristic was preferred to fast or variable ones.

Keywords: human-robot coexistence, human-robot interaction, HRI, performance, trust, acceptance, distraction, visual behavior

1   INTRODUCTION

Current issues in industrial production call for the development of HRI applications. Industrialized countries are faced with demographic changes in the work force, including a significant decrease in workers' physical abilities (Gaudart,

2000). Production planning must create new strategies to combat this problem. Additionally, ergonomists will still face critical situations where the traditional ergonomic workplace design is not sufficient to generate healthy working conditions. The HRI concept seems to be a suitable solution to these problems (Bortot et al., 2012).

Regarding the development process, system designers must understand how to design safe, ergonomic and accepted systems while increasing efficiency. Acceptance is a complex construct, influenced by several factors. When HRI is present, the following parameters, among others, have a major impact on the human worker: robot velocity/acceleration, size of the robot, distance and minimum distance between the robot and the human, visual impression of the robot, acoustic stimulus caused by the robot, and robot trajectory planning. All of these can and should be manipulated such that human workers do not perceive the robot as distracting but rather as a good interaction partner. Experiments are one possibility for determining the right settings for these factors. Some of the factors have already been investigated in various earlier empirical studies, but in other application fields mostly (e.g. HRI with humanoids in everyday life applications). A few selected results are presented in the following paragraph.

Two important factors that significantly influence the acceptance of HRI systems are the distance between the robot and its human user as well as the velocity of the robot movements. Various studies looked at both of these parameters, mainly in the field of humanoid robots. Butler and Agah varied both of the factors when conducting a study with a humanoid robot in 2001. The results of the study revealed that, for the factor speed, human subjects were comfortable with all speeds except the fast approach speeds. The most popular choice was slower than walking speed. The authors believed this was because of the subjects' unfamiliarity with the robot (Butler and Agah, 2001). Other authors found that speed of a robotic partner should be adjustable in order to adapt its capacity to the learning curve of first-time users. In a questionnaire study, Khan found that humans preferred adaptable robot velocities (Khan, 1998). In addition to many studies with humanoid robots, some experiments dealt with industrial robots as well. Kulic and Croft used physiological measurements to describe perceived safety in an HRI scenario. Fast robot motions tended to reliably elicit a strong, measurable arousal response, which could be due to a consciously experienced affective state such as anxiety or surprise (Kulić and Croft, 2007). Nonaka et al. confirmed the increase of the discomfort felt with higher robot velocities (Nonaka et al., 2004).

Hanajima et al. (2005) and Yamada et al. (1999) detected a dependency between robot velocity and the distance between the two entities. Participants in the studies accepted higher velocities as their distance to the robot increased. This indicates that TTC (time to collision) could be a layout parameter instead of pure distance.

Previous studies conducted by the authors of this paper revealed that the noise caused by an industrial robot can cause distraction and must be analyzed as well. Beauchamp and Stobbe (1990) considered not only robot velocity, but also various other factors such as workplace illumination, luminance contrast, task demands, and noise. For noise, the following two conditions were applied: 65 dbA and 100 dbA.

The results of the study do not show any differences between those two conditions; in other words, the environmental noise was not found to influence human performance. It should be mentioned, however, that both conditions were composed of normal industrial noises; the acoustic stimulus caused by the robot was not considered explicitly. Bartneck et al. (2009) found that humans felt uncomfortable if the robot was operating behind them, even if this in an extreme layout for HRI. So their perception of the robot was limited to acoustic stimulus. In the study by Hanajima et al. (2004), subjects were faced with three different conditions. The robotic device provided them with either visual or auditory stimulus, or a combination of both modalities. In both cases where only one channel was available, the Skin Potential Response (SPR), and thereby arousal, increased.

The following study aims to detect the most influential factor(s) on human acceptance for a scenario in which humans must collaborate with a medium-sized industrial robot (RV20-16, Reis robotics, Fig. 1).

2 METHOD

2.1 Framework conditions of the study

The setup of the study involved human coexistence with an industrial robot. Participants sat at a workbench and had to solve a continuous two-dimensional tracking task by controlling a joystick. During this exercise, the robot simulated handling maneuvers close to the human (Fig. 1).

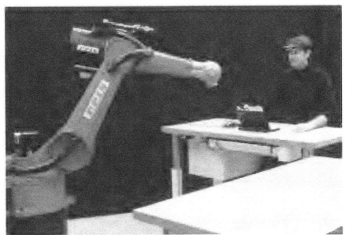

Figure 1 Experimental setup: Analyzing the participants' visual behavior and performance in a two-dimensional tracking task while they were coexisting with an industrial robot.

The study was conducted with 30 participants. Three of them were female, 27 male. The average age of the subject group was 28.17 (SD = 7.66) and therefore very young for typical manufacturing scenarios. Fig. 2 shows that most of the participants indicated at least an above-average trust in technical systems and had little or no previous experience with industrial robots.

256

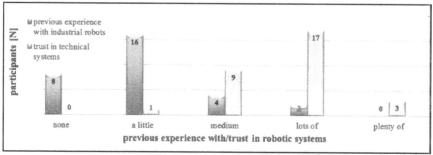

Figure 2   Characterization of the subject group.

Beyond that, at the beginning of the study, participants were asked if they had ever experienced any kind of accident with technical systems before. Out of 30 subjects 26 hadn't so, i.e. four participants had undergone such an experience. This supposes a generally positive attitude towards coexistence with the robot.

2.2   Procedure

Participants were asked to perform a two-dimensional tracking task as accurately as possible, which was displayed on the screen (10.1″) of a *Lenovo S10-3T* netbook. With the help of a joystick, they had to position a small dynamic green cross-hair on top of a larger (red) cross-hair. The continuous character of the task demanded the subjects to focus, and accordingly to control the cross-hair, constantly. Contemporaneously, the robot moved around in the proximity of the subjects. Each of the subjects completed seven runs, which were distinguished according to the following parameters: robot velocity, working area of the robot and hearing abilities of the participants (Chapter 2.4). All these independent variables were randomized to prevent effects of sequence. Fig. 3 illustrates the procedure that every participant went through.

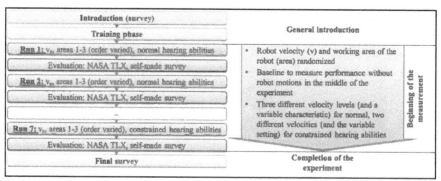

Figure 3   Experimental procedure: Every participant was confronted with seven different characteristics for the setting. Each was followed by an evaluation step.

2.3   Apparatus/dependent variables

The tracking task provided objective parameters to assess the deteriorating effect of the robot on human performance. It measured the deviation of the two cross-hairs (in pixels) and created the error-indicator root mean square error (RMSE) for three different dimensions (x, y, and overall = two-dimensional). The results of the tracking task are assumed to give an impression of the level of intrusion caused by the human-robot coexistence, as larger deviations are probably caused by reduced concentration on the task (because of the distraction caused by the robot). During the test execution and the follow-up data analysis, it became apparent that the design of the tracking task needed to be questioned. From time to time, the deflection of the small dynamic cross-hairs in the x direction was exceptionally large, such that the participant was unable to control the system even though he or she intended to do so. This phenomenon could not be observed in every test, and therefore reliability was no longer ensured. As a consequence, the organizers of the study decided to use deflection in the y direction only and disregard both the results in the x direction and the two-dimensional ones.

Visual behavior was analyzed as a second objective parameter to determine the distraction caused by the robot. Eye tracking systems can be used to illuminate various research areas, such as information perception for displays, the distracting effect of secondary tasks, and strategies concerning visual behavior. For this study, a head-mounted Dikablis eye tracking system (Lange et al, 2010) was used to analyze visual behavior, and thereby determine the time of gaze deviation (distraction caused by the robot). Visual behavior was defined as a dichotomous variable, i.e. gazes could be directed either to the display of the netbook or the robot. All videos were coded manually, assigning every single frame to either one of the regions. The results of the coding process were two time intervals for each of the regions, in other words a gaze distribution for the duration of every single run. The analysis conformed to the ISO 15007 standard (ISO 15007, 2002).

In addition to objective parameters, subjective evaluation methods were applied in order to comprehensively determine the participants' workload. After each experimental run (Fig. 3) they had to fill out the NASA-TLX evaluation sheet (Hart and Staveland, 1988) as well as a short survey created by the study organizers. The subjectively estimated workloads were evaluated after the subjects were exposed to the following conditions: one robot velocity characteristic with either normal or constrained hearing abilities, but in all three of the different working areas. Additionally, the NASA-TLX index was determined after the baseline (no robot motions), which led to eight different indices over the course of the experiment.

2.4   Independent variables

Three different independent variables were systematically varied in the course of the study. Altogether, four different velocity characteristics of the robot were applied: The robot moved slowly, medium-fast, fast, or in a combination of these three velocities (variably).

258

The distance between the two entities was varied by keeping the Tool Center Point of the robot within certain working areas (at different heights). Three different areas were defined (Fig. 4).

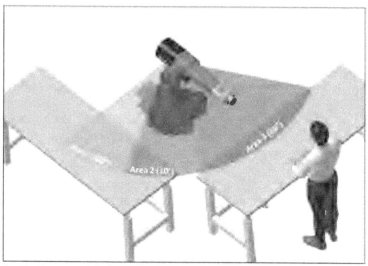

Figure 4   Working areas 1 to 3 were varied systematically to avoid errors caused by effects of sequence.

During the first half of the experiment, participants had normal hearing abilities, while the second half was conducted with constrained hearing abilities as the subjects had to wear earmuffs. As a result, the surrounding noise was no longer perceived. Nonetheless, the motions of the robot, at least those at the highest velocity level, could still be heard (depending on the individual hearing abilities of each participant).

3   RESULTS

3.1   Procedure of the statistical analysis

Based on the results of Kolmogorov-Smirnov tests, i.e. based on whether raw data followed a normal distribution or not, either different kinds of ANOVAs (results fulfilled the condition of normal distribution) or Friedmann tests (raw data was not distributed normally) were applied. Degrees of freedom were corrected using the Greenhouse-Geisser correction factor if the criterion of sphericity was not met. For all analyses, the significance level was set to 0.05.

3.2   Tracking task

Three different ANOVAs were applied in order to detect statistically significant influence factors (Tab. 1).

Table 1 ANOVAs applied to detect statistically significant influence factors (number of characteristics in brackets)

ANOVA type	Constant factor	Factor 1	Factor 2	Factor 3
three-factorial	-	robot velocity (3)	working area of the robot (3)	hearing abilities (2)
two-factorial	normal hearing abilities	robot velocity (4)	working area of the robot (3)	-
two-factorial	constrained hearing abilities	robot velocity (3)	working area of the robot (3)	-

Under the condition 'constrained hearing abilities', only three different robot velocity characteristics were applied (the medium-fast velocity level was dropped). Neither of the two-factorial ANOVAs reveals any statistically significant results. As no interaction effects of the first or the second order appear for the three-factorial ANOVA, the resulting main effects can be considered separately. Robot velocity appears to be a statistically significant influencing factor $(F(1.397, 40.517) = 3.682, p < .05, \eta^2=.113, 1-\beta=.546)$, as the variable velocity characteristic leads to a significantly worse statistical performance by the subjects compared to the 'slow velocity level' condition. Additionally, participants performed better with constrained hearing abilities compared to normal hearing abilities $(F(1, 29) = 40.29, p < .001, \eta^2=.581, 1-\beta=1)$.

3.3 Eye tracking

As the tracking task was defined as the primary task that had to be solved as well as possible, some of the subjects did not look at the robot at all throughout the entire experiment. This led to the result of 0 frames (seconds) where they were looking at the robot (distribution: 100% on the screen, 0% on the robot). As a consequence, the eye tracking data was not distributed normally and ANOVAS had to be substituted by ranked analyses of variances. If the chosen Friedmann test delivered significant results, Wilcoxon tests followed as a single paired post-hoc test. Depending on the number of single paired post-hoc tests, the significance level was adjusted (lowered) using the Bonferroni correction factor.

Under certain circumstances, various working areas of the TCP have statistically significant influences. If the robot moves with one particular velocity characteristic and is not acoustically perceived as loudly as under normal circumstances (constrained hearing abilities), the TCP's working area has no effect on the visual behavior. That changes the moment the subjects coexist with the robot with normal hearing abilities (Tab. 2).

Table 2 With normal hearing abilities of the subjects, the TCP's working area influences the visual behavior for every velocity characteristic (*p<.05, **p<.01, ***p<0.167)

Hearing abilities	Robot velocity	Result	Wilcoxon tests (significant)
normal	slow	$\chi^2(2) = 7.259$*	-
normal	medium-fast	$\chi^2(2) = 8.930$*	Area 3 vs. Area 1*** Area 3 vs. Area 2***
normal	fast	$\chi^2(2) = 11.405$**	Area 3 vs. Area 2***
normal	variable	$\chi^2(2) = 6.318$*	-

In all of the three significant post-hoc tests, Area 3 has the highest duration of glances at the robot, i.e. Area 3 causes the largest distraction.

Finally, the general influence of the participants' hearing abilities is investigated. Three out of nine paired comparisons (three robot velocity characteristics*three working areas) reveal statistically significant results. Differences in visual behavior occur in Area 3 for the slow ($\chi^2(1) = 9.308$, p = .002; T = 9, r = -0.329) and the fast ($\chi^2(1) = 10$, p = .002; T = 0, r = -0.362) velocities, as well as in Area 2 for the slow velocity level ($\chi^2(1) = 8$, p = .005; T = 0, r = -0.325). All in all, the influence of hearing abilities on visual behavior occurs mainly in Area 3, and in one case in Area 2, whereas it did not have any effect when the robot moved in Area 1. Every significant difference was caused because of more and longer glances at the robot when the subjects had normal hearing abilities.

3.4 NASA-TLX

Three different statistical tests clarify the estimated workload. For both conditions, normal and constrained hearing abilities, one-factorial ANOVAs are carried out (factor: robot velocity). Furthermore, the influence of the additional factor of hearing abilities is considered using a two-factorial ANOVA. The latter reveals a statistically significant influence for the hearing abilities ($F(1, 29)=1.974$, p<.01, $\eta^2=.240$, 1-β=.832). Participants in the study estimated their workload as being higher with the condition of normal hearing abilities.

Neither of the one-factorial ANOVAs showed significant results (normal hearing abilities: $F(2.363, 68.540)=2.078$, p=.125; constrained hearing abilities: $F(2, 58)=0.719$, p=.491). Thus for robot velocity, as well as for the interaction between robot velocity*hearing abilities, no significant influences can be determined.

3.5 Overall evaluation

The results of the tracking task, NASA-TLX evaluation and the eye tracking data show an influence by the different characteristics on participants' hearing abilities. This effect is supported by statistically significant results for all three of the parameters. Participants not only subjectively preferred constrained hearing abilities (i.e. not perceiving the robot as loudly as under normal conditions), they also performed the tracking task better and were less distracted by the motions of the robot.

No significant influence was detected for robot velocity on gaze behavior. Consequently, the results of the tracking task lead to the conclusion that robot velocity has a significant influence on human performance. Participants preferred slow robotic motions as opposed to fast or variable velocities.

The eye tracking data shows a partial significance for the robot's working area, with Area 3 (the closest area) the most distracting. Differences can be observed among the parameters for "normal hearing abilities of the participants" and medium/fast robot velocities. Nevertheless, only three out of 21 paired comparisons are significant.

In sum, participants' hearing abilities can be shown to have a distinct influence on human performance within HRI. Some effects were detected for both of the other variables, but not to the same extent as for the hearing abilities.

4   SUMMARY AND DISCUSSION

The results of the study reveal that the auditory stimulus of an industrial robot has a major influence on the distraction/performance of humans in HRI applications. Further experiments should focus on the independent variable 'noise' and examine the influence of different noise levels thoroughly.

As explained before, some of the participants did not look at the robot at all. This was due to the fact that the experimental design allowed participants to stay focused on the tracking task at all times. In most cases, gazes on the robot were a consequence of strong distractions, and minor ones did not cause the subjects to look at the robot. This calls visual behavior into question as an adequate parameter for assessing human distraction. In any case, under different circumstances, e.g. interaction scenarios where the participants necessarily have to look at the robot every once in a while, visual behavior still seems to be a good parameter for describing the human perception of a robot in an coexisting/interaction scenario.

The fact that the study was conducted with young, inexperienced participants who trusted technical systems (Fig. 2) must be kept in mind. It must be assumed that operators in production halls have a more negative mindset towards industrial robots because their attention was, and still is, continuously drawn to the hazards that arise from such machines. Thus the distraction of participants who work in industrial production should be greater than that of the investigated participants. This suggests that visual behavior can serve as a suitable factor for measuring human distraction/performance.

262

ACKNOWLEDGMENTS

The authors would like to acknowledge the extraordinarily good cooperation with their partners in the EsIMiP project (AZ-852-08), which is funded by the Bayerische Forschungsstiftung.

REFERENCES

Bartneck, C., D. Kulić, E. Croft and S. Zoghbi. 2009. Measurement Instruments for the Anthropomorphism, Animacy, Likeability, Perceived Intelligence, and Perceived Safety of Robots. *International Journal of Social Robotics* Vol. 1, 71–81.

Beauchamp, Y. and T. J. Stobbe. 1990. The effects of factors on human performance in the event of an unexpected robot motion. *Journal of Safety Research* 21 (3), 83–96.

Bortot, D., Ding, H., Antonopolous, A., Bengler, K. 2012. Human motion behavior while interacting with an industrial robot. *Work: A Journal of Prevention, Assessment and Rehabilitation* 41 (Supplement 1/2012), S. 1699–1707.

Butler, J. T. and A. Agah. 2001. Psychological Effects of Behavior Patterns of a Mobil Robot. *Autonomous Robots* 10. 185-202.

Gaudart, C. 2000. Conditions for maintaining ageing operators at work - a case study conducted at an automobile manufacturing plant. *Applied Ergonomics* (31), 453–462.

Hanajima, N., M. Fujimoto, H. Hikita, and M. Yamashita (Ed.). 2004. Influence of Auditory and Visual modalities on Skin Potential Response to Robot Motions. *Proceedings of the 2004 IEEE/RSJ International Conference on Intelligent Robots and Systems (IROS)*, 1226-1231.

Hanajima, N., Y. Ohta, H. Hikita and M. Yamashita. 2005. Investigation of impressions for approach motion of a mobile robot based on psychophysiological analysis. *2005 IEEE International Workshop on Robots and Human Interactive Communication*, 79–84.

Hart, S. G. and L. E. Staveland. 1988. Development of NASA-TLX (Task Load Index): Results of empirical and theoretical research. P. A. Hancock and N. Meshkati (Eds.): *Human Mental Workload*. Amsterdam: North Holland Press.

ISO 15007-1 2002. Road vehicles - Measurement of driver visual behaviour with respect to transport information and control systems - Definitions and parameters.

Khan, Z. 1998. Attitudes towards intelligent service robots. Technical report TRITA-NA-P9821. NADA.

Kulić, D. and E. Croft. 2007. Physiological and subjective responses to articulated robot motion. *Robotica* 25, 13-27.

Lange, C., R. Spies, H. Bubb, and K. Bengler. 2010. Automated Analysis of Eye-Tracking Data for the Evaluation of Driver Information Systems According to EN ISO 15007-1 and ISO/TS 15007-2. *Advances in Ergonomics Modeling and Usability Evaluation.* 147-152.

Nonaka , S., K. Inoue, T. Arai and Y. Mae. 2004. Evaluation of human sense of security for coexisting robots using virtual reality. 1st report: Evaluation of pick and place motion of humanoid robots. *Proceedings of the 2004 IEEE International Conference on Robotics & Automation.* 2770-2775.

Yamada, Y., Y. Umetani and Y. Hirawawa. 1999. Proposal of a psychophysiological experiment system applying the reaction of human pupillary dilation to frightening robot motions. *Proceedings of IEEE International Conference on Systems, Man and Cybernetics.* 1052-1057.

CHAPTER 28

The Effect of Anthropomorphic Movements of Assembly Robots on Human Prediction

Sinem Kuz, Antje Heinicke, Dominik Schwichtenhoevel, Marcel Ph. Mayer, Christopher M. Schlick

Institute of Industrial Engineering and Ergonomics of RWTH Aachen University
Aachen, Germany
{s.kuz; a.heinicke; d.schwichtenhoevel; m.mayer; c.schlick}@iaw.rwth-aachen.de

ABSTRACT

From a user centered point of view an important basic requirement to enable human-robot cooperation is to achieve conformity with operator's expectations of robot behavior. Therefore, this study focuses on the question, whether anthropomorphic robot movement trajectories can lead to an improved anticipation of the robot's behavior. Based on a virtual simulation environment a robotized assembly cell consisting of the assembly robot and the actual workplace was considered. In order to be able to simulate anthropomorphic movements, the human wrist trajectories of defined pick and place movements were obtained using an infrared motion capture system. The captured data were used to navigate the virtual assembly robot. Within the experiment anthropomorphic and robotic trajectories were distinguished. During the experiment, the main task of the participants was to predict the movement's destination as quickly as possible. Thus, the corresponding reaction value was analyzed to investigate the influence of anthropomorphic robot movements on human prediction in industrial environments.

Keywords: human-robot cooperation, anthropomorphism, industrial settings

1. INTRODUCTION

One important issue in the design of socio-technical systems especially regarding so called intelligent systems is the perception of such systems by the human operator as intelligent systems. Anthropomorphism e.g. by assigning humanlike features to technical systems is one promising approach to help humans apprehend systems as intelligent entities, be it by appearance or actions.

Concerning the first aspect Krach et al. (2008) investigated the effect of increasing anthropomorphism in appearance on human perception, without varying the inherent "intelligence" of the system. Actually, the basic system used within the study was a simple computer game. Anthropomorphism was varied from a simple laptop, over a humanlike robot to an actual human sitting behind the laptop and acting as if being the opponent. The results of this survey show a positive correlation between anthropomorphism of a non human-entity and it's attribution with intelligence.

Within the Cluster of Excellence "Integrative Production Technology for High-Wage Countries" (Brecher, 2011) at the Faculty of Mechanical Engineering of RWTH Aachen University, a research project has been established to study cognitively automated assembly cells (Mayer et al., 2011). These systems inherit certain intelligence by means of cognitive functions such as decision making or problem solving, and learning. In order to be able to use the full potential of cognitive automation, cognitively automated systems have to be perceived as intelligent systems.

In an industrial environment however, transferring anthropomorphism in appearance might be difficult to do. Hence, current investigations focus on the question, whether ascribing an industrial robot with anthropomorphic action patterns can lead to a better anticipation of its behavior by the human operator.

In order to investigate this idea, a laboratory experiment was conducted on the basis of a simulation environment consisting of an assembly robot and a working place. The simulation environment is based on the experimental assembly cell as introduced e.g. by Kempf et al. (2008). This paper focuses on the results of this empirical study regarding the conformity with the operator's expectations.

2. ANTHROPOMORPHISM AND ROBOTS

Research of the last years indicates that the acceptance of non-human agents like robots can be increased by anthropomorphic attributes (Kupferberg et al., 2011; Duffy, 2003). Hence, especially studies in the field of physical interaction of humans and robots focus more and more on building robots with different degrees of humanlike characteristics. The anthropomorphism of a robot can include adjustments in the robot's physical design as well as its movement or both. Usually anthropomorphism is used in the area of social robotics because there is almost no physical separation between humans and robots. The adaptation of human features to an industrial robot is only rarely investigated because efficiency and robustness are the most important aspects of a robot in industrial settings. Nevertheless, direct physical interaction between humans and robots in industrial environments might enable the useful combination of the specific skills of the human worker with the abilities of an industrial robot. Thus, the advantages of anthropomorphism could also be applied for cooperative work of humans and robots in industrial environments. However, a basic requirement for joint action is to understand the other's actions and intentions. Concerning the perception and understanding actions, neuroscientific researchers discovered special brain areas (mirror neurons) in humans and non-human primates that are activated both by action generation and observation of other humans' actions (Rizzolatti et al., 1996a; 1996b). Regarding this fact, some studies have tried to investigate human mirror neuron activation during the observation of robot actions and concluded different approaches. Tai et al. (2004) examined movement sequences that were repeated identical but could not prove any significant mirror neuron activation of the humans during monitoring an industrial robot. However, Gazzola et al. (2007) concentrated on different sequences of movement and could show an activation of the human mirror neuron system by observing industrial robot actions. Accordingly based on the findings by Gazzola et al., the question is whether augmenting industrial robot movements with anthropomorphic features would achieve a better anticipation of the robot's behavior by the human worker.

3. EMPIRICAL STUDY

The laboratory study conducted in this work aims at the investigation of the influence of anthropomorphism in industrial settings on the conformity with operator's expectations of system behavior. The results include the comparison between the effect of the anthropomorphic and robotic movement trajectories of predefined pick and place movements on the prediction accuracy of the human.

A total number of 24 male subjects, who are either taking part in an engineering bachelor/master program or have finished their studies, participated in the laboratory study. The participants were aged between 20 and 33 years (mean 25.21 years, SD 3.799). 83.33% of the participants had already experience with 3D simulation environments. Almost 38% of the subjects had already worked with robotic environments and even 33% of them for several months.

In order to analyze and organize humanlike pick and place movements, the human hand trajectories were measured using an infrared optical tracking system.

266

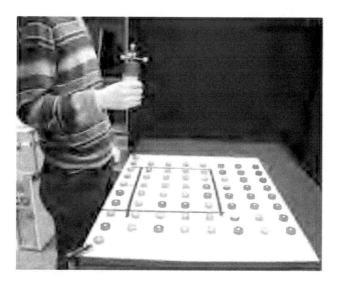

Figure 1 Posture during wrist height trajectory tracking

As shown in figure 1, the human always started in the same posture in front of the grid (80 cm × 60 cm) and placed the tracked cylinder with markers on every round in the marked area (36 cm × 28.8 cm) on the grid. The black marked area on the grid corresponds to the workplace of the robot in the virtual simulation environment. The 3D marker data recorded by the optical motion capture system describes the human wrist position during the execution. Beside the hand trajectories, the duration of the complete movement from the start position until placing the cylinder was also recorded and examined. Figure 2 represents the human wrist and the robot end-effector trajectories for the same start and endpoints in comparison. As the robot's trajectory proceeds on a relatively straight course, the human's trajectory appears more curved.

After recording and analyzing the position data for all fields of the grid, the experimental setup shown in figure 1 should be transformed in a virtual scenario. Hence, a C++ simulation environment that was already developed and used within different works of the Cluster of Excellence (Odenthal et al., 2009) was adapted through a few changes and employed as the virtual presentation of the described scenario. The 3D elements were realized by using the OpenGL library for C++. The developed virtual scene consists of an assembly robot with a cylinder and a grid with an adjustable number of fields. Within this work the virtual grid consists of 20 fields and represents the working place of the robot. Finally, the tracked 3D coordinates of the human wrist were organized to drive the model of the virtual assembly robot.

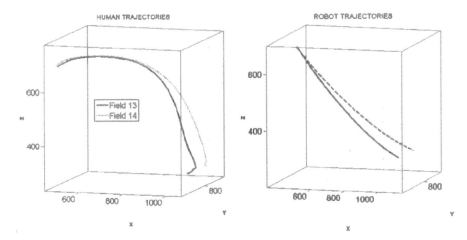

Figure 2 Trajectories of the human (left) and the robot (right)

The main task of this study was to monitor the assembly robot performing different placing movements and to predict the target point. Furthermore subjects had no knowledge about the fact that the robot would execute two different types of movement trajectories. They were seated in front of a TFT LCD 28″ monitor, and asked to observe the virtual scene (the industrial robot and its workplace), which was presented from the lateral perspective. During the execution, the robot placed the cylinder on every field of the grid. Each of these fields was approached in a human and a robot kind of way whereby the order of the motion sequences were permuted. The participants should monitor each of these 40 motion sequences and select a field by using a computer mouse, where the robot would place the cylinder (Figure 3). Subjects were asked to select the target field as early as possible within the movement. After the selection the scenery was interrupted and the next motion sequence could be started either by the participants or was started after predefined seconds.

Before the main part of the study the participants had to fill in information such as the personal data, experiences with computers, with virtual reality systems as well as with robots and industrial environments. A visual acuity test, a stereo vision test and a color vision and emmetropia were carried out by means of the vision test device. All of them fulfilled the requirements concerning a stereo vision, color vision and emmetropia according to DIN 58220 (1997).

As independent variables the two movement types (human and robot) and the different field positions were considered.

As dependent variables two types of reaction values of the participants were analyzed. The reaction value was given as the elapsed time in milliseconds from the beginning of a movement until the reaction of the participant as well as the covered way on the movement sequence. This paper will only focus on the results regarding the covered way on the corresponding movement sequence.

The statistical analysis in this work was calculated using the statistical software package SPSS Version 19.0. The determined data have been investigated with the help of a repeated-measures analysis of variance. The chosen level of significance for each analysis was α=0.05.

Figure 3 The virtual robot and the workplace

4. RESULTS AND DISCUSSION

In this section the determined results regarding 'the reaction value' as the dependent variable will be presented. This value corresponds to the trajectory elapsed during the experiment from the start of a movement sequence until the subject selected a field and is independent of the result of the prediction. Therefore the end-effector positions of the robot during a movement phase were traced till participants marked a field. These coordinates describe the behavior of the respective movement trajectory until the moment of the selection. Hence it was possible to set this tracked sequence in relation to the relevant trajectory and calculate a normalized value which can be regarded as the participant's reaction value. We analyzed the effect of the two main factors movement type and the position on the grid.

Results show significant effects of the movement type ($F(1,23)=6.647$, $p=0.017$) as well as the position ($F(19,437)=7.449$, $p=0.000$). Findings also show a significant disordinal interaction effect between the movement type and the position ($F(19, 209.683)=7.036$, $p=0.000$). Based on this interaction type it is not possible to make statements about the main effects (Field, 2009). Nevertheless on closer inspection of the mean reaction values as the way from the beginning of the movement until the reaction of the participant, there are still discernible tendencies regarding the position. The comparisons show that the covered way on the movement sequence for fields that are closer to the virtual robot are shorter for humanlike movements than for robotic ones. On the other hand findings for the top series of the workplace generally show the same for robotic movements. Accordingly, in short distances participants rather had the feeling that they have recognized the target field for humanlike and at longer distances for robotic movements (Figure 4).

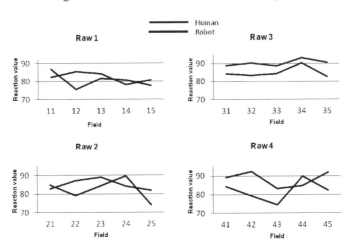

Figure 4 Reaction values in comparison for each row on the workplace

Furthermore, the number of correct predictions was also examined. Results show that only 24 (human movement trajectories=2 and robotic movement trajectories=22) of 960 predictions are indeed correct. According to this the small number of correct predictions clearly shows that it was in general not easy for the participant to recognize the right target field under both conditions. The reason for this might be due to the lack of depth information in the visualization of the simulation environment. This fact was also partially remarked by the subjects during or after the experiment.

5. SUMMARY AND CONCLUSION

The focus of this work was to investigate whether enhancing industrial robot movements with humanlike characteristics increase the prediction accuracy of the humans. In order to investigate this idea, a laboratory experiment was conducted. In order to be able to simulate anthropomorphic movements, the hand trajectories of defined pick and place human movements were obtained using an infrared motion capture system. The captured data were analyzed and converted to drive the model of the virtual assembly robot. Therefore the study distinguished two different kinds of movement trajectories which are the anthropomorphic and the typical trajectories of industrial robots. During the experiment the participants were asked to make predictions of the movement destination. Within the evaluation phase the trajectories until the reaction moment of participants were analyzed regarding the movement type and the fields on the grid. Due to the disordinal significant interaction effects, the interpretation of the two main effects is not possible. Nevertheless, there are tendencies regarding the distance and movement type. In general reaction values were better at short distances when the movement trajectory was humanlike and at longer distances for robotic movement trajectories.

Additionally correct predictions were not analyzed further because of the very small number considering the actual prediction iterations. The probable reason for this might be the lack of depth information of the visualization of the simulation environment.

Another dependent variable in the study was the reaction time, which corresponds to the elapsed time in milliseconds from the beginning of a movement sequence until the reaction of the participant. Although the tested virtual movement trajectories were identical for all subjects, there can probably appear minimal differences in the duration of the same sequence of movements due to the developed simulation environment. For this reason, based on several simulations, an average value was determined for each of the 40 movement trajectories. Hence, it was possible to compute normalized reaction time values for each selection of the participants. Detailed analysis of this value and other outcomes are still under construction and will be introduced in further work.

In conclusion the tested robot characteristics within this work can be seen as a low degree of anthropomorphism because only trajectories and not physical characteristics of human pick-and-place movements were considered. Hence, an interesting avenue for future work would be to increase the degree of anthropomorphism such as adjustments in appearance of the industrial robot.

AKNOWLEDGEMENT

The authors would like to thank the German Research Foundation DFG for the kind support within the Cluster of Excellence "Integrative Production Technology for High-Wage Countries".

REFERENCES

Brecher, Ch. 2011. Integrative Production Technology for High-Wage Countries, Springer, Berlin ISBN 978-3-642-21066-2

DIN 58220. 1997. Sehschärfebestimmung. Berlin: Beuth. (in German)

Duffy, B. 2003. Anthropomorphism and The Social Robot. Special Issue on Socially Interactive Robots, Robotics and Autonomous Systems 42 (3-4).

Field, A. 2009. Discovering Statistics Using SPSS. London: Sage Publications.

Gazzola, V., G. Rizzolatti, B. Wicker and C. Keysers. 2007. The anthropomorphic brain: The mirror neuron system responds to human and robotic actions. Journal of Neuroimage, 35(4), 1674-84

Kempf, T., W. Herfs, and Ch. Brecher. 2008. Cognitive Control Technology for a Self-Optimizing Robot Based Assembly Cell. Proceedings of the ASME 2008 International Design Engineering Technical Conferences & Computers and Information in Engineering Conference, American Society of Mechanical Engineers

Krach, S., F. Hegel, B. Wrede, G. Sagerer, F. Binkofski, and T. Kircher. 2008. Can machines think? Interaction and perspective taking with robots investigated via fMRI. PLoS ONE, 3(7), e2597.

Kupferberg, A., S. Glasauer, M. Huber, M. Rickert, A. Knoll, and T. Brandt. 2011. Biological movement increases acceptance of humanoid robots as human partners in motor interaction. Journal of AI & Society, 26(4), 339-345.

Mayer, M., C. Schlick, D. Ewert, D. Behnen, S. Kuz, B. Odenthal, and B. Kausch. 2011. Automation of robotic assembly processes on the basis of an architecture of human cognition, In: Production Engineering Research and Development, 5 4, ISSN 0944-6524, 423-431

Odenthal, B., M. Ph. Mayer, W. Kabuß, B. Kausch, and C. M. Schlick. 2009. Investigation of Error Detection in Assembled Workpieces Using an Augmented Vision System. In: Proceedings of the IEA 17th World Congress on Ergonomics (CD-ROM) Beijing, China, 1-9

Rizzolatti, G., L. Fadiga, V. Gallese, and L. Fogassi. 1996a. Premotor cortex and the recognition of motor actions. Cognitive Brain Research, 3, 131–141.

Rizzolatti, G., L. Fadiga, M. Matelli, V. Bettinardi, E. Paulesu, D. Perani, and F. Fazio. 1996b. Localization of grasp representations in humans by PET: 1. Observation versus execution Experimental Brain Research, 111, 246–252.

Tai; Y.F., C. Scherfler, D.J. Brooks, N. Sawamoto & U. Castiello. 2004. The human premotor cortex is 'mirror' only for biological actions. Journal of Current Biology, 14, 117–120.

CHAPTER 29

Design and Implementation of a Comprehensible Cognitive Assembly System

Christian Brecher[1], Simon Müller[1], Marco Faber[2], Werner Herfs[1]

[1]Laboratory for Machine Tools and Production Engineering, RWTH Aachen
{c.brecher, s.mueller, w.herfs}@wzl.rwth-aachen.de
[2]Institute of Industrial Engineering and Ergonomics, RWTH Aachen
m.faber@iaw.rwth-aachen.de

ABSTRACT

The commissioning and programming of complex robotized handling operations requires extensive planning efforts, which do not contribute directly to the added value. A cognitive control system is able to generate essential sequences of actions at run time on the foundation of model-based knowledge regarding task, system and process flow and in consideration of current environmental information. A software framework on the basis of the cognitive architecture Soar, which allows for the creation of intelligent robotized handling processes, has been designed. In order for a human operator to supervise and interact with such a cognitively controlled system, a promising approach is to design the behavior of the system in a cognitive-ergonomic way. Thus, this paper describes the design and implementation of Methods-Time Measurement 1 (MTM-1) key elements as cognitive-ergonomic process logic to control an industrial robot in an assembly task.

Keywords: cognition, MTM, robotics

1   INTRODUCTION

Due to shorter innovation cycles and more complex products, one of the key factors for the success of current businesses exists in the challenge to find the best strategy and layer of granularity for planning. In regard to production systems two different approaches can be distinguished: planning orientation and value orientation (Schmitt et al. 2011).

Planning orientated approaches are based on the notion that specific changes in the environment can be foreseen. These approaches usually require large models with lots of parameters. A planner utilizing this method does not have to wait for a change to happen, but is able to decide based on an expected trend or a specific scenario. This approach is practical and efficient in static environments, but fails for unforeseen and unpredictable changes (Gausemeier et al. 2009, Schuh et al. 2011).

In order to adapt to fast changing market requirements and shorter product lifecycles, the value oriented approach adjusts activities according to the value stream of an enterprise. Thus it reduces the central planning efforts by decentralizing the decision-making powers. However, the full potential for optimization can only be reached, if all decentralized systems know every effect on a change of their internal parameters to the global optimum. Since this is usually hardly possible or even impossible to accomplish, this presents one of the main disadvantages for these systems (Orilski and Schuh 2007).

Most current businesses position themselves somewhere in the dilemma of planning orientation and value orientation. In order to be successful in the future, however, it is necessary to resolve that conflict. For technical systems, this task requires decentralized systems that are able to generate an internally optimal behavior for external goals from a central superior planning system, even under the influence of disturbances. The shifting of planning processes into the lower level decentralized parts requires the ability of self-adjustment regarding new tasks or changed external conditions. One category of systems that fulfills this criterion are the so called "self-optimizing systems". Self-optimizing systems repeatedly execute the three steps: "analysis of the current situation", "determination of goals" and "adaption of system behavior" (Adelt et al. 2009).

Especially for assembly systems, the requirements regarding dynamics, speed and diversity have grown significantly in the last years. Most of today's assembly systems lack means and strategies to adapt to these increasing challenges. A plausible approach to generate goals within these systems is to supply them with "intelligence" (Kempf 2010). This paper describes the design and implementation of such a system.

2   COGNITIVE CONTROL UNIT (CCU)

It is apparent that intelligence of technical systems does not necessarily have to be comparable to human intelligence, since the area of application at hand and the human brain's area of operation are significantly distinct, i.e. the human can generally solve tasks that are much more complex. Rather, it is desired to alter the static and preprogrammed flow of a system resulting in a more dynamic and autonomous character, hence enabling a decision making according to the current status of the environment. In order to realize such a system, knowledge has to be prepared, formalized and made available to executing units (Kempf, Herfs and Brecher 2009).

For technical systems, cognitive architectures that attempt to model human behavior and reasoning provide cognitive functionalities in form of software implementations. In this project, the cognitive architecture Soar has been chosen to simulate cognitive functions. Soar's internal knowledge base is structured in the form of production rules (if-then rules) and thus has the advantage over emergent systems of not needing time-consuming and potentially unreliable preconditioning.

In order to allow cognitively automated systems to interact with their environment, it becomes apparent, that these systems need to be able to continually analyze, fuse, interpret and evaluate data. To create some abstraction between the data and knowledge, an architecture linking the cognitive system and the controlled automation system is necessary. A common architecture for robotic applications is the three layer model by Russell and Norvig comprised of a planning layer, a coordinating layer and a reactive layer (Russell and Norvig 2003). This model also supports one of the key factors in regard to automation: an abstraction between both code and knowledge, and requirements regarding time. However, this structure does not satisfy the requirements of common production systems, since they usually have to provide more interfaces as, for instance, for human operators.

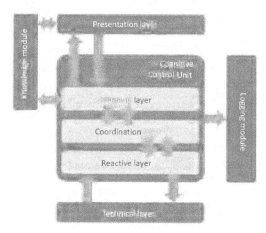

Figure 1: Model of the Cognitive Control Unit (CCU) (Mayer et al. 2011)

Therefore, the cognitive assembly system presented in this paper extends the aforementioned three layer model by a presentation layer and a technical layer. It embodies the CCU (cp. Figure 1) and allows for an integration and enhancement through distributed software modules. The upmost layer (presentation layer) incorporates the human machine interface and an interface for defining an explicit knowledge base. The planning layer serves as the deliberative layer, in which the actual decision for the next action in the production process is made. The services provided by the coordination layer can be invoked by the planning layer to start the execution of specific actions, e.g., the coordination or sequencing of multiple technical actuators. The reactive layer is responsible for a low response time of the whole system, e.g., in order to effectively and efficiently respond to real time control situations. Planning, coordination and reactive layer comprise the CCU, whereas the simulated cognitive functionality is located in the planning layer. The technical layer provides the interface for actuating the machines. The layer model is additionally extended by two modules, namely the knowledge module and the logging module. The knowledge module contains the necessary domain knowledge of the system, in this case in terms of if-then rules (production rules). The logging module records all actions of the system. This architecture serves as a general design framework for modeling the actual automation and production system.

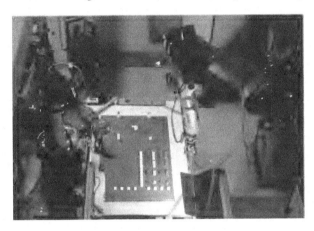

Figure 2: Assembly cell as a validation scenario

The main functionality and capability of the CCU has been developed for the control of a robotic cell for part handling (cp. Figure 2). Hence, the input of the system is a model of an assembly group, which is interpreted and built autonomously by the assembly cell. Thus, the system requires the appropriate means to interpret a model of the assembly presented as a 3D-CAD model and to generate an executable plan, which contains the necessary actions for assembling the target CAD model a real product. Additionally, the system must have knowledge about the actuators accessible to execute the constructed plan. The area of operation, which was chosen to demonstrate the skills of the cognitive assembly system effectively and efficiently, consists of an assembly task using Hubelino® bricks. These bricks are in shape and color very similar to Lego Duplo®. One of the advantages of the Hubelino bricks is that they only vary little regarding the size of the parts. Modular assemblies can easily be pictured using different colors. Little variance as well as the different shapes also simplify the detection of parts using technical solutions. Even by utilizing only a small selection of parts, one can construct very simple as well as rather complex models. By breaking down the selection of parts, it is possible to even lower the requirements for the technical system. This is specifically interesting for image recognition systems, in order to achieve a high detection rate for moving objects. Additionally, the Hubelino models can be scaled geometrically, thus enabling for a transformation of small assemblies into bigger ones, while maintaining the abstract design. A graphical user interface can be used to create the models, convert them to CAD models and present them to the interpreting system. The applicability of this system's abilities has been demonstrated in an industrial scenario for the assembly of switch-cabinet systems (Mayer et al. 2010).

In addition to the technical components a human operator can join in solving the assembly task. This also allows for the system to solve even more complex tasks, like gripping small or hardly accessible objects. For situations that are identified by the system as not solvable a human operator can be included as a mediator, giving him the opportunity to assess the situation, recommend proceedings or resolve the problem.

To sum up, the system needs to be able to interpret a 3D data model and to coordinate the assembly. Thus, a sequence of actions has to be generated based on the assembly's construction. In order to construct the assembly, a sequence of actions has to be generated on-the-fly, whereupon all of the available components have to work together synchronously. The traditional construction of a complex system usually binds lots of experienced developers in pursuing activities without value creation. Hence the shifting of these tasks into an entity that delivers the same results without human interaction has the potential to increase the overall productivity of assembly processes. This applies for both the interaction and the planning in these systems.

In order to allow a human operator to supervise or participate in an assembly task, it is essential, that he is able to understand the actions performed by the system. The methodology of MTM-1 is able to establish this transparency of the system.

3   DESIGN OF THE COMPREHENSIBLE ASSEMBLY SYSTEM

Planning of assembly processes traditionally requires programming programmable logic controllers (PLCs), robot controllers (RCs) or other control units by the human operator, who hence creates a mental non-formalized model of the resulting process. With a cognitively controlled system, these activities are reduced to the layer of granularity that the system can handle by itself. The sequence of actions carried out by a cognitively controlled system and the sequence designed by a human, have the same goal, but might, e.g. be ordered differently. This leads to differences and incompatibilities between the mental model of the process in the mind of the human operator and the process carried out by the technical system. Since these systems also vary in the representation and processing of information, it is crucial for the performance of the man-machine cooperation to find a way to overcome these inconsistencies (Mayer et al. 2011).

3.1 Application of MTM for robotized handling

By modeling the knowledge regarding the process in a cognitive-ergonomic manner it resembles the structures in the mind of the human operator. Thus, the resulting behavior should be similar or even equal to the behavior expected by human operators. Therefore, the process logic for the control of articulated robots performing assembly tasks has been designed cognitive-ergonomically by means of key elements of Methods-Time Measurement 1 (MTM-1) (Maynard, Stegmerten, and Schwab 1948).

MTM-1 is a world-wide used and standardized system for decomposing motion-sequences of assembly tasks into basic movements, whereupon each basic movement is linked with a predetermined time value. Thus, complete assembly tasks can be analyzed yielding a total time value specifying the time that is needed by a trained worker for performing the assembly. This analysis can be used to optimize the conditions within the assembly area. According to the most typical action sequences, MTM-1 distinguishes five basic movements: REACH, GRASP, MOVE, POSITION and RELEASE. Every assembly action can be decomposed into a sequence of these actions. Furthermore, they have been empirically validated and, as they are expectation conformal, they can be easily understood, learned and optimized by a human.

Since this research is not focused on the time-based optimization of processes but rather on the general evaluation of the concept of a transparent assembly process, the time concept of MTM-1 has not been considered here. Instead, the cognitive assembly system takes advantage of the easily understandable decomposition of assembly sequences into the basic movements. They have been utilized to define production rules, which can be interpreted by the processing unit of the cognitive architecture (Mayer et al. 2008). In detail, each MTM-1 operator has been reproduced by an own production rule. The rule for the operator REACH is, for instance, proposed if any required component has been conveyed to the system and the rule for the operator GRASP reacts if the robot is at the position of a required component. However, the production rules are independent from each other, i.e. each action can be followed by any other action as long as its preconditions are satisfied. Hence, the cognitive system is not restricted to strict assembly sequences, but is able to react to unforeseen situations as, for instance, a dynamic change of the assembly. In order to support the rules for the basic movements, there are additional production rules for managing the behavior of the cognitive system. These rules are, for instance, responsible for monitoring the destination system or setting preferences between multiple possibilities of similar operations. The latter is important for distinguishing between actions on the queue and the buffer, since the former ones are favored.

3.2 Demonstration scenario

In order to demonstrate and verify the capabilities of the presented system a simple scenario has been created (cp. Figure 3). The scenario contains three different areas, namely supply area, assembly area and buffer area, which can be accessed by the robot. All available parts are at the beginning in the supply area, whereupon the type and color of a brick at a certain position is random. In order to determine, which parts are present, they need to be detected sequentially. The

278

construction of the assembly group takes place in the assembly area. Due to the random nature of the part supply and the algorithmic and physical restrictions, it is possible that a part is needed for an assembly, but not at the time, when it is detected. To cope with this restriction, such parts can be stored in the buffer area for future use. At last, parts, which are never needed, can be disposed.

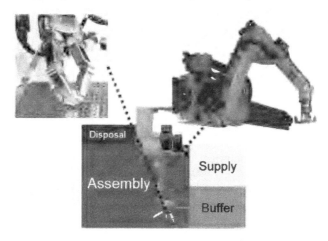

Figure 3: Simple scenario with Kuka KR 30 Jet (right) and SCHUNK SDH-2(left)

The technical components of this simple scenario are an articulated robot (Kuka KR 30 Jet) and a three finger gripper hand with haptic sensors (SCHUNK Dexterous Hand 2, SDH-2). These components are accessed on the technical layer of the CCU architecture and thus have to supply methods and services for their technical capabilities. For the KR30 Jet all of the basic movement commands like Point-to-Point (PTP) and Linear (LIN) as well as more sophisticated movements and interactions are provided. Additionally, the Robot Sensor Interface (RSI) mode can be activated, which allows for the control of the robot in its interpolation cycle (12 ms). This mode is useful for situations requiring real time access with more advanced sensory input, e.g., image recognition procedures. For such situations, a RGB camera is attached to the flange of the robot. In regard to the SDH-2 several gripping commands and positions can be accessed. Additionally, the tactile sensor information in each of the three fingers can be utilized, e.g., to create advanced positioning operations or to react in erroneous situations.

As previously described, the assembly of a certain target assembly group can begin, once the target CAD model has been created and submitted to the CCU. Since at this point, none of the parts located in the supply area is known to the CCU, the only possible action sequence is the one to detect color and shape of another part in this area. After that, there are three possible scenarios:

- The part is never needed: The part is disposed.
- The part is needed either immediately (assembly) or later (buffer): The control of the assembly system is handed over to the MTM-1 rules (see below).

- It is unknown if the part is needed: This results in an internal buffering of the information of that part and the detection of another part in the supply area.

3.3 Implementation of MTM

In order to deliver the control of the assembly system to the MTM-1 rules, there has to exist a mapping between the technical behavior and the basic MTM-1 operations. Thus, the CCU architecture translates the operations into explicit control commands:

- REACH: The reach command is translated into a PTP command to a position, which is slightly above the brick that needs to be gripped. The position of this brick is known, since the detection of this part has taken place previously.
- GRASP: Short LIN move in order to position the gripper, followed by the actual gripping operation.
- MOVE: Short LIN move to detach the brick from the surface, followed by a PTP movement to a point that is slightly above the target position. This target position can be either in the assembly or buffer area.
- POSITION: For the current scenario this is the most complex command. The brick needs to be assembled at its target position. A simple approach could be to just assemble the part with a vertical movement towards the mounting surface. Due to uncertainties that are inherent in both the robot and the gripper, extensive testing has shown that this approach usually fails. In order to create a more resilient process, the movement has been designed similar to a human assembly process, e.g., assembling the part sideways. Additionally, the tactile sensors of the gripper hand can be used to detect pressure and slipping effects on a brick. Brecher et al. provides a comprehensive overview of the resulting strategies and algorithms (Brecher et al. 2012).
- RELEASE: Detaching of the gripper and short LIN command in order to position the gripper above the assembled brick.

If none of the MTM operators is applicable any more, for instance because there are no more bricks in the internal and physical buffer, a finishing command re-enables the other units of the CCU, leading to e.g., the detection of another part in the supply area. The sequence of these operations is not predetermined, as they have initially equal weights. However, further rules may manipulate these preferences according to the particular situation. Since all matching production rules fire at the same time, the resulting sequence could be different each time.

As a result, all assembly operations can be carried out by basic MTM-1 operations. An example for a generated MTM-1 sequence is given in **Fehler! Verweisquelle konnte nicht gefunden werden.**. Aside from POSITION the operations are built upon basic robot movement commands. The gripper geometry is responsible for most of the physical restrictions of the assembly. Even though these restrictions need to be considered in both the cognitive architecture and the gripping planner, all of the resulting actions only affect the technical layer of the CCU. The complexity and requirements of the technical implementation are, however, somewhat dependent on the scenario.

4   CONCLUSIONS

The commissioning and programming of complex robotized handling operations requires extensive planning efforts, which do not contribute directly to the added value. Time-consuming imperative coding processes result in inflexible programs that are difficult to adapt to varying requirements. In order to adjust to the shorter innovation cycles for current products, methods to reduce these efforts by automatically adapting to new environments and new objectives are required. A cognitive control system is able to generate essential action-sequences at run time on the foundation of model-based knowledge regarding task, system and process flow (e.g., in a robot cell) and in consideration of current environmental information.

Hence, a software framework on the basis of a cognitive architecture, which allows for the creation of intelligent robotized handling processes, has been created. The Cognitive Control Unit (CCU) interprets a 3D-CAD model and coordinates the manufacturing of an assembly group for randomly supplied parts. Thus, possible assembly sequences need to be calculated and form the basis for the action flow generation. All components involved (e.g., robots, grippers, cameras) synchronously participate in the assembly process. The capabilities and interactions of miscellaneous components are key aspects, since only a rich reactive system enables a rich cognitive system. Reactive grasping and perception of image information assure the efficiency of the overall system.

The benefits of the CCU are contrasted by a possible loss of control on the overall system. In order to allow for an effective and safe supervision of the assembly process by the human operator, the behavior of the corresponding machines must be transparent for him. For common control devices and machines this is usually rather easy, as they demonstrate a deterministic, preprogrammed and verifiable behavior. For cognitively controlled units, however, the decisions made within the cognitive architecture need to conform to the operator's expectations. In regard to assembly sequences, the most effective way for machines – and especially for robots – to indicate their objectives, is to work according to the mental model of the human operator. Thus a concept was developed and implemented, that creates a systematic behavior according to the expectations of a skilled worker by using a human centered modeling approach on the basis of the MTM-1 system.

The implementation abstracts from the technical implementation of a robot's behavior by providing interfaces for the common MTM-1 commands REACH, GRASP, MOVE, POSITION and RELEASE. The software architecture of the CCU consists of multiple layers using distributed software modules. In order to work in a real production environment it is for example necessary to differentiate on the subordinate MTM behavior for different situations, depending on the current environment, e.g., to apply different POSITION procedures depending on the surroundings of an assembly part.

ACKNOWLEDGMENTS

The authors would like to thank the German Research Foundation DFG for the support of the depicted research within the Cluster of Excellence "Integrative Production Technology for High-Wage Countries".

REFERENCES

Adelt, P., et al., 2009, Selbstoptimierende Systeme des Maschinenbaus - Definitionen, Anwendungen, Konzepte. In: J. Gausemeier, F.J. Rammig, and W. Schäfer, eds. HNI-Verlagsschriftenreihe. Paderborn: W. V. Westfalia Druck GmbH.
Brecher, C., et al., 2012. Assembly Motion Planning using Controlled Collisions. In: 4th International Conference on Applied Human Factors and Ergonomics 2012, (submitted)
Calvo, P., T. Gomila, and A. Gomila, 2008. Handbook of cognitive science: an embodied approach: Elsevier Science.
Gausemeier, J., C. Plass, and C. Wenzelmann, 2009. Zukunftsorientierte Unternehmensgestaltung: Strategien, Geschäftsprozesse und IT-Systeme für die Produktion von morgen. München: Hanser Fachbuchverlag.
Kempf, T., W. Herfs and C. Brecher, 2009. SOAR-based Sequence Control for a Flexible Assembly Cell, In: Proceedings of the 2009 IEEE Conference on Emerging Technologies and Factory Automation, 22.-26. September 2009, Palma de Mallorca, Hrsg.: Grau, A.; Campos, J.; Oliver, G., IEEE Palma de Mallorca 2009, ISBN 978-1-4244-2728-4
Kempf, T., 2010. Ein kognitives Steuerungsframework für robotergestützte Handhabungsaufgaben. 1st ed. Aachen: Apprimus-Verlag.
Mayer, M., et al., 2008. Task-Oriented Process Planning for Cognitive Production Systems using MTM, In: Proceedings of the 2nd International Conference on Applied Human Factors and Ergonomic (AHFE) 14.-17. July 2008, Las Vegas, Nevada, USA,
Mayer, M., et al., 2011. Selbstoptimierende Montagesysteme auf Basis kognitiver Technologien. In: C. Brecher, ed. Integrative Produktionstechnik für Hochlohnländer. Heidelberg: Springer Verlag, 83–255.
Maynard, H.B., G.J. Stegemerten ; J.L. Schwab, 1948. Methods-time measurement. New York: McGraw-Hill.
Orilski, S. and G. Schuh, 2007. Roadmapping for competitiveness of high wage countries. In: Proceedings XVIII ISPIM Conference: Innovation for Growth - the Challenges for East and West. Warsaw.
Russell, S.J. and P. Norvig, 2003. Artificial Intelligence: A Modern Approach. 2nd ed. Upper Saddle River, New Jersey: Prentice Hall.
Schmitt, R., et al., 2011. Selbstoptimierende Produktionssysteme. In: C. Brecher, ed. Integrative Produktionstechnik für Hochlohnländer. Heidelberg: Springer Verlag, 747–1057.
Schuh, G., et al., 2011. Individualisierte Produktion. In: C. Brecher, ed. Integrative Produktionstechnik für Hochlohnländer. Heidelberg: Springer Verlag, 83–255.

CHAPTER 30

Navigation as a Key for Self-Optimizing Assembly Processes

R. Schmitt, P. Jatzkowski, A. Schönberg

Laboratory of Machine Tools and
Production Engineering WZL
RWTH Aachen University
R.Schmitt@wzl.rwth-aachen.de

ABSTRACT

Assembly relies strongly on human factors in high wage countries. It needs to integrate collaborative components in today's production systems, which are required to perform various cooperative tasks with multiple entities rather than monolithic solutions. Modular concepts introduce a high amount of planning and complexity in volatile environments with unforeseen requirements. Self-optimization is mainly addressed in the virtual domain, today. Economic assembly however needs versatility on both the virtual planning and the real assembly side. Reconfigurable multi robot systems with high absolute pose accuracy and synchronization are extended by large-volume measurement systems as well as local sensors to provide virtually aligned state estimations of multiple end effectors in a global reference system. Assembly systems are then enabled to navigate in the virtually planned machine state by socio-technical realizations of self-organization; they autonomously form a complete self-optimizing assembly. Implementations are shown for cooperating robots and a collaborative process for alignment of fuselage structures.

Keywords: automation, metrology, global reference systems

1   INTRODUCTION

In contrast to specialized automation systems, multiple robots may be used to adjust processes without further or only minimal hardware changes. The rapid propagation of new technologies, offensive competition, closer networking goods and flows of capital as well as fragmenting and dynamic new configuration of the value added exposes today's industry to unknown challenges [1]. Flexible systems suffer economic drawbacks in uncertain scenarios due to their need to compromise for a broad range of expected requirements. Recent developments focus on versatile systems that are capable to work at technically and economically optimized working points [1]. This approach opposes the flexibility concept, as it does not assume to meet all expected requirements, but tries to be optimized until a change need is identified and then adopted.

Self-optimizing planning tools have become very powerful in the last years; however lack the feedback from the real world to adjust their internal models. Instead of implementing fuzzy elements for planning conformity, shifting the focus to develop conforming real world elements is another option being researched on. The reduction of discrepancies between the virtual planning and the real world production line is the key to self-optimizing assembly. Cognitive abilities raise the production system's awareness and self-influence. The vital role of metrology in this context may lead to better production concepts as it supplies the ability to react efficiently to changes. The focus of metrology will shift from the work-piece to the production system itself [2].

2   SELF-OPTIMIZING PLANNING

Versatility takes place largely in the digital planning world, today. Enablers of versatility address scalability, mobility, universality, compatibility and modularity [1] that mainly eases the (automated) planning. Self-optimizing systems partly address the execution of planned stages by integrating cognitive capabilities to the processes that enable the machines to act in uncertain scenarios (e.g. [3] [4]). While the planning with ideal digital models is widely assumed to be existent in this context, the alignment with the real assembly system is still a major working topic. The effects of change are clearly visible in ideally modeled scenarios and may therefore be judged easily for benefits and drawbacks. The result of "versatile planning" is a changed digital state for the assembly system, which is then used to guide the machines to change to new working point, forming a "versatile production" [2]. Virtual and real assembly may be aligned with two approaches:

- The first approach extends the ideal models of the planning
  environment with uncertainty factors and compensation capabilities;
- The second approach allows the assembly entities to navigate
  on the planned state at runtime.

Both approaches need information about their behavior to concurrently navigate in the worlds of virtual planning and real assembly. Production navigation works analogue to traffic navigation by using a global referencing system in combination with ideal world planning data and local sensing to control local effects. With highly interacting systems, control hierarchies introduce exponential complexity because of unforeseen interdependencies by using direct control mechanisms like path correction [5] or adjustment of measurement results [6]. This drawback of the navigation approach needs to be addressed by a more scalable and robust control approach.

3   SELF-ORGANIZATION FOR NAVIGATION IN SOCIO-TECHNOLOGICAL PRODUCTION SYSTEMS

The emergent behavior of a self-organizing system shows good characteristics concerning scaling and robustness in relation to influences of noise or changes of parameters [7]. Successful approaches to deal with self-organization in management science [8] switches the technical perspective to the social perspective of strongly interacting systems. The main criteria to describe self-organization are redundancy, autonomy, complexity and self-reference [7].

In self-organizing systems no separation between organizing, arranging or steering parts takes place. All parts of the system represent potential actors and impose a high degree of redundancy and autonomy. The parts are interlaced by mutual, permanently changing relations, which can change likewise at any time and form a complex system. Self-referencing exhibits an operational closeness. That is each action of the system retroacts on itself and becomes the starting point for further behavior [7]. Operational closeness acts not due to external environmental influences, but independently and solely responsible out of itself. Self-reference represents however no contradiction in relation to the openness of systems [7].

The basic mechanism of self-organization is the exploration of different regions in the state space until an attractor is reached, a configuration that closes in on itself. This exploration may be performed by deterministic or stochastic variations of the dynamic system. Therefore self-organization depends strongly on noise in the system and may be accelerated or deepened by imposing more noise [9]. Production processes generally contain a significant amount of noise to support the exploration [10]. In the surrounding of an attractor further variation outside the attractor is precluded, and thus freedom of the system components is restricted to behave independently. This is equivalent to the increase of coherence, or decrease of statistical entropy, that defines self-organization [9]. Although present organizations and societies as well as technological concepts incorporate many aspects of self-organization, the lack of understanding of self-organization makes it difficult to introduce radical implementations.

The advantages of using self-organization in complex socio-economic systems, rather than centralized planning and control, may also be adapted to the socio-technological systems of assembly. Self-organizing systems may be suitable as a paradigm for future complex technical systems [11]. The use of self-organizing approaches may lead to more stable systems in changing environments with multiple interacting entities. In the past years first approaches to use self-organization were developed [12] and established highly adaptive systems. The original concept of self-organization however is often only partly addressed in manufacturing science [13].

4   GLOBAL REFERENCE SYSTEMS IN INTERACTING ASSEMBLY SCENARIOS

The terrestrial global positioning system provides universal navigation information about geometry and time on a planetary scale. In combination with "calibrated" maps, traffic participants are able to plan their route or automatically stay on a programmed path. On a facility scale, no monolithic reference system exists, however may be established by combining the internal state of the machines and large-volume metrology systems as a dimensional reference. The digital shop-floor usually provides communication capabilities in today's production systems to gather a holistic machine state. The availability of a holistic machine state extends the cognitive systems to enable self-organizing behavior of the machines to align with a virtually planned state. Large-volume metrology systems may be used to provide valid data for the integration of metrological attractors to compensate systematic deviations in a global reference system (GRS). Numerous alternatives for measuring pose accuracy in complete production environments, such as laser trackers, multilateration [14], vision systems [15] or iGPS exist.

Most production machines contain internal sensing to work properly, especially (multiple) robots however suffer from systematic deviations especially in changing environment. The concept of GRS adds external sensing to the cognitive system of production entities by using a standardized interface to provide deviations from the machine state in the real world with a virtual state existent in the planning world. The self-referencing entities act as a response to the data, which is internally interpreted to represent the own behavior and therefore induces e.g. a control action to form an attractor as required by self-organization

To allow the gathering of an aligned virtual/real machine state, several key requirements are used to define GRS [2]:

- Metrology systems are a part of the infrastructure
  → there is no direct link to machines
- Common coordinate system in both production and simulation
- Evaluation of spatial and temporal relations

Interfacing is the key issue for GRS. Production and metrological entities generally use vendor specific interfaces to interact. The entities may be separated in the groups 'non-manipulating' and 'manipulating'. Metrology systems mostly act as a data-provider and manipulators as a data-receiver in the assembly communication network, which establishes the GRS. The concept needs to be detailed for the implementation of GRS. All entities of the GRS are required to align their information to the virtually planned state and to publish it in the network.

4.1 Cooperating robots

Standardized interfaces are currently not existent to allow true self-organization on assembly entities like robots and therefore require special solutions to implement GRS in cooperating applications (machine-machine interaction). Generic clients for tool/base/timeline transformations and calculation of deviations to the virtual machine state are used with an offline-calibration phase prior to production. At runtime, data from the metrology systems are aligned to the programmed virtual machine state and may be used to calculate the deviation between virtual and real world for alignment. The estimated deviations are inserted in the cognition system of the robots consisting of the machine-variables for Reis or Fanuc robots or special sensor-options like the robot sensor interface from Kuka (Fig. 1).

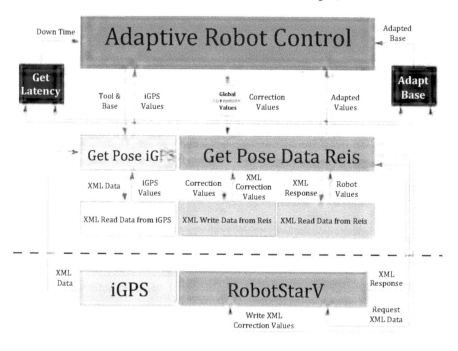

Figure 1: Self-referencing system layout for Reis robots and iGPS

The problem of achieving absolute accuracy in the spatiotemporal positioning and movement of cooperating robots are evaluated in the iGPS based robot cell at the Laboratory for Machine Tools and Production Engineering WZL of the RWTH

Aachen University (Fig. 2). Nikon Metrology iGPS receivers are attached to the robot end-effectors to provide spatiotemporal information on the robot poses in six degrees-of-freedom. The kinematics and error budget predict accuracy achievable for the robots to be at least within 0.5 mm for measurements at a rate of up to 40 measurements per seconds. iGPS provides a spatiotemporal base to form the GRS.

Figure 2: iGPS based robot park at WZL with cooperating Kuka and Reis robots

For Kuka robots iGPS controlled positioning and movement has been shown in [5] and has been converted to the GRS-approach without performance changes. Reis robots use machine-variables to implement feedback control. The modular adaptive robot control uses the XML-ethernet option for reading and writing machine-variables ('Get Pose Data Reis' in Fig. 1). This option has a cycle time of about 1ms and therefore allows the exchange of ARP and RPK within interpolation cycles as supported by the existing adaptive robot control. Only changes for the robot-interface and the setup of the sensor-option are needed for controlling the Kuka and Reis robots at positioning and during movements. For positioning only, tool and base offset options exist for Kuka, Reis and Fanuc robots.

The performance of iGPS-controlled positioning and movement is almost identical for Kuka [5] and Reis, which is expected because of the identical setup for most parts of the self-referencing system except the interfacing to transfer ARP and RPK. This change has no significant effect, as both interfaces are able to update within the interpolation cycle of the robot controller (Kuka: 12ms, Reis: 10ms). On the lift in Fig. 3 the positioning of the Reis RV130 is shown with a weight added to the TCP at the first mark and the start of the control at the second mark. On the right the robot moves in x-direction and deviations are shown for the z-direction. At the mark, controlling is enabled and the robot aligns to the programmed path.

288

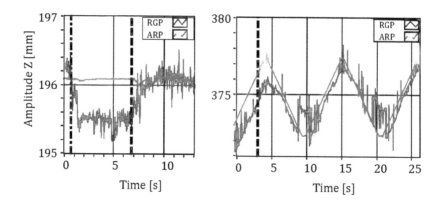

Figure 3: Self-referencing positioning with a weight added (left) and controlled movement (right, robot moves in x-direction) for Reis RV130

4.2 Collaborative Inspection

The iGPS based robot cell is a development platform for GRS-based interacting entities, mainly used for evaluating cooperative and collaborative issues in simple scenarios as a technology enabler (e.g. efficient assembly, fixtureless welding). Industry-related scenarios like an automated alignment of fuselage structures [2] show the capabilities to meet process requirements and to form an automated system in several measurement tasks (Fig. 4). Collaborative (man-machine interaction) aspects are widely existent in aerospace manufacturing due to a high variety of tasks to be performed during the production of large complex structures.

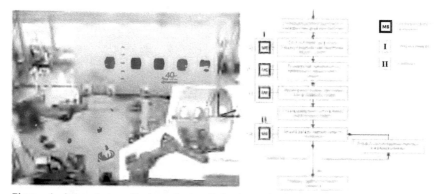

Figure 4: Automated positioning of aircraft fuselage structures at the Instituto Tecnologica Aeronatica (ITA) in Sao Jose dos Campos in Brasil in the test setup to compare Nikon Metrology iGPS, K-series and laser radar

Once delivered, in scenario 1 a new fuselage part has to be inspected for applying a best-fit estimation for the alignment phase. The result of a first scenario is a need for laser radar for geometric inspection prior to aligning it based on the performance to meet the following measurement requirements.

- Inspection of given tolerances
- Reliability: Measurement accuracy
- Reliability: Measurement repeatability and reproducibility
- Process time
- Process costs
- Process complexity
- Environmental conditions
- Safety at work
- Admissible workspace volume

In scenario 2 the alignment is controlled locally at the joining circle of the cylindrical parts and globally to allow the assembly process after alignment using best-fit mechanisms. iGPS is the most favourable measurement system for the alignment task, however may be extended by laser radar for certain tasks especially in the tracking-preparation phase. This scenario's scoring system is based on the following requirements:

- Fuselage segment tracking preparation
- Fuselage segment tracking
- Measuring accuracy
- Measuring repeatability
- Process automation
- Measurement convenience
- Process costs
- Environmental conditions

In summary it can be stated that a single iGPS does not represent an acceptable solution for the overall process at hand, while a single Laser Radar may be acceptable with significant drawbacks. The strong points of both systems cover the entity of both scenarios' requirements. The iGPS's comparatively high measurement uncertainty during alignment control can be compensated for with averaging multiple measurements, which is acceptable due to high frame rates. The possibility to provide Laser Radar measurements of the iGPS monument receivers shows the compatibility of these two instruments. The more exact the iGPS' monument receiver positions are registered in the system configuration, the higher is the measurement performance for the iGPS. The significant increase in investment costs remains the combination's only weakness which has to be put into correlation with the conditions for a single system strategy with the Laser Radar. The GRS-implementation's strengths are especially existent, if

- maximal alignment time reduction is important
- a prompt implementation of the automated alignment process is a priority
- future scenarios requiring dynamic control are supposed to be covered, too

5 CONCLUSIONS

Self-referencing assembly using GRS autonomously forms a best-fit between the digital planning and the real assembly world. This is assumed to be the key for versatility as an answer to meet a broad range of potentially unexpected market requirements. Versatile planning is only partly influenced by this approach as the assembly entities expect an ideally modeled process. Self-optimized planning may be focused on with low requirements on taking real world complexity into account. Assembly scenarios at different scales are evaluated for collaborative and metrological aspects to provide a validation framework for self-organizing versatile assembly in this contribution.

Stabilization at distinct working points may be described by self-organization and is currently used as an explanation model mainly in socio-economical sciences. It may be extended to production systems, provided a universal data provider for cognitive interfaces of self-referencing production entities exists. GRS evaluate the spatiotemporal deviations between a virtual planning and the real world production using Large-Volume metrology to compensate systematic deviations.

Especially for cooperative applications the standardized feedback of global reference systems are fed into the cognitive systems of the assembly to allow self-referencing. The adaptive robot control implementation uses a modular concept to switch between different measurements systems and vendors for manipulators. Performance for absolute positioning and movements of single robots show promising results. With the extension of temporal synchronization, a control system for multiple cooperating robots raise production capacities by parallelization to meet versatility demands instead of reaching limits of single machines. Humans are the most unpredictable entity in a complex assembly scenario. In order to allow a smooth collaboration for several tasks, GRS are capable to provide the input for the cognitive systems of the assembly entities to understand the human actions. On the other side, globally referenced information reduces complexity for the human participants in today's productions systems, too. Self-optimizing assembly gains robustness in both the planning and the execution stage to retain competiveness in volatile markets especially for high-wage countries.

ACKNOWLEDGEMENTS

The authors would like to acknowledge the german research foundation (DFG) for supporting the excellence cluster titled: "Integrative Production Technology for High-Wage Countries" as well as the german ministry of education and research (BMBF) for funding the project with the title "Changing Production Systems actively (ProAktiW)".

REFERENCES

[1] Nyhuis, P., Reinhart, G., Abele, E., 2008, Wandlungsfähige Produktionssysteme – Heute die Industrie von morgen gestalten, PZH Produktionstechni-sches Zentrum, Garbsen, Germany.

[2] Demeester, F.,Dresselhaus, M.,Essel, I.,Jatzkowski, P.,Nau, M.,Pause, B.,Plapper, P.,Schmitt, R.,Schönberg, A.,Voss, H., 2011, Referenzsysteme für wandlungsfähige Produktion, Wettbewerbsfaktor Produktionstechnik, Shaker, 449-477

[3] Loosen, P. et al., 2011. Self-optimizing assembly of laser systems. Production Engineering, 5(4), 443-451

[4] Schmitt, R. et al., 2011. Multiagent-based approach for the automation and quality assurance of the small series production. Emerging Technologies & Factory Automation. Toulouse: IEEE, 1-8

[5] Schmitt, R., Schönberg, A., Damm, B., 2010, Indoor-GPS based robots as a key technology for versatile production, Proceedings for the joint conference of ISR 2010 (41st International Symposium on Robotics) and ROBOT-IK 2010 (6th German Conference on Robotics), 199-205.

[6] Schmitt, R., Schönberg, A., 2010, Kompensation von Abweichungen bei Absolutbewegungen von Industrierobotern durch Indoor-GPS, Automation 2010 - Branchentreff der Mess- und Automatisierungstechnik, 411-414.

[7] Probst, G.J., 1987, Selbstorganisation - Ordnungsprozesse in sozialen Systemen aus ganzheitlicher Sicht, Paul Parey Berlin, Germany.

[8] Hülsmann, M. et al., 2007, Self-Organization in Management Science, Understanding Autonomous Cooperation and Control in Logistics, 169–192.

[9] Heylighen, F., 1999, The science of self-organization and adaptivity, Knowledge Management, Organizational Intelligence and Learning, and Complexity, The Encyclopedia of Life Support Systems (EOLSS), Eolss Publishers, 253-280.

[10] Schwenke, H., Knapp, W., Haitjema, H.,Weckenmann, A., Schmitt, R., Delbressine, F., 2008, Geometric error measurement and compensation of machines, CIRP Annals - Manufacturing Technology, 57/2, 660–675.

[11] Sedlacek, K. D., 2010, Emergenz: Strukturen der Selbstorganisation in Natur und Technik, BoD – Books on Demand, 33-34

[12] Leitão, P., 2008, Self-Organization in Manufacturing Systems: Challenges and Opportunities, Second IEEE International Conference on Self-Adaptive and Self-Organizing Systems, Ieee, 174-179.

[13] Fuchs-Kittowski, K., 2009, Selbstorganisation und Gestaltung informationeller Systeme in sozialer Organisation, Selbstorganisation in Wissenschaft und Technik, Wissenschaftlicher Verlag Berlin, 122-184.

[14] Cuypers, W, Van Gestel, N, Voet, et al., 2009, Optical measurement techniques for mobile and large-scale dimensional metrology, Optics and Lasers in Engineering, 47 , 292-300.

[15] Jeon, S., Tomizuka, M., Katou, T., 2009, Kinematic Kalman Filter (KKF) for Robot End-Effector Sensing, Journal of Dynamic Systems, Measurement, and Control, 131/2, 021010-1-8.

CHAPTER 31

Assembly Motion Planning Using Controlled Collisions

Christian Brecher, Daniel Behnen, Thomas Breitbach, Werner Herfs

Laboratory for Machine Tools and Production Engineering
RWTH Aachen University
{c.brecher, d.behnen, th.breitbach, w.herfs}@wzl.rwth-aachen.de

ABSTRACT

In order to utilize industrial robots for filigree assembly tasks new handling and assembly skills for automated assembly systems are required. This paper presents an approach using a combination of knowledge, planning and "intelligent" gripper behavior to cope with a broad spectrum of fast-changing products and with assembly system limitations such as fuzzy tolerances caused by either product parts or system components.

Keywords: assembly, planning, cognition

1   INTRODUCTION

The production industry in countries with high labor costs, such as Germany, faces global competition and is concerned with losing more and more branches of production to low wage countries. Currently the production of goods requiring assembly of multiple components, like the automotive industry, is located in both high and low wage countries whereas the production and assembly of electronic devices like mobile phones is almost solely located in low wage countries. In general, production processes, which can be supported by automation, are more likely to be economically accomplishable in high wage countries due to less involvement of human resources than processes with a low degree of automation (Brecher, 2011, Chapter 6).

In the field of product assembly current systems provide a high degree of accuracy and repeatability, but cannot compete with human labor due to a lack of flexibility of both the kinematics and the control software. Current assembly systems are suited for well-programmed assembly movements and must be reprogrammed costly and tested extensively if new products or product variants need to be assembled.

An approach towards flexible assembly systems realizes "intelligent" assembly systems, thus decreasing the manual effort of planning and programming. The minimization of planning tasks demands for some degree of autonomy within production systems. In other words the production system is expected to behave similarly to intelligent and flexible humans (Kempf, 2010).

This paper describes an approach towards a flexible and autonomous assembly cell which uses a cognitive architecture to plan and to conduct the assembly of model products, using only a model of target assembly structure (i.e. the final product) as task description. Besides the overall system architecture of the assembly cell an approach towards motion planning and motion control of the effector inspired by human assembly behavior is described.

2   AUTOMATED ASSEMBLY USING INDUSTRIAL ROBOTS

To be able to implement and verify a cognitive assembly control architecture and software, a flexible assembly cell has been designed and built. The cell was designed to be able to handle varying products and was built with various control components to be able to verify a complex control architecture. As the challenge for conventional assembly processes often lies in handling and joining, an assemblable model product family, namely LEGO Duplo, was chosen. The automated and human-assisted assembly of LEGO Duplo models has proven to be challenging and analogies with real-world industrial assembly processes, like control box assembly, have been identified and investigated (Brecher, 2011, Chapter 6).

An assembly process to be carried out by the cell is started by an operator who graphically designs a target assembly group comprised of a number of single parts (bricks). The task of the assembly cell is to autonomously find a possible assembly sequence using the goal description and to execute the task using the given technical components (Kempf, 2010).

The setup comprises four main modules (see figure 1):
- an articulated robot (five degrees of freedom, sitting on a linear track), which is equipped with a camera system and a flexible, human-inspired gripper hand used for assembly,
- an articulated robot (six degrees of freedom) with a 3D camera,
- a conveyor system equipped with six individually controlled transfer lines, pneumatic track switches and a couple of light barriers,
- a pallet handling system with a gripper.

Furthermore a number of locations need to be distinguished:
- a storage pallet, which is initially equipped with a number of bricks,
- a buffer for bricks to wait for further treatment,
- an assembly area for automatic assembly,
- two disposals to be reached either by the conveyor or a robot.

294

Figure 1 shows the overall assembly cell for the scenario. The concept of the setup ensures that not only the objectives change, but also the sequence of supplied raw parts. Initially the parts are randomly placed onto the conveyor by the pallet handling system. While circulating on the conveyor, they need to be grabbed from the moving belt previous to assembly. Therefore an optical tracking system, which uses the camera mounted on the assembling robot, determines the speed and orientation of the parts that need to be grabbed.

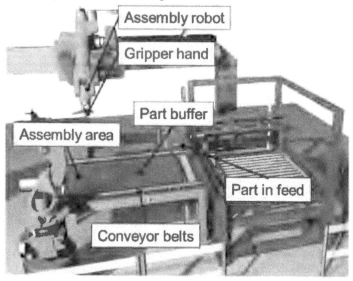

Figure 1 -Assembly cell installation with two robots, a gripper, conveyor belts, part feed in, buffer and assembly area.

The overall assembly problem is composed of a number of steps each of which is highly relevant in regard to industrial handling applications:

1. The system has to be able to retrieve part information from a part geometry database and an offline engineering tool (CAD) and autonomously create an assembly sequence.
2. This sequence has to be transformed into a synchronized sequence of machinery operations including any number of subsystems (action plan).
3. The generation and control of robot movements (path plan) needs to be performed.
4. A valid gripping strategy (gripping plan) needs to be found.

All of these planning steps usually take a lot of programming expertise of a number of persons, especially since multiple stages of the traditional engineering process and various subsystems are affected.

In the following chapter the generation of "cognitive" behavior is described and the aspects 1 and 2 are addressed. In Chapter 4.2 an approach towards a solution for the third aspect, namely the generation and control of robot movements is presented.

3   COGNITIVE PLANNING AND CONTROL

The overall layered architecture of the control unit is shown in figure 2. The planning layer operates on a high abstraction level with symbolic problem definitions and must satisfy only "soft real-time" demands. The reactive layer, on the other hand, has to monitor machine-dependent control loops in "hard real-time". The coordination layer mediates between these two layers. It transforms abstract instructions from the planning layer into precise machine control commands. In the other direction, the information from the various sensors is aggregated to form an overall picture of the situation and is transmitted to the planning layer as a basis for decisions. In order to satisfy the demands of a holistic consideration of the human-machine system, the classic three-layer architecture was expanded to include further layers and modules. The presentation layer forms the interface to the human operator. The interactive goal definition and description of the task, as well as the presentation of the current internal state of the control unit is realized by the human-machine interface. In addition, external data formats such as CAD data of the product to be assembled, are transmitted in internal representation forms and made available to the other components. The planning layer can send enquiries to the knowledge module by transmitting the system state received from the coordination layer. The knowledge module then analyzes the objects contained in this state and can derive further information by means of reasoning with the available knowledge and transmits the results to the planning layer. In the logging module, all of the data generated during system operation is persistently stored so that in the event of an error, the cause can be reconstructed if necessary. To support system transparency, all the data stored in the logging module during operation is accessible to the user via the presentation layer.

This framework is arranged around the cognitive architecture SOAR ("Soar Homepage", 2012). SOAR operates in sense-plan-act cycles on a symbolic representation of static domain knowledge and actual system states within a discrete system model. Thus far, SOAR has been implemented e.g. in psychological, (military) aviation and business applications but not yet for anything similar to a production plant.

In SOAR knowledge about the domain is situated in the procedural memory, expressed in "IF...THEN"-like rules (productions) which are stored in the long-term memory. The problem space consists of several states, described by features, and operators which can be applied to states. SOAR navigates through this problem space using the rule base in order to reach a certain goal state. This navigation can happen either offline (planning) or online and is performed in the working memory. Here all knowledge including environment information which is relevant in the current situation (be it online or offline) is brought together. The architecture hereby implements all of the necessary tasks for a cognitive system: In each decision cycle input from the environment is considered (perception), all relevant knowledge is evaluated (elaboration), one operator is chosen (decision), productions fire accordingly (application) and information to the environment is generated (output), which eventually leads to a new state. That means, in contrast to classic planning tools SOAR applies all relevant knowledge useful in a certain situation which makes it more flexible and responsive (Kempf, Herfs, and Brecher, 2009). Unlike purely reactive systems SOAR is able to evaluate certain actions virtually thus performing planning functionality.

Although the overall behavior of the assembly cell is planned and controlled by SOAR, the geometric reasoning required to determine feasible assembly sequences for given product configurations can hardly be realized with SOAR's rule based reasoning (Brecher, 2011, Chapter 6). Hence, a specialized planner has been developed to find feasible assembly sequences given the restrictions posed by the gripper's abilities and extensive knowledge about the accuracies of sensors and robots. The planner is located within the soft real-time planning layer and is directly used by the cognitive processor, e.g. SOAR, to aid in decision making and to provide motion path information for the reactive layer.

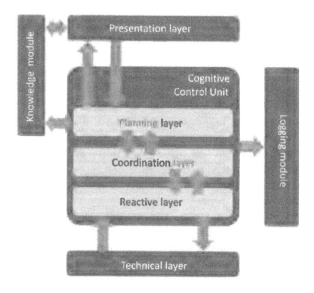

Figure 2 Layered control architecture of the cognitive assembly cell (Brecher, 2011, Chapter 6).

4   GEOMETRICAL ASSEMBLY PLANNING

In order to coordinate the assembly sequences, all possible sequences are pre-calculated by the control unit and represented as a graph (Brecher, 2011, Chapter 6). For the generation of this assembly graph the planner has to determine which parts can be mounted at different assembly stages considering the state and handling abilities of the assembly cell, the shape of the gripper tool and the forces and their directions in the assembly process. Furthermore, grabbing parts from the conveyor belt using optical tracking introduces positioning deviations, which need to be taken into consideration when planning the assembly motions for narrow placements.

To be able to plan feasible assembly sequences, geometrical restrictions need to be taken into account for the following offline planning operations:

1. Generation of a feasible assembly sequence
2. Movement path for the gripper tool and the handled parts
3. Grip pose planning

The planning operations are entangled: For the generation of the assembly sequence it is crucial to consider the gripper's abilities, which also influences possible movement paths for the gripper tool. Hence, the three planning operations need to be performed using the same knowledge on parts and resources. Since all three planning operations need reasoning based on geometric data, the planner for the assembly cell is implemented as one component, located within the planning layer (see figure 2).

4.1 Assembly restrictions for model bricks

In regard to geometrical restrictions the assembly of LEGO bricks allows for some classifications, which simplify decisions based on discrete and distinguishable assembly situations.

The LEGO model system poses restrictions on the buildable models. For this scenario only a limited number of different brick types is used. These bricks have the common restriction that they can only be attached to each other at the top or the bottom where the nubs fit. In most situations it is not possible to assemble one brick beneath an already assembled one. Consequently, a strict bottom up assembly is implied. Surveys and experiments have shown that humans tend to build complex structures layer by layer (see (Mayer, Schlick, and Ewert, 2011)). It is desirable to adapt the assembly planning algorithm to this behavior. This also reduces the complexity of that algorithm because path planning throughout many partly assembled layers is avoided but not strictly prohibited.

Another design element of this scenario is the discrete coordinate system used for LEGO parts. Each brick's dimensions are multiples of 0.4 mm. Furthermore, the distances of the nubs are fixed, thus the number of joining positions is limited.

When connecting two parts, the forces applied to other already connected parts have to be considered. Due to the adherence of all parts, any force causing torque has to be considered as being destructive to the already assembled structure. Hence, when mounting parts, no free mounting positions may lay in the direction of the force vector. A simplified rule states that each part has to be mounted on top of another part or on the base plate. This rule again restricts the assembly process to strict bottom up assembly.

The previous restrictions for the assembly of bricks are independent of the assembly system. Although the assembly cell in this scenario contains state-of-the-art industrial robots with high accuracy, additional restrictions are posed by tolerances induced by various components of the assembly system, like the optical tracking system.

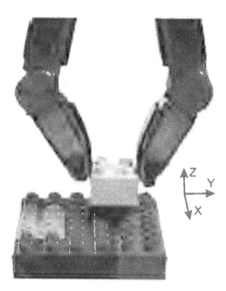

Figure 3 Gripper tool assembling a brick. High tolerances apply in the X-direction.

As parts are grabbed from a moving conveyor belt using a CCD camera and image processing, the position of the grabbed brick between the two fingers of the gripper hand has a tolerance of 3 mm (see figure 3, the tolerance applies only to the X-position of the brick). This tolerance cannot be reduced significantly due to the resolution of the CCD camera and the resulting accuracy of the detected position of the brick on the conveyor. Hence, when assembling a brick, these tolerances have to be taken into account.

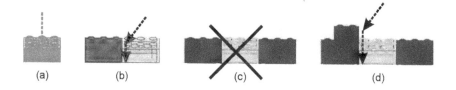

| (a) | (b) | (c) | (d) |

Figure 4 Classification of assembly situations.

In order to cope with the described instabilities, geometrical properties of the scenario are utilized. Therefore different states of the assembly group need to be considered (see figure 4). The following situations occur when a brick is assembled:

1. Free positioning (figure 4 (a)): No surrounding bricks are within the range of tolerance
2. One-side contact (figure 4 (b)): The brick has to be placed adjacent to one other brick
3. Opposing-side contact (figure 4 (c), (d)): Two and only two opposing sides of the brick to be mounted are adjacent to other bricks

4. Multi-side contact (not shown in figure 4): Two non-opposing sides or more than two sides are adjacent to other bricks

In the first situation no further restrictions have to be considered when planning the brick placement, given that the movement path is free of obstacles. The second situation inevitably appears when coherent models are assembled. A one-side contact has to be considered when planning the assembly motion since it poses a collision risk for the moving part and the adjacent assembled part. If the assembly motion path does not restrict the assembly, a one-side contact is considered to be desirable when planning an assembly sequence because it allows the safe assembly of a part. In opposing-side contact situations a valid placement is not always possible as described in the subchapter 4.2. When planning possible assembly sequences, opposing-side contact situations are only feasible, if the model is not assembled layer by layer. The planner has to make sure that no multi-side contact situations arise in valid plans as the gripper tool used in the assembly cell is only capable of gripping a brick on opposing sides, which must not be blocked by adjacent bricks in the assembly position.

4.2   Motion planning

For the planning of assembly motions a feasible path for the gripper tool and the attached brick has to be found. Motion control is based solely on position target values and force control is not applied. A variety of algorithms for collision-free movement planning has been published (LaValle, 2006), (de Berg, 2008). A simple solution to account for tolerances is to use the expected tolerances to enlarge a bounding box containing the brick that has to be assembled and use this bounding box for path planning. However, given the restrictions described in Chapter 4.1, a motion planning algorithm cannot find a collision free motion path for the situations 2-4 if the brick is considered to be in the center of the bounding box. Therefore, a probabilistic motion planning algorithm is used to plan a collision free path which ends where the first non-avoidable collision near the goal destination is detected. For situation 1 the algorithm finds the complete motion path including the assembly movement. However, for the other described situations, a situation specific approach is needed. Depending on the detected situation, motion planning for the final assembly move follows predefined patterns:

In situation 2, a movement path sketched in figure 4 (b) is chosen. Following this path, the moving brick intentionally collides with the adjacent brick. The collision occurs on the side where the position is highly influenced by the (possible) derivations induced by the optical. Since the gripper applies only enough force to the grabbed brick to keep it attached, applying force to overcome the nonpositive connection of the brick and the gripper finger makes to brick slip through the fingers when the collision occurs, thus enabling one degree of freedom regarding finger and brick. Depending on the actual position within the tolerance range, the brick can slip up to the full distance of that range. When the gripper finally moves down to assemble the grabbed brick, no further collision can occur. Hence, the adjacent brick serves as a rail for the moving brick. The described system behavior is inspired by human behavior where deficiencies in accuracy are overcome by utilizing mechanical aids, such as supporting the work piece on stable elements of the environment.

For situation 3, the described approach has to be modified. The motion path planner detects two collisions – one for each adjacent brick – and cannot use the motion pattern of situation 2 because the brick in motion would certainly collide with both adjacent bricks, twist and jam the assembly. A solution is to determine if there is an assembly situation similar to situation 2 above (see figure 4 (d)). When there is another brick above only one of the bricks adjacent to the assembly position, another "rail" is found and can be used, otherwise no feasible solution can be used in the plan (see figure 4 (c)). The described procedure can be executed recursively to find assembly motions for larger model structures.

4.3   Generalization of the approach

The described approach heavily utilizes the properties of LEGO models. Although LEGO models are toys, their assembly shows analogies to assembly processes in industry applications. The plastic casings of clamps used in control boxes for instance have similar properties as LEGO bricks. In order to generalize the described approach for industry applications like control box assembly, the information used to determine the described patterns and situations need to be modeled. The probabilistic motion planning algorithm uses only shapes to determine a 3D path to the destination and thus already uses information represented in geometrical models. Generalizing the presented approach using the case-based motion pattern requires modeling the implicitly used information on the geometrical shapes of LEGO bricks. Utilizing the surface of one brick as a rail for another one is possible in the described model assembly system because the plastic surface of the bricks has three important properties: it has a low friction, is endurable enough to absorb the collision with other bricks and is "scratch-proof". Besides the described surface properties, it is important that there is no significant cleavage between two bricks, since this could cause the moving brick to twist.

The next step towards finding assembly motion paths for a more universal assembly cell is to develop geometric reasoning algorithms, which are able to handle more than the described situations. These algorithms need to reason with less simplified models and incorporate more knowledge about parts that need to be assembled. The required information, like surface structures, weights and stiffness is often available in CAD models and can be used to create dexterous assembly systems.

5   CONCLUSIONS

Using industrial robots for filigree assembly tasks poses requirements on the dexterity of the robot's effector. Although a broad spectrum of assembly tasks are highly automated today, industry's requirements on the assembly of close-tolerance components have not lead to pervasive use of automated assembly tools. Today's need for changeable production systems as a response to a varying spectrum of products contradicts specialized highly productive assembly solutions for single tasks. In order to cope with a broad spectrum of possible products, fuzzy tolerances and position derivations induced by either product parts, the assembly system or sensors, an assembly system using a combination of knowledge and "intelligent" motion planning has been presented. Given an assembly order describing a feasible sequence of parts and assembly target positions, the task of an assembly system is to find and control movements which place a part attached to a gripper at the correct location. For assembly tasks beyond the complexity of "pick&place", a suitable grip and a collision-free movement path for both effector and part has to be found. Given accurate models of parts and grippers, path planning algorithms can be used to find optimal paths. However, in real-world assembly tasks tolerances of parts and gripper position as well as the spatial correlation of part and gripper (grip pose) lead to collisions. In narrow modules these collisions cannot be avoided without increasing the positioning accuracy of both robot arm and gripper tool as well as the hand eye coordination beyond construction limits. A feasible solution for an enhanced path planning algorithm besides increasing the sensor accuracy to compensate position derivations is to adapt human assembly behavior to avoid collisions leading to jammed up parts within the module. Instead, collisions can be forced to allow already assembled parts with plane surfaces to "guide" parts moving to their final assembly position. The presented path planning algorithm finds possible collisions and determines possible motion paths exploiting solicited collisions.

REFERENCES

de Berg, M., O. Cheong, and M. van Kreveld, et al. 2008. *Computational Geometry: Algorithms and Applications.* Heidelberg: Springer-Verlag.

Brecher, C. 2011. *Integrative Production Technology for High-Wage Countries.* Berlin Heidelberg: Springer-Verlag.

Kempf, T., W. Herfs, and C. Brecher. 2009. SOAR-based Squence Control for a Flexible Assembly Cell. *Proceedings of the 2009 IEEE Conference on Emerging Technologies and Factory Automation.* Palma de Mallorca: IEEE

Kempf, T. 2010. *Ein kognitives Steuerungsframework für robotergestützte Handhabungsaufgaben.* Aachen: Apprimus Verlag.

LaValle, S. M. 2006. *Planning Algorithms.* Cambridge: Cambridge University Press.

Mayer, M., C. Schlick, and C. Ewert, et al. 2011. Automation of robotic assembly processes on the basis of an architecture of human cognition. *Production Engineering Research and Development.* Plublished online. ISSN 0944-6524.

"Soar Homepage". Accessed February, 2012, http://sitemaker.umich.edu/soar/home

Improving Operator's Conformity with Expectations in a Cognitively Automated Assembly Cell Using Human Heuristics

Marcel Ph. Mayer, Christopher C. Schlick

Institute of Industrial Engineering and Ergonomics of RWTH Aachen University
Aachen, Germany
{m.mayer; c.schlick}@iaw.rwth-aachen.de

ABSTRACT

Cognitive automation of assembly processes can be seen as promising approach to overcome the vicious circle of automation. On the basis of the cognitive architecture SOAR a cognitive control unit (CCU) was developed which is able to simulate human information processing at a rule-based level of cognitive control. Thus, the CCU can plan assembly processes autonomously and can react to changes in assembly processes due to increasing number of products that have to be assembled in a large variety in production space as well as changing or uncertain conditions. Towards a "Humanoid-Mode" for automated assembly systems comparable to the horse-metaphor for automated vehicles human assembly strategies where identified in empirical investigations and formulated as production rules to be included in the knowledgebase of the CCU. Within this contribution exemplary results regarding the effect of such an automation approach on operator's conformity with expectations regarding system behavior in a supervisory task are presented.

Keywords: cognitive automation, assembly, SOAR, Humanoid Mode

1   INTRODUCTION

Today one must conclude that especially in high-wage countries the level of automation of many production systems has already been taken far without paying sufficient attention to the specific knowledge, skills and abilities of the human operator. According to the law of diminishing returns, this kind of naive increase in automation will likely not lead to a significant increase in productivity but can also have adverse effects (Figure 1). According to Kinkel et al. (2008) the amount of process errors is on average significantly reduced by automation, but the severity of potential consequences of a single error increases disproportionately. These *"ironies of automation"* (Bainbridge, 1987) can be considered a vicious circle (Onken & Schulte, 2009), where a function that was allocated to a human operator due to poor human reliability is automated.

The novel concept of cognitive automation by means of simulation of human cognition aims at breaking this vicious circle. Based on simulated cognitive functions, technical systems shall not only be able to (semi-) autonomously carry out manufacturing planning, adapt to changing supply conditions and be able to learn from experience but also to simulate goal-directed human behavior on a rule-based level (Rasmussen, 1986) and therefore significantly increase the conformity with operator expectations.

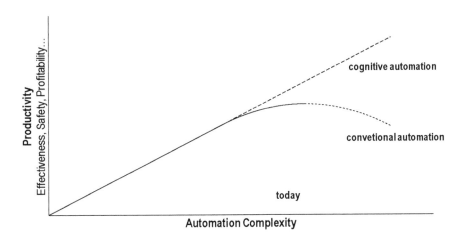

Figure 1   The complexity of cognitive automation in comparison to conventional automation regarding productivity

In order to be able to use the full potential of cognitive automation, one has to expand the focus from a traditional human-machine system to joint cognitive systems (Hollnagel & Woods, 2005). In these systems both the human operator and the cognitive technical system cooperate safely and effectively at different levels of cognitive control to achieve a maximum of human-machine compatibility.

2   PREVIOUS WORK

Comparing the role of the human operator to Sheridan's classic supervisory control model (Sheridan, 2002) the functions of detailed process planning and programming are allocated to the functions of the cognitively automated systems. Additionally, during manufacturing execution the process is now continuously monitored by cognitive automation technology. As a consequence, in cognitive automation the amount of planning effort is shifted to a different layer. The task of monitoring the system however remains one of the duties of the human operator. On the basis of the preprocessed signals and signs he/she has to monitor the system and its environment for a possible violation of boundary conditions or dissatisfied constraints and intervene accordingly. Therefore, the human operator needs to have detailed knowledge about the state of the machine and the state of the process to make decisions. Hence, an important basic requirement to the user-centered design of cognitively automated systems is the conformity of operator's expectations with system behavior. (Mayer et al., 2008, 2009)

Under the assumption that a numerical control of a technical system based on human decision making will lead to a better understanding by the human operator regarding the intention of the technical system, the cognitive architecture SOAR (Lehman et al., 2006) was chosen. SOAR as a computational model does not require initial training – in comparison to emergent systems like artificial neuronal networks. Hence, changes to the knowledge base can be performed quickly, which is of great interest especially for an application in an industrial environment.

An experimental assembly cell (Kempf et al., 2008) was developed to investigate the novel concept of cognitive automation. Within the experimental cell two robots carry out a certain repertoire of coordinated pick and place operations with small work pieces. One robot is stationary the other robot is sitting on a linear track. A conveyor system of four individually controlled transfer lines, pneumatic track switches and light barriers is arranged so that the parts can circle around the workplace. First, the stationary robot takes the bricks from a pallet and puts them on a conveyor. The second robot, waiting on the linear track for the part, has to identify the part with respect to color and shape. If the part is included in the final state of the product to be assembled, it picks it from the conveyor and puts it on the working area either at the corresponding position in the assembly or in a buffer area for further processing. (Mayer et al., 2010, 2012a)

Focusing a human centered description to bridge the gap between process knowledge and mental model one promising approach in this special scenario is the use of motion descriptions as action primitives. Motions that are familiar to human operators from manually performed tasks might be easier to anticipate than complex programming code. Since in production systems complex handling tasks have to be broken down into fundamental elements, we use the methods time measurement system (MTM) as a library of fundamental movements, neglecting the underlying time information in a first step. The MTM motion descriptors that were mapped onto production rules in the knowledge base are equivalent. They do not necessarily have to be applied in a periodic sequence. These production rules correspond to the MTM-1 basic motion descriptors REACH, GRASP, MOVE, POSITION and RELEASE. In addition, other rules are included that represent the physical constraints depending on the parts to be assembled (e.g. joining-angle or conditions required for positioning an element) and include information about whether a supplied part can be installed directly or must be stored in the buffer for later assembly. (Mayer et al., 2009, 2010)

To validate the cognitive automation concept, simulation studies investigating the influence of various factors such as part size, queue length and feeding regime on the processing time as well as the number of action primitives were carried out. The most interesting and somewhat counter intuitive finding of these studies was the effect of the feeding regime. The following two variations of the feeding regime were investigated: deterministic supply of needed parts and random supply including unneeded parts. The simulation results unambiguously show a disproportional increase in processing time with increasing part size and queue length for deterministic part feed. Conversely, a decrease in processing time over the queue length was observed for a stochastic part feed. (Mayer et al., 2011)

Even though the desired target assemblies were assembled correctly and within the expected number of simulation cycles, it must be acknowledged that the variance of the observed assembly sequences was immense. Despite the application of an anthropocentric taxonomy in the form of MTM-1, the question is whether the described approach is adequate to ensure the conformity of the system behavior with operator's expectations. From an anthropocentric perspective, the results appear to be inadequate, as they may not be compatible with human expectations due to their high procedural variance. One possible conclusion is that based on Marshall's Schema Model (Marshall, 2008) the elaboration knowledge of the cognitive simulation model is underrepresented. (Mayer et al., 2009, 2010)

To improve elaboration knowledge two independent experimental trials with a total of 36 subjects were carried out to identify human buildup heuristics that could be included in the knowledgebase of the cognitive system. Based on the data three fundamental assembly heuristics could be identified and validated (Mayer et al., 2010): (1) humans begin an assembly at edge positions of the working area; (2) humans prefer to build in the vicinity of neighboring objects; (3) humans prefer to assemble in layers. To develop a "humanoid mode" for cognitively automated assembly systems similar to the "horse-metaphor" for automated vehicles (Flemisch et al., 2003), the identified assembly heuristics where formulated as production rules. When the reasoning component is enriched with these rules, a significant increase in the predictability of the robot when assembling the products could be achieved. (Mayer et al., 2012b)

Even though the presented results show the potential of cognitive automation in a production environment, the positive effect of a "humanoid mode" on operator's expectations yet has to be proven.

3   METHOD

In a supervisory task the human operator continually monitors the activities of a system and compares theses activities with his/her mental model. Based on the mental model, expectations for following activities can be formulated and compared. The hypothesis for the study presented in the following can be formulated as follows: If the knowledge base of the cognitive system is extended by production rules based on human heuristics, the system's buildup sequence can be better anticipated by the human operator for it is compatible with his/her mental model of the assembly process.

To verify the hypothesis, prediction experiments similar to the studies of Shannon (1951) were carried out. In our studies a human operator had to predict the next expected part to be positioned based on an observed buildup sequence. For the experimental setup a total of five differing cognitive simulation models were used to generate assembly sequences. The basic simulation model contained just the previously mentioned MTM-1 rules (Model 1). Hence this simulation model is capable of performing all physically possible assembly sequences. The other four simulation models were derived by extending the first simulation model by the rule regarding the vicinity of neighboring parts (Model 2), the rule regarding the buildup in layers (Model 3), a linear combination of these rules (Model 4) as well as a combination which allows only neighboring parts within layers (Model 5). Additionally, assembly sequences performed by human operators were integrated in the experimental setup (Model H).

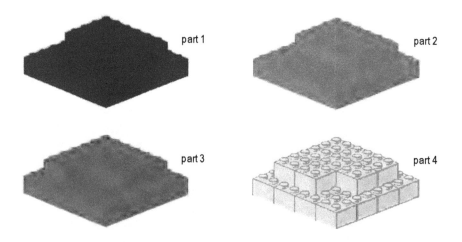

Figure 2 Assembly objects used within the experiment

The experimental task is described in the following: The target state of the product to be assembled, a pyramid of 30 identical cubical parts, was positioned within the field of view of the subject. The buildup sequence up to a predefined current state was visualized on a table mounted display. Each subject had to perform a total of 48 predictions, which were provided in randomized order. After the presentation of the assembly sequence the subject had to position the next part on an assembly object (four different objects, see figure 3) corresponding to the predefined current state positioned in front of him/her. After each prediction the subject had to rate his/her mental fatigue as well as his/her performance (adapted from the NASA TLX).

The buildup sequence was visualized on a 28" TFT screen. The assembly area with the corresponding assembly object was positioned in front of the subject. The target state was positioned slightly left. The assembly sequence, the corresponding assembly object as well as the target state were positioned in close proximity. The part to be used for the prediction of the next step was positioned on a tableau on the right. To light barriers were used to record time information. To record the position of the predicted part a video camera was used. The apparatus used within the experiment is shown in figure 3.

Figure 3 Apparatus with display to present the assembly sequence, target state (left), current state (lower center) and tableau providing the part to be positioned (right)

308

A total of 27 subjects (18 male, 9 female) participated in the experiment. The age of the subjects ranged between 19 and 31 years (mean = 25.19; SD = 2.96). 25 of the subjects were right-handed, the remaining were left-handed. Regarding their experience in assembly, the mean value was 2.15 on a scale from 0 (no experience) to 5 (experienced). None of the subjects was dealing with assembly in his/her everyday life. All subjects were unpaid volunteers.

As independent variables the cognitive simulation models used to generate the assembly sequences (Model 1, Model 2, Model 3, Model 4, Model 5 and human assembly sequence – Model H) as well as the four current states were varied. In addition, each combination of factors was repeated, resulting in the previously mentioned 48 predictions to be performed by the subjects.

As dependent variables the time to perform the prediction as an objective measure as well as the subjective rating of mental fatigue and performance were taken into account. In the framework of this paper only the results regarding mental fatigue are presented.

4   RESULTS

The subjective rating regarding fatigue was investigated regarding chronological effects of the experiment. For that reason each of the 48 predictions within the order of predictions performed was regarded as factor levels of a factor time. Since not all data is normally distributed, the Friedman-Test on a significance level of $\alpha = 0{,}05$ was performed as a nonparametric version of a repeated measures ANOVA.

The results show a significant effect of the factor time on the subjective rating of mental fatigue ($p<0.001$). In figure 4 the collected data regarding mental fatigue are plotted over the factor levels of the factor time. As expected, subjective rating increases with the number of predictions performed.

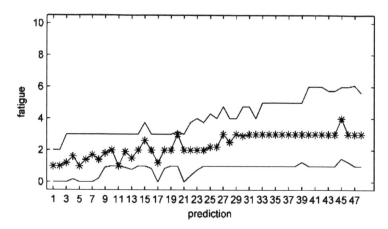

Figure 4 Subjective rating of mental fatigue (lower quartile, median, upper quartile) as a function of the chronological order of predictions.

Additionally, the subjective ratings were analyzed regarding the influence of the cognitive simulation models as well as the current states. Since the preconditions for the performance of an ANOVA are violated, an adapted version of the Scheirer-Ray-Hare-Test (SRH; Scheirer et al., 1976) as a nonparametric version of a completely repeated measures ANOVA was performed on a significance level of $\alpha = 0,05$.

A significant effect on the subjective rating of mental fatigue was found for the factor cognitive simulation model ($H = 3.291$, $p = 0.002$). A post-hoc comparison according to Bonferroni on a significance level of $\alpha = 0.05$ shows a significant difference between Model 3 and Model H. The median of mental fatigue as well as the upper and lower quartile are plotted over the levels of the independent factor simulation model (figure 5).

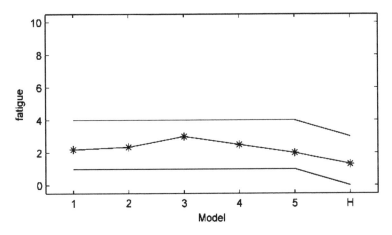

Figure 5: Subjective rating of mental fatigue (lower quartile, median, upper quartile) as a function of the cognitive simulation models

The results regarding the influence of the cognitive simulation models are remarkable in so far as the experiments were designed comparable to double-blind trials. Neither the subjects nor the experimenter knew which cognitive simulation model was underlying the assembly sequence. Therefore, based on the presented results we can draw the conclusion, that cognitive automation when extended by human heuristics can reduce load on the human operators.

5   SUMMARY AND OUTLOOK

Regarding highly automated manufacturing systems that shall produce customer-specific products, an increase in conventional automation will not necessarily lead to a significant increase in productivity. Novel concepts towards proactive, agile and versatile manufacturing systems have to be developed to solve the so called polylemma of production (Schuh & Orilski, 2009). Cognitive automation is a promising approach to improve proactivity and agility. Despite this novel automation approach the experienced machining operator will always play a key architectural role as a solver for complex planning and diagnosis problems. Moreover, he/she is supported by cognitive simulation models which can solve algorithmic problems on a rule-based level of cognitive control (Rasmussen, 1986) quickly, efficiently and reliably and take over dull and dangerous tasks.

To develop a "Humanoid-Mode" for cognitively automated assembly systems similar to the horse-metaphor for automated vehicles (Flemisch et al., 2003) human assembly strategies where formulated as production rules and integrated in the knowledgebase of the cognitively automated system. These production rules underlying human heuristics could increase the predictability of the robot when assembling various products.

The empirical study presented in this contribution aimed at answering the question, whether these production rules can lead to a better conformity of operator's expectations with system behavior in a prediction experiment. The results show, that using an anthropocentric, cognitively automation approach the subjective rating of mental fatigue can be reduced. However, further analysis of the dependant variable time as well as the subjective rating of performance has to be carried out.

ACKNOWLEDGMENTS

The authors would like to thank the German Research Foundation DFG for the kind support within the Cluster of Excellence "Integrative Production Technology for High-Wage Countries".

REFERENCES

Bainbridge, L. 1987. Ironies of Automation. In. *New Technology and Human Error*, eds. J. Rasmussen, K. Duncan, J. Leplat. Chichester: Wiley

Flemisch, F. O., C. A. Adams, S. R. Conway, K. H. Goodrich, M. T. Palmer, P. C. Schutte 2003. *The H-Metaphor as a Guideline for Vehicle Automation and Interaction*, NASA/TM—2003-212672

Hollnagel, E., D. D. Woods 2005. *Joint Cognitive Systems: Foundations of Cognitive Systems Engineering*, Boca Raton: Taylor & Francis Group

Kinkel, S., M. Friedwald, B. Hüsing, G. Lay, R. Lindner 2008. *Arbeiten in der Zukunft, Strukturen und Trends der Industriearbeit. Studien des Büros für Technikfolgen-Abschätzung beim Deutschen Bundestag* – 27. Berlin: edition sigma (in German)

Lehman J. F., J. Laird, P. Rosenbloom 2006. *A Gentle Introduction to Soar, An Architecture for Human Cognition: 2006 Update*, Accessed October 7, 2011, http://ai.eecs.umich.edu/soar/sitemaker/docs/misc/GentleIntroduction-2006.pdf

Mayer, M, B. Odenthal, M. Faber, C. Schlick 2012a. Cognitively automated assembly processes: a simulation based evaluation of performance. *Work: A Journal of Prevention, Assessment and Rehabilitation* 41(Supplement 1):3449-3454

Mayer, M., B. Odenthal, D. Ewert, T. Kempf, D. Behnen, C. Büscher, S. Kuz, S. Müller, E. Hauck, B. Kausch, D. Schilberg, W. Herfs, C. Schlick, S. Jeschke, C. Brecher 2012b.

Self-optimising Assembly Systems Based on Cognitive Technology. In: *Integrative Production Technology for High-Wage Countries*, ed. C. Brecher. Berlin: Springer

Mayer M, C. Schlick, D. Ewert, D. Behnen, S. Kuz, B. Odenthal, B. Kausch 2011. Automation of robotic assembly processes on the basis of an architecture of human cognition. *Production Engineering Research and Development* 5(4):423-431

Mayer, M., B. Odenthal, M. Faber, W. Kabuß, N. Jochems, C. Schlick 2010. Cognitive Engineering for Self-Optimizing Assembly Systems. In. *Advances in Human Factors, Ergonomics, and Safety in Manufacturing and Service Industries*, eds. W. Karwowski, G. Salvendy. Boca Raton: Taylor & Francis

Mayer, M., B. Odenthal, M. Faber, J. Neuhöfer, W. Kabuß, B. Kausch, C. Schlick 2009. Cognitive Engineering for Direct Human-Robot Cooperation in Self-optimizing Assembly Cells. In: *Human Centered Design, Lecture Notes in Computer Science 5619*, ed. M. Kurosu. Heidelberg: Springer

Mayer, M., B. Odenthal, M. Grandt, C. Schlick 2008. Task-Oriented Process Planning for Cognitive Production Systems using MTM. In. *Proceedings of the 2nd International Conference on Applied Human Factors and Ergonomic (AHFE) 14.-17. July 2008*, eds. W. Karwowski, G Salvendy. Las Vegas: USA Publishing

Onken, R., A. Schulte 2010. *System-ergonomic design of cognitive automation. Studies in Computational Intelligence 235*. Berlin: Springer

Rasmussen J. 1986. *Information Processing and Human-Machine Interaction. An Approach to Cognitive Engineering*. New York: North-Holland

Scheirer, J., W. S. Ray, N. Hare 1976. The Analysis of Ranked Data Derived from Completely Randomized Factorial Designs. *Biometrics* 32(2):429-434

Schuh, G., S. Orilski 2007. Roadmapping for Competitiveness of High Wage Countries. *Proceedings of the XVIII. ISPIM Conference.*

CHAPTER 33

Cooperative Disassembly of Human and Robot Using an Augmented Vision System

Barbara Odenthal, Marcel Ph. Mayer, Wolfgang Kabuß,
Christopher M. Schlick

RWTH Aachen University
Insitute of Industrial Engineering, and Ergonomics,
Aachen, Germany
{b.odenthal; m.mayer; w.kabuss; c.schlick}@iaw.rwth-aachen.de

ABSTRACT

Within a research project, a numerical control unit based on a cognitive architecture and its ergonomic human-machine interface are developed for a robotized production unit. In order to cope with novel systems, the human operator will have to meet new challenges regarding the work requirements. Therefore, based on a first prototype of an augmented vision system (AVS) assisting the human operator in error detection the system is enhanced in order to deal with the task of a cooperatively error correction of the human operator and the cognitive controlled technical system. A laboratory experiment was carried out to compare two different participant groups (different professional education) and to investigate different decision support modes. A total of 44 participants participated in the experiments.

Keywords: dis-/assembly, cognition, decision support, augmented reality

1   INTRODUCTION

The research project "Integrative production technologies for high-wage countries" (Brecher, 2011) has been initiated to study self-optimizing assembly cells (Mayer et al., 2011) and to design innovative ergonomic human-machine interfaces of the cell's numerical control. Self-optimizing assembly cells are an especially promising approach to improve the competitiveness of manufacturing companies in high-wage countries (Klocke, 2009, Gausemeier, 2008) through advanced methods of modeling and optimization and to support the associated development of skills and knowledge of experienced machining operators (Luczak, Reuth, and Schmidt, 2003) on the shop floor. A novel architecture of the cell's numerical control based on the cognitive architecture SOAR (State, Operator Apply Result, see Leiden et al., 2001) forms the technological basis of this approach and is the foundation for human-machine cooperation on a rule-based level of cognitive control (Rasmussen, 1986). The cell's numerical control is termed a cognitive control unit (CCU), which is able to process procedural "knowledge" encoded in production rules. The experienced machining operator (German: Facharbeiter) plays a key architectural role in the concept of self-optimization.

To study human-machine interaction in self-optimizing assembly systems, an experimental assembly cell was designed where a robot carries out a certain repertoire of coordinated pick and place operations with small work pieces (Kempf, Herfs, and Brecher, 2008). In order to support the human operator while intervening in the case of assembly errors, an augmented vision system (AVS) was developed and implemented focusing on the presentation of the assembly information. The aim of using this system was to place the human operator in a position to detect the assembly errors in a fast and reliable way. In order to cooperatively correct the detected error the AVS was extended to support the decision making of the human operator. In this case, the development of a cooperative disassembly strategy by the human operator is especially important. The execution of the disassembly shall be carried out in cooperation between the human operator and the CCU controlled robot. A detailed description of the functional design of the CCU can be found in (Mayer et al., 2010, Ewert et al., 2010). Normally, after a representation of the goal state (e.g. 3D-assembly object) is handed over to the CCU by the human operator, the CCU itself works out an assembly plan and executes its action according to it. During execution, an error occurs that the CCU is only qualitatively able to detect. So, the CCU passes information to the operator that an error has occurred. In this situation the error must be identified and corrected by the human operator.

A high-resolution stereoscopic head-mounted display (HMD) in see-through mode is used to accurately superimpose synthetic assembly information in the field of view of the operator. To do so, an infrared motion tracking system is used by the CCU to determine the position and orientation of the operator's head in real time. To investigate ergonomic advantages and disadvantages of this so termed head-mounted Augmented Vision System (AVS; Odenthal et al., 2009) for supervisory control of the CCU, a comparative study was carried out. Based on a first prototype of the AVS (Odenthal et al., 2009), additional functionalities have been designed and implemented in order to support the decision making of the operator regarding the disassembly of an assembly group. According to Schlick, Bruder, and Luczak (2010), the non-manual work distinguishes three different phases: the information perception, information processing and information output. After detecting the stimulus, the human has to realize the meaning of the stimulus, has to identify essential characteristics and has to choose one acting alternative. Hence; the AVS

314

was designed to support the experienced machining operators during this particular process in choosing one acting alternative. The task of the operator is to plan if and which parts of the assembly group shall be disassembled by him or if the task of planning and executing the disassembly shall be handed over to the robot. In order to support the development of action alternatives and the selection of one, the AVS calculates situationally which assembly parts are removable by the human operator and/or the robot and which parts are not removable. Based on these information, different decision support modes were designed to display additional geometric information supporting this cooperative disassembly task under the restriction that human intervention in the running process shall be focused on situations in which it is essential. A laboratory study was carried out to find - from an ergonomic point of view - good solutions. The study is described in the following.

2   LABORATORY STUDY

2.1   Participants

A total of 44 participants participated in the laboratory study. All of them satisfied the requirements concerning stereo vision, color vision and emmetropia according to DIN 58 220 (1997). The participants were divided into two groups:

Group 1 "engineers": This group consists of 28 male (future) engineers who are either in a bachelor/master program at university or have finished their studies. The age of the participants was between 21 and 35 years (mean 26.1 years, SD 4.2). The average experience in assembling was 2.6 and in LEGO assembly 3.5 on a scale ranging from 0 (low) to 5 (high). One person stated problems in the past after working with electronic information displays (back pain).

Group 2 "technicians": The second group "technicians" consists of 16 male participants who did a technical apprenticeship (e.g. in the field of assembly or as an (industrial) mechanic). Most participants of this group attend currently the technical school of the army. The age of the participants was between 23 and 37 years (mean 27.9 years, SD 3.7). The average experience in assembling was 3.7 and in LEGO assembly 4.1 on a scale ranging from 0 (low) to 5 (high). One person stated problems in the past after working with electronic information displays (burning eyes).

2.2   Task

Caused by the trend of an increasing level of production automation in high-wage countries and due to occupational safety, the human interaction in the running process shall be focused on situations in which the human intervention is essential. Changes between the robot and the human operator are time consuming and error-prone operations. Hence, the following restrictions were set during the laboratory study: 1) The human operator has to disassemble as few parts as possible; 2) The assembly object should be disassembled with a minimum of handovers between human and robot. During the experimental phase the human operator had to virtually disassemble eight models in cooperation with a virtual robot by means of the AVS. Regarding the cooperative disassembly, there are two possibilities to remove parts. 1) The human operator marks a assembly part by means of an input device and removes it by a voice command, or 2) the human operator decides that the robot has to remove parts (so called human-robot change (HRC) function). The HRC function is also activated by a voice command. The CCU controlled robot removes as much parts as possible until no more parts can be removed autonomously (see the definition of parts removable by the robot). Eight distinct assembly groups are developed according to the following criteria: All models consist of a total of 27 cubic parts. A minimum of three parts must be removed by the human operator.

2.3   Experimental design

The experimental design distinguishes two factors: 1) the so-called decision support mode (DSM) with four levels: no support, Human, Robot, and Robot & Human; 2) the group with two levels: engineers, and technicians.

The decision support mode (DSM) were distinguished, see Figure 1:

- *No support*: No further information than the assembly group that shall be disassembled are presented.
- *Human*: Within this mode, the AVS determines - based on the current situation - the assembly parts that are removable by the human operator and marks them by color. The definition for removability by the human operator is as follows: The whole top surface and at minimum one of the lateral surfaces of the assembly part are completely exposed.
- *Robot*: Within this mode, the AVS determines the assembly parts that are removable by the robot and marks them by color. The definition for removability by the robot is that the whole top surface and two opposing lateral surfaces of the assembly part are completely exposed. Therefore, all objects removable by the robot can also be disassembled by the human operator but not vice versa.
- *Robot & Human*: The AVS determines the assembly parts that are removable by either the robot or the human. To distinguish between the two kinds first the parts removable by the robot are marked in one color and additionally the parts removable solely by the human are marked in a different color.

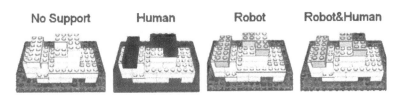

Figure 1: Different decision support modes

2.4 Apparatus

The apparatus is shown in Figure 2. The position and orientation of the HMD were tracked by the infrared real time tracking system smARTtrack of A.R.T. GmbH. A binocular optical see-through HMD (nVisorST by NVIS Inc.) was used to display the synthetic assembly information.

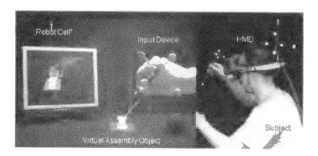

Figure 2: Main components of the AVS; view through the HMD (left)

A self-developed marker pen was used by the participants to virtually mark the parts in order to disassemble the assembly group. The manipulation of the virtual models was performed by voice control (VoCon2.1 by Philips). The robot cell was simulated and virtually presented on an off-the-shelf TFT flat screen. The cell consists of a robot that carried out the commanded pick and place operations for the disassembly by the robot in the virtual working area.

2.5 Procedure

The procedure was divided into four phases:

- Pretests: First, the personal data were collected. Furthermore, a visual acuity test, a stereo vision test and a color test are processed. Because of the fact, that the system is controlled by voice, a calibration of the voice recognition system had to be carried out in advance.
- Training under experimental conditions: After completing the pretests, the participants had some minutes to get familiar with the system.
  Data acquisition: The participant started with filling in a questionnaire about visual fatigue. Each participant conducted eight trials because each decision support mode was tested by two different assembly groups. Each participant started with the mode "no support" as baseline. The execution time started with the beginning of the system variant. The task was to completely disassemble a virtual assembly group in cooperation with the robot. With the infrared optical tracking system the position and the orientation of the HMD was measured in real time and the user's viewing direction was calculated. In case the HRC function within the disassembly was called, the presentation of the virtual model disappeared on the HMD and the simulation of the robot cell started on the table

mounted display. After removing all possible assembly parts of the model by the robot, the presentation of the (partly) disassembled object appeared again on the HMD. After the complete disassembly of a model, the next model followed. After each trial the participant had to fill in the NASA Task Load Index.

- Post tests: Finally, within the post test (after the fourth trial) the participant had to take off the HMD and had to fill in the questionnaire about visual fatigue.

The total time-on-task was approximately two hours for each participant.

2.6 Dependent variables

The following dependent variables were examined:

- Human execution time (T_{human}):
  The human execution time (T_{human}) represents the time elapsed from the trial start until the virtual object is disassembled completely reduced by the disassembly time of the robot.
- Distance to optimal Performance (*DoP*):
  The Distance to optimal Performance (*DoP*) is a measure for the deviation of optimal performance and a function of the ratio r_{BH} which is defined as the number of disassembled parts by the human operator N_{BH} divided by the minimal number of assembly parts N_{BHmin} and the normalized human cycle (HC_{norm}) which is defined as the number of human cycle N_{HC} during the experiment divided by the minimal number of cycles N_{HCopt}. Because of their linear independence, *DoP* is defined as

$$DoP = \sqrt{(r_{BH} - 1)^2 + (HC_{norm} - 1)^2}$$

$$\text{with } r_{StH} = \frac{N_B}{N_{BH\,min}} = \frac{N_B}{3} \text{ and } HC_{norm} = \frac{N_{HC}}{N_{HCopt}} .$$

With the optimal value of the ratio r_{BH} (= 1) and the optimal value of HC_{norm} (= 1) the optimum value of *DoP* is 0.

3 RESULTS

The statistical analyses were carried out on the basis of the whole data set from both groups. For the not normally distributed data, an analysis of variance by ranks with data alignment was calculated (Bortz et al., 2008). The significance level was set to $a = 0.05$.

3.1 Human Execution Time

The means and the 95% confidence intervals of the human execution time under the different experimental conditions are shown in Figure 3.

The decision support mode (DSM) significantly influenced the human execution time (H=4.101; p=0.043), hence the null hypothesis H_{01} was rejected. When using the decision support mode "Human", the average T_{human} was 10.4% shorter compared with the T_{human} of the condition "no support". When comparing the decision support mode "Robot" and "no support", the average T_{human} of the robot condition was 11.4% shorter. In the case of the comparison between the condition "Robot & Human" and "no support", the "Robot & Human" condition led to an average T_{human} that was 5.2% shorter than under the "no support" condition.

In group 2 (technicians) only the "Robot" mode compared with the "no support" condition led to an efficiency improvement (7.3%). In contrast to that, the „Robot & Human" mode led to an increase of the T_{human} (+10.7%) Within the group of engineers, the experimental conditions with decision support information (compared with the „no support" mode) led to a reduction of the T_{human} (at minimum 13.9%).

The factor "Group" showed a significant influence on the human execution time (H=9.237; p=0.026). Therefore the null hypothesis H_{02} could be rejected as well. The average T_{human} of the "technicians" was 24.7% longer than the T_{human} of the engineers. Statistically significant interactions were not found (DSM*Group: H=5.687; p=0.128).

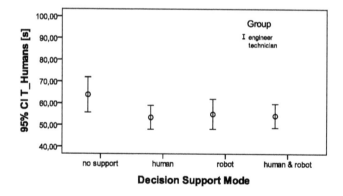

Figure 3 Human Execution Time [s] under the experimental conditions

3.2   Distance to optimal Performance (DoP)

The means and the 95% confidence intervals of the distance to optimal performance under the different experimental conditions are shown Fig. 4.

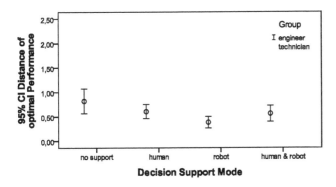

Figure 4: Distance of optimal Performance under the experimental conditions

The Decision Support Mode significantly influenced the difference of Distance to optimal Performance ($H = 9.216$, $p = 0.0266$), so that null hypothesis was rejected. The between-subject factor „Group" showed no significant influence of the *DoP* ($H = 0.238$, $p = 0.6257$), so that null hypothesis could not be rejected. Statistically significant interactions were not found (SM*Group: $H = 1,111$, $p = 0.7744$). When using the supporting mode "Human" (*DoP*=0.70), the average *DoP* was 4.8% lower compared with the *DoP* of the condition "no support" (*DoP*=0.74). When comparing the supporting mode "Robot" and "no support", the average *DoP* of the robot condition was 40.8% lower (DoP_{Robot}=0.44). In the case of the a comparison between the condition "Robot & Human" and "no support", the "Robot & Human" condition led to an average *DoP* that was 8.9% lower than under the "no support" condition ($DoP_{Robot\&Human}$=0.67).

In group 2 (technicians) only the "Robot"(*DoP*=0.53) mode compared with the "no support" "(*DoP*=0.60) condition led to an efficiency improvement (-11.4%). In contrast to that, the "Human", and the „Robot & Human" mode led to an increase of the *DoP* (DoP_{Human}=0.875; $DoP_{Robot\&Human}$=0.869). Within the group of engineers, the experimental conditions with decision support information (compared with the „no support" mode) led to a reduction of the *DoP* (at minimum 26.2%).

3.3   Results of the NASA-TLX

As pointed out, the data of the NASA Task Load Index (Hart, and Staveland, 1988) were collected after each trial. In general, there was no significant difference between the weighted load regarding the decision support mode or the different groups. Regarding the items, it can be detected that the technicians evaluated averagely the *performance* better than the engineers. This can be a hint, why there was the time difference between the two groups. Maybe the technicians took more time to achieve a good task execution. Regarding the *mental demand*, the highest average value of the technicians could be found for the mode „robot & human". This higher *mental demand* could be one reason for the higher execution time in this mode. It can be shown, that the *frustration* is much lower in the three modes with additional decision support information than in the mode "no support".

320

3.4 Results of the Visual Fatigue Questionnaire

As pointed out, the data of the visual fatigue questionnaire were collected before and after the last (fourth) trial. In Group 1 (engineers) the largest increases in visual fatigue are 1.45 for the item "neck pain", 0.96 for the item "headache", and 0.8 for the item "mental fatigue. In Group 2 (technicians) the largest increase is 0.83 for "neck pain". The acquired fatigue values are much smaller than expected. It is to be expected that longer load cycles would significantly increase visual fatigue (Pfendler, and Schlick, 2007).

4. SUMMARY & CONCLUSION

In this paper, the results of a laboratory study that deals with the cooperative disassembly of assembly groups by a human and a robot were presented. Therefore, different decision support modes (DSM) and two participant groups who differ in their professional education were evaluated regarding the human execution time (T_{human}) and the distance to optimal performance (DoP). Regarding the T_{human} both groups differ significantly. Figure 3 shows that in group 2 (technicians) only the "Robot" mode compared with the "no support" condition led to an efficiency improvement. In contrast to that, the „Robot & Human" mode led to an increase of the T_{human}. In the group of engineers additional information led to a decreasing T_{human}. Regarding the distance of optimal performance, the experiment had shown that the mode "Robot" achieved the best results regarding the average DoP compared with the "no support" mode (-41%). A reason for that could be that the "Robot" condition is directly linked to the fulfillment of the task. Especially in the group of the technicians, the modes with the determined and marked parts that are removable by the human operator led to a higher average DoP compared with the "no support" condition. It seems that in modes with highlighted parts that are removable by the human operator, the participants tend to disassemble parts rather than to hand the disassembly over to the robot. In the case of a cooperative disassembly under the aforementioned conditions and on the basis of the empirical findings, the following design recommendations can be given. In a group of engineers it doesn't matter which DSM with additional information is presented. But there is a trend to the "Robot" mode. In a group of technicians, there is no mode to prefer but the "Robot & Human" mode seems to be worst choice because of the increasing execution time and the increasing distance to optimal performance.

ACKNOWLEDGEMENT

The authors would like to thank the German Research Foundation DFG for the kind support within the Cluster of Excellence "Integrative Production Technology for High-Wage Countries".

REFERENCES

Bortz, J., G.A. Lienert, K. Boehnke 2008. *Verteilungsfreie Methoden in der Biostatistik.* Heidelberg: Springer Medizin Verlag (in German)

Brecher, Ch 2011 *Integrative Production Technology for High-Wage Countries.* Berlin: Springer

DIN 58220 (1997) *Sehschärfebestimmung.* Berlin: Beuth. (in German)

Ewert, D., M. Mayer, S. Kuz, D. Schilberg, S. Jeschke 2010 A hybrid approach to cognitive production systems. In: *Proceedings of the International Conference Advances in Production Management Systems*, eds. M. Garetti. Cernobbio

Gausemeier, J. 2008. From Mechatronics to Self-Optimizing Systems. In: *Proceedings of the 7th International Heinz Nixdorf Symposium on Self-optimizing Mechatronic Systems: Design the Future*, eds. J. Gausemeier, F.J. Rammig, W. Schäfer. Paderborn: HNI-Verlagsschriftenreihe.

Hart, S.G., L.E. Staveland 1988. Development of NASA-TLX (Task Load Index): Results of empirical and theoretical research. In: *Human Mental Workload*, eds. P.A. Hancock, N. Meshkati. Amsterdam: North Holland Press.

Kempf, T., W. Herfs, C. Brecher 2008. Cognitive Control Technology for a Self-Optimizing Robot Based Assembly Cell. In: *Proceedings of the ASME 2008 International Design Engineering Technical Conferences & Computers and Information in Engineering Conference*, American Society of Mechanical Engineers

Klocke, F. 2009. Production Technology in High-Wage Countries – From Ideas of Today to Products of Tomorrow. In: *Industrial Engineering and Ergonomics*, ed. C. Schlick. Berlin: Springer

Leiden, K., K.R. Laughery, J. Keller, J. French, W. Warwick, S.D. Wood 2001. *A Review of Human Performance Models for the Prediction of Human Error. Prepared for National Aeronautics and Space Administration System-Wide Accident Prevention Program.* Ames Research Center, Moffet Field, CA.

Liepmann, D., A. Beauducel, B. Brocke, R. Amthauer 2007. *IST 2000 R- Intelligenz Struktur Test 2000 R Manual.* (in German)

Luczak, H., R. Reuth, L. Schmidt 2003. Development of error-compensating UI for autonomous prodction cells. *Ergonomics* 46:19-40.

Mayer, M., B. Odenthal, M. Faber, W. Kabuß, N. Jochems, C. Schlick 2010. Cognitive Engineering for Self-Optimizing Assembly Systems. In: *Advances in Human Factors, Ergonomics, and Safety in Manufacturing and Service Industries*, eds. W. Karwowski, G. Salvendy. Boca Raton: Taylor & Francis

Mayer, M., C. Schlick, D. Ewert, D. Behnen, S. Kuz, B. Odenthal, B. Kausch 2011. Automation of robotic assembly processes on the basis of an architecture of human cognition. *Production Engineering Research and Development*, 5:423-431

Odenthal, B., M. Mayer, W. Kabuß, B. Kausch, C. Schlick 2009. Investigation of Error Detection in Assembled Workpieces Using an Augmented Vision System. In: *Proceedings of the IEA2009 - 17th World Congress on Ergonomics.* Beijing: China

Pfendler, C., C. Schlick 2007. A comparative study of mobile map displays in a geographic orientation task. *Behaviour & Information Technology* 26:455-463

Rasmussen, J. 1986. *Information Processing and Human-Machine Interaction. An Approach to Cognitive Engineering.* New York: Elsevier Science Inc.

Schlick, C., R. Bruder, H. Luczak 2010. *Arbeitswissenschaft.* Berlin: Springer (in German)

CHAPTER 34

Artificial Intelligence in Optimizing Environmental Factors towards Human Performance

[1]Ahmad Rasdan Ismail, [2]Hj Baba Md. Deros, [2]Mohd Yusri Mohd Yusoff

[1]Universiti Malaysia Pahang, 26600 Pekan, Pahang, Malaysia
[2]Universiti Kebangsaan Malaysia, 43600 Bangi, Selangor, Malaysia
arasdan@gmail.com

ABSTRACT

Worker's performance in the automotive industry especially from the components assembly department are believed will be affected by the environmental factor. The intention of this study is to investigate the effect of environmental factor such as illuminance, Wet Bulb Globe Temperature (WBGT), and relative humidity to the worker productivity at automotive industry and optimize these factors by using Response Surface Method and Artificial Neural Network (ANN). The study was carried out at a room with area of 17m^2 which equipped with environmental parameter control system. Subjects were performed the task of manual assembly operation. The information and data collection is done using the measurement instruments of QuestTemp°36 and Heavy Duty Light Meter. Experimental result shows that there is the relationship of these three environmental factors on the worker's productivity. The analysis from Response Surface Method indicated that the productivity was predicted to be 1.0 under the condition of WBGT is 25.9°C and relative humidity 40% with illuminance of 441.51 lux. Meanwhile through the ANN, the data were trained to react as linear relationship. The linear relationship from ANN revealed that the optimum value of production (value≈ 1) can be achieved with the temperature (WBGT) is 22.7°C. The findings from the study can be proof nearly equal and therefore the authors believed that both methods can be used to propose the optimum environmental factors towards human productivity.

Keywords: Environmental factors, Productivity, Optimization, RSM, ANN

1 INTRODUCTION

Human capital is the momentum and resources to the innovation and can be determined from the full benefit of labor productivity (Dzinkowski, 2000). Productivity is an important matter which involves the manufacturing industry. The automotive industry is an important industry to the economy of Malaysia. Quality of work, management, and working conditions were generally identified able to increase the productivity of workers (Prokopenko, 1992). Environmental factors considered in this context are illuminance, relative humidity and WBGT. These factors largely determine the safety, health, and performance in the workplace and daily life (Dul and Weerdmeester, 2008). Optimum comfort will help to form the optimal productive workers. According Seppa'nen et al. (2003), an alleged relationship with performance decline 2% per °C increase in temperature in the range between 25°C to 32°C. In fact, some studies explain that the parameters such as temperature increase workplace can provide a drastic impact on productivity in the workplace (Croasmun, 2004). Response Surface Method and Artificial Neural Network (ANN) analysis were used to optimize the environmental factors in order improve the productivity in this study. Myers and Montgomery (1995) states that the response surface method is an empirical model to determine the relationship between various parameters and response to various criteria and find the desired parameter of interest to the response. This is a successive strategy to build and optimize empirical model.

2 METHODOLOGY

In this study, a total of 3 adults of both gender (2 males and 1 female) were used as subjetcs. The subjects does not have experience in this task and they were trained for a week before perform the actual study. They performed their tasks and considered to be undergoing the same physical effects. Employees were given salary without looking at their performance in order to avoid the factors influencing their motivation during the task performed. The demographic information of subjects obtained through questionnaire. QuestTemp°36 used to observe the WBGT and relative humidity level in the working environment whereas the illuminance level was measured by Heavy Duty Light Meter. The environmental factors were observed and recorded for every 10 minutes to ensure that all the parameters controlled in a certain range. Figure 1 describes 3D illustration of the arrangement of instruments and observation devices. The device control parameters, such as adjustable lighting, air conditioning and dehumidifier system is shown in the figure.

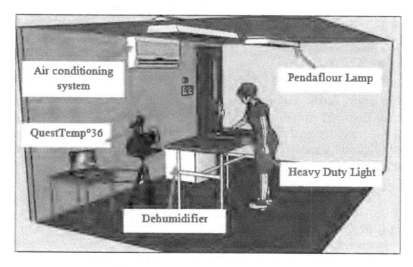

Figure 1: 3D illustration of testing chamber.

Before beginning the process of data acquisition in the field of study, the tools of observation calibrated. After the data observed and recorded, the data undergo the process of ANN analysis for each of the line graph data. This analysis carried out repeatedly to get the best reading. After obtaining the data on variations of the best, the optimal value determined. This is important to determine the accuracy of the data. On other hand, the data also were analyzed using Response Surface Method via MINITAB software in order to study the relationship of each factor towards the productivity of workers. Further optimization of environmental factors is carried out through MINITAB to find the optimum environmental factors that can produce productivity of 1.0. This study conducted for 45 days with each of the parameters of environmental conditions were repeated three times with different operator. The target time for a completed unit is 1.8 minutes. Productivity of workers is the ratio of actual output to targeted output. Therefore, productivity is calculated by using Equation 1 as below.

$$\text{Productivity} = \frac{\text{Actual output}}{18} \tag{1}$$

3   RESULTS AND DISCUSSION

The observed and recorded data were analyzed using both ANN analysis and RSM method to find the optimum environmental factors value in order to achieve the optimum productivity (value ≈ 1).

3.1   Artificial Neural Network (ANN) Analysis

WBGT have been adjusted and the average production rate for each employee analyzed using software available Neurosolution to show the relationship between production rate and the overall factors involved. According ANN analysis, the factors involved were fixed and independent variables for the rate of production were trained repeatedly in the ANN analysis. For this ANN analysis is using a hidden layer for the training data which is Levenberg-Marquardt (LM) algorithm used and data absorbed at random to be trained. Since the data trained by ANN is a random, different value for the rate of production of the ANN analysis is obtained even using the same factor analysis. The results of the analysis showed that uniform data were generated from the relationship of each factor to the rate of output. Regression (R^2) value represents level of optimization and the higher percentage reflects the precision of the graph. Figure 2 shows the graph of productivity versus temperature (WBGT). From the graph can be seen that the higher the value of the WBGT temperature causes the output (productivity) value decreased.

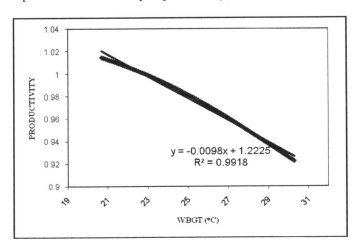

Figure 2: Network Output for Various Inputs WBGT (°C)

The optimum WBGT temperature value for optimum production rate of 1.0 for each set of assembly is 22.7°C using the equation 2:

$$y = -0.0098x + 1.2225 \tag{2}$$

3.2   Response Surface Method (RSM) Analysis

A mathematical model was developed to relate the relationship of each factor on workers' productivity. All coefficients in the model is then converted into a mathematical equation 3 in real terms:

$$y = 1.25743 - 0.009192311x_1 - 0.00265556x_2 + 0.000197083x_3 \tag{3}$$

where y is the response, x_1 is WBGT (°C), x_2 is relative humidity (%), and x_3 is lighting level (lux).

Table 1 Predicted regression coefficient for first order model

Term	Coefficient	SE Coefficient	T	P
Constant	0.9952	0.0124	80.514	0.000
WBGT	-0.0598	0.0170	-3.530	0.001
Relative humidity	-0.0398	0.0169	-2.353	0.023
Illuminance	0.0788	0.0169	4.658	0.000
S=0.0829188	R-Sq= 49.19%	R-Sq (adj)= 45.47%		

From Table 1, it was found that the coefficient of illuminance (p=0.000), coefficient of WBGT (p=0.001), and coefficient of relative humidity (p=0.023) where the coefficient of illuminance showed most significant effect on the worker's productivity. This indicates that any changes in the value of the WBGT, relative humidity, or illuminance will cause the significance changes in the productivity. Meanwhile, if viewed in terms of coefficient of multiple determination, R^2, which is 0.49. This value indicates the extent to which the validity of the model built with existing model in determine the dependent variable. Given the value of R^2 is less than 0.5, it is not really suitable to be adopted. Since this is the preliminary study, so the analysis must be proceeded to the development of second order model for more accurate findings.

Optimal value of productivity can be achieved with combination of WBGT 25.9°C, relative humidity 40%, and illuminance 441.51 lux. Optimization of environmental factors study done by Ismail et al. (2009) also found that WBGT is around 26°C and this achieves equality with the study results. According to Juslén et al. (2007), illuminance level of 500 lux is the minimum requirement for the assembly line at eletrical industry. So, the illuminance level of 441.51 lux is adequate for the component assemby line at automotive industry.

4   CONCLUSIONS

In conclusion, the optimal rate of environmental factors for temperature, relative humidity, and illuminance to obtain the optimum production rate (value ≈1.0) for a set of manual production lines in the automotive industry in Malaysia has been successfully achieved. Through RSM method, a mathematical model was developed to relate the relationship of each factor on workers productivity. Environmental factors that can provide optimal productivity value of 1.0 are WBGT 25.9°C, relative humidity 40%, and illuminance 441.51 lux. Using ANN analysis, the optimum value of production (value 1) can be achieved when the temperature (WBGT) is 22.7°C. Although the others research get the optimum temperature in average of 24°C in accordance with regulations issued by the ISO 7730, which provides the comfortable temperature is between the value of 24°C to 27°C. It is because others studies get the response within single factors without the influence of other environmental factors. In this study, two others factors (relative humidity and illuminance) considered beside temperature (WBGT). Therefore, the result which is optimum value of the temperature in industry is in average of 22.7°C is accepted. Since the findings equal, it is believed that both methods can be used to propose the optimum environmental factors towards human productivity.

ACKNOWLEDGMENTS

The authors would like to acknowledge Ministry of Higher Education of Malaysia for support through Research University Grant (No. UKM-GUP-TK-08-16-059). Also appreciation to the research assistants and the Metal Industries Company for the assistance and cooperations given to complete this study.

REFERENCES

Bommel, W.J.M.van., Beld, G.J.van. & Ooyen, M.H.F. van. 2002. Industrial lighting and productivity. Philips Lighting, The Netherlands.

Croasmun, J. 2004. Comfort Means Productivity for Office Workers. Accessed August 12, 2007. http://www.ergoweb.com/news.

Dul, J. & Weerdmeester, B.A. 2008. Ergonomics for Beginners. Third Edition, Taylor & Francis Group.

Dzinkowski, R. 2000. The value of intellectual capital. *Journal of Business Strategy* 21/4 (July-August): 3-4.

Ismail, A.R., Rani, M.R.A., Makhbul, Z.K.M., Sopian, K., Tahir, M.M., Ghani, J.A. & Meier, C. 2009. Optimization of environment factors: a study at Malaysian automotive industry. *Journal of Scientific Research*, 27(4): 500-509.

Ismail, A.R., Rani, M. R. A., Makhbul, Z. K. M., Nor, M. J. M. and Rahman, M. N. A. 2009. A Study of Relationship between WBGT and Relative Humidity to Worker Performance. *World Academy of Science, Engineering and Technology* 51.

Juslén, H.T., Verbossen, J. & Wouters, M.C.H.M. 2007. Appreciation of localized task lighting in shift work- a field study in the food industry. *International Journal Of Industrial Ergonomics*, 37(5): 433-443.

Lan, L. & Lian, Z.W. 2009. Neurobehavioral tests to evaluate the effects of indoor environment quality on productivity. *Building and Environment*, 44(11): 2208-2217.

Myers, R.H. & Montgomery, D.H. 1995. Response Surface Methodology. John Wiley & Sons, USA.

Myers, R.H., Montgomery, D.C., Cook, C.M.A. 2009. Response Surface Methodology: Process and Product Optimization Using Designed Experiments. Third Edition, John Wiley & Sons, Inc, New Jersey.

Prokopenko, J. 1987. Productivity management: A practical handbook. International Labour Organisation. Switzerland.

Seppa'nen O.A., Fisk W.J., Mendell M.J.1999. Association of ventilation rates and CO2-concentrations with health and other responses in commercial and institutional buildings. *Indoor Air*, 9: 252-74.

Tsutsumi, H., Tanabe, S., Harigaya, J., Iguchi, Y. and Nakamura, G. 2007. Effect of humidity on human comfort and productivity after step changes from warm and humid environment. *Building and Environment*, 42: 4034-4042.

CHAPTER 35

Perceptions of Traffic Enforcers on Road Traffic Noise and Its Potential Non-Auditory Effects

A.C. Matias, K. M. Baraquel, M. A. Barrios, L. C. Millare, C. N. Ocampo

University of the Philippines
Quezon City, PHILIPPINES
aura.matias@coe.upd.edu.ph

ABSTRACT

Non-auditory effects of noise is defined as 'all those effects on health and well-being caused by exposure to noise, with the exclusion of effects on the hearing organ and the effects due to the masking of auditory information (Smith, 1992). The study primarily aims to obtain information on the perceptions on road traffic noise and subjective non-auditory responses of traffic enforcers who are exposed to a time weighted average noise of 85 decibels along a main highway in Metro Manila, Philippines. In analyzing the data, correlation tests were performed to determine possible interrelationships between demographic variables age, number of years worked, annoyance level, and the potential non-auditory effects of noise sleep, weariness, stress, irritability and depression which were used suggesting recommendations on how the working conditions of the enforcers can be improved. Correlation coefficients suggest that age is significantly associated with level of annoyance and level of annoyance is significantly associated with irritability.

1   INTRODUCTION

Metro Manila, being the Philippine center of commerce and trade experiences heavy traffic everyday. Jeepneys, buses and private vehicles which serve as the primary modes of public transport congest the city streets during rush hour (6:00-

9:00 AM, 4:00-8:00 PM). Road traffic noise typically constitutes interminable honking and whirring of cars, exhaust, engines, defective parts and wheels. Currently, there are no available data to determine the magnitude of the problem in the Philippines -- a clear indicator that noise as a great hazard to health has yet to be acknowledged in the country. Impact of noise assessment has been limited to noise perception surveys administered by the NCTS (National Center for Transportation Studies) and other concerned institutions (Fajardo, 2009). Traffic enforcers, being exposed to high levels of road traffic noise during their eight-hour shift, are subject to its harmful effects on human health. Negative effects on health of environmental noise (such as road traffic noise) are not limited to hearing (Berglund, Berglund and Lindvall, 1975). Prolonged exposure to noise can cause various non-auditory effects of noise such as sleep disturbance, impaired performance, cardiovascular disease, endocrine responses such as stress and psychological symptoms. It is therefore important that non-auditory effects of road traffic noise also be identified to determine the measures necessary to ensure the safety and protection of the traffic enforcers who play a vital role in maintaining the smooth flow of traffic.

2   METHODOLOGY

A total of 70 traffic enforcers were interviewed using a survey questionnaire composed of two parts. Questions in part 1 obtained personal information from the respondents such as age, gender; shift schedule, and perceptual responses of awareness of health hazards of noise exposure, perceived seriousness of its effects and annoyance level. Questions in part 2 obtained information on the frequency of occurrence among traffic enforcers of some of the non-auditory effects of noise, specifically, Difficulty to Sleep, Weariness, Stress and Depression. A 5-point verbal scale corresponding to the frequencies was used. Figures 1 and 2 illustrate how the respondents were segmented.

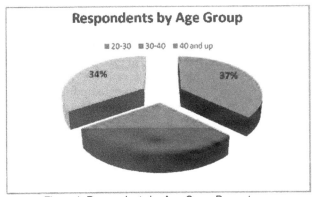

Figure 1. Respondents by Age Group Percentages

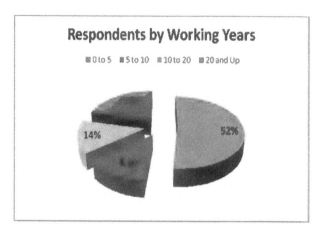

Figure 2. Respondents by Working Years Percentages

3   RESULTS AND DISCUSSION

Results from Part 1 of the survey give information on the general perception of the MMDA traffic enforcers on road traffic noise. The following results were obtained:

- 81.43% of the respondents are aware of the health hazards of being exposed to noise
- 70% realize the seriousness of the effects of noise
- 52.85% of the traffic enforcers do nothing to protect themselves from traffic noise
- 67.14% consider horn-blowing as the most annoying source of road traffic noise
- 32.26% answered 7 as the annoyance level out of a possible 10
- 75.71% believe that discipline is the key to noise reduction.

Although the traffic enforcers realize the hazardous effects of noise on health, majority of them fail to protect themselves. The result that the blowing of horns is the most annoying source of noise can be related to the subjects' discernment that discipline is the key to noise reduction. Blowing of horns may be reduced if drivers follow traffic rules and practice the responsible use of horns.

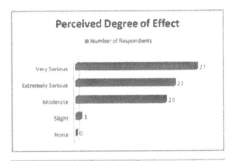

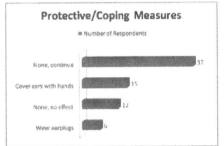

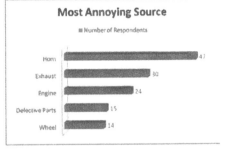

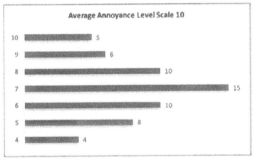

Figure 3. Summary of Results of Survey Part 1

Analysis for the second part of the survey seeks for the interrelationships between the variables age, working years and shift to annoyance and potential effects sleep/ weariness/stress/irritability/depression. The data gathered were tested using Statfit2 and it was established that it follows the normal distribution. ANOVA analyses were done to determine association between the variables. Because the said variables are ordinal (having three or more levels with a natural ordering, such as never, sometimes, moderately, often and always), measures of concordance and correlation coefficients for ordinal categories were also computed using Minitab. Concordance is used to describe the relationship between sets of paired observations. A pair is concordant if the pair of observations is in the same direction. A pair is discordant if the pair of observations is in opposite directions.

3.1 Age/Working Years/Shift versus Annoyance

Table 1 and 2 give the summary of the ANOVA analysis of annoyance level versus age, number of working years and shift. F test statistics being greater than the F critical for Age and Working Years imply that annoyance level of the traffic enforcers differ significantly between age groups and number of working years. On the other hand, F less than the critical value for Annoyance Level versus Shift suggests that annoyance level is not significantly affected by the time of shift.

Table 1 Age/Working Years/ Shift Versus Annoyance Level

	Correlation Coefficients for Ordinal Categories		
	Goodman &Kruskal's Gamma	Kendall's Tau-b	P-Value
Age	0.4159	0.2614	0.0082
Working Years	0.4268	0.2486	0.0122

Table 2 Correlation Of Age, Working Years Versus Annoyance Level

	Analysis of Variance				
	F		F critical	Spearman's Rho	Pearson's Coefficient
Age	6.2803	>	3.1338	0.2867	0.2515
Working Years	5.0223	>	2.7437	0.1848	0.1386
Shift	0.5048	<	4.1393		

The Goodman &Kruskal's Gamma coefficients indicate association between age/number of working years and level of annoyance. Tau-b coefficients, on the other hand tell us that annoyance level increases as the ranking of age/number of working years increases. The null hypothesis for the test of concordance is that the probability of concordance equals the probability of discordance. Assuming α-level as 0.05 for the test, the null hypothesis is rejected, thus it can be concluded that there is enough evidence to conclude that the variables (age vs. annoyance, working years vs. annoyance) improve in the same direction as the correlation coefficients suggest that there are more concordant pairs than discordant ones.

3.2 Age and Working Years versus Potential Non-Auditory Effects

Table 3 shows that irritability differs between age groups and working years while depression differs with working years. Examining the p-values, it can be said that at an $\alpha = 0.05$, there is enough evidence to reject the null hypothesis that the probability of concordance is equal to the probability of discordance for age-irritability, age-depression and working years-depression.

Table 3 Anova Analyses For Age/Working Years Versus Irritability and Depression

	Age			Working Years		
Irritability	6.2715	>	3.1337	3.5445	>	2.7437
Depression				3.5815	>	2.7437

	Age		Working Years	
	Pearson's r	Spearman's rho	Pearson's r	Spearman's rho
Irritability	0.3594	0.3561		
Depression			0.2750	0.2917

	Age			Working Years		
	Goodman & Kruskal's Gamma	Kendall's Tau-b	P-Value	Goodman & Kruskal's Gamma	Kendall's Tau-b	P-Value
Irritability	0.4743	0.3237	0.0014			
Depression				0.4024	0.2600	0.0081

4 CONCLUSION

Although the traffic enforcers realize the hazardous effects of noise on health, majority of them do nothing to protect themselves. Results also show that the enforcers consider the blowing of horns as the most annoying source of noise and that discipline is the key to noise reduction. Results of Part 2 of the survey indicate a significant association between 1) age and level of annoyance 2) age and irritability 3) number of working years and level of annoyance 4) number of working years and depression and 5) level of annoyance and irritability. For all of these relationships, Minitab outputs gave out positive correlation coefficients however, they are not large enough to conclude a strong positive association between the said variables. Mapping out the results as illustrated in Fig. 4, however suggests that age being significantly associated with level of annoyance, and level

334

of annoyance being significantly associated with irritability might be a pathway by which noise affects the traffic enforcers. It suggests that older enforcers tend to be more annoyed to the noise, and their level of annoyance is associated with their feeling of being irritated.

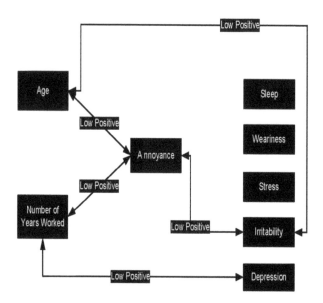

Figure 4. Summary of Results for Part II of Survey

5   RECOMMENDATION

Based on the results of this study, a number of recommendations were made in order to improve the working environment of the traffic enforcers. Since majority of the traffic enforcers deem the sound of the vehicles' horn as the greatest contributor to noise, it is recommended that an anti-noise resolution be enforced to prohibit unnecessary blowing of horns. Realizing the negative physical effects of noise, the resolution prohibits honking of horns when driving near residential areas, churches, schools and hospitals. Only public service vehicles such as ambulances, fire trucks and police patrol cars are exempt from this regulation. Violating drivers would incur a penalty per offense. A stricter implementation and further dissemination of this regulation to prevent the possible effects of noise in Metro Manila is recommended to help minimize road traffic noise caused by the irresponsible honking of horns.

REFERENCES

Berglund, B., U. Berglund, and T. Lindvall "Scaling loudness noisiness and annoyance of aircraft noise," Journal of the Acoustical Society of America, 57, 930-934 (1975).

Berglund, B., U. Berglund, and T. Lindvall, "Scaling loudness noisiness and annoyance of community noises", Journal of the Acoustical Society of America, 60 1119-1125 (1976).

Fajardo, B., "A Study on the Individual Perceptions of Road Traffic Noise," Masters dissertation, School of Urban and Rural Planning, Univ. of the Philippines, Diliman, March 2009.

Job, R.F.S. , J. Hatfield, N.L. Carter, P. Peploe, R. Taylor, and S. Morrell, " General scales of community reaction to noise (dissatisfaction and perceived affectedness) are more reliable than scales of annoyance". Journal of the Acoustical Society of America, 110-2, 939-944 (2001).

Kuwano, S., T. Mizunami, S. Namba, and M. Morinaga, "The effect of different kinds of noise on the quality of sleep under the controlled conditions" Journal of Sound and Vibration, 250-1, 83-90, (2002).

Smith, A.P. and D.E. Broadbent, "Non-auditory Effects of Noise at Work": A Review of the Literature. HSE Contract Research Report No 30, London: HMSO, 1992.

Stephen, A.S and M.P. Matheson, "Noise pollution: non-auditory effects on health", Department of Psychiatry, Medical Sciences Building, Queen Mary, University of London, London, UK.

CHAPTER 36

Temporal Strategy and Performance during a Short-cycle Fatiguing Repetitive Task

T. Bosch, S.E. Mathiassen , D. Hallman, M.P de Looze, E. Lyskov, B. Visser, J.H van Dieën

TNO Healthy Living
Hoofddorp, The Netherlands
Tim.Bosch@TNO.nl

ABSTRACT

This study investigated temporal changes in movement strategy and performance during fatiguing short-cycle work. Eighteen participants performed six 7-minutes work blocks with repetitive reaching movements at 0.5 Hz, each followed by a 3-minute rest break for a total duration of one hour. Electromyography (EMG) was collected continuously from the upper trapezius muscle, the temporal movement strategy and timing errors were obtained on a cycle-to-cycle basis, and perceived fatigue was rated before and after each work block. Clear signs of fatigue according to subjective ratings and EMG manifestations developed within each work block, as well as during the entire hour. For most participants, timing errors gradually increased, as did the waiting time at the near target. The results showed that the movement strategy may change, possibly for the purpose of alleviating fatigue, but also that the change may not be sufficient to eliminate fatigue.

Keywords: repetitive work, temporal strategy, performance, fatigue, EMG

1   INTRODUCTION

More than 60% of the European working population reports to perform repetitive hand or arm movements during more than 25% of their working time

(Eurofound, 2010). Muscle fatigue may develop during repetitive work requiring low or moderate force (e.g. Nussbaum, 2001), and has been proposed to be an important precursor to the development of neck and shoulder muscle disorders (e.g. Takala, 2002). Therefore, studies of the development and effects of fatigue during repetitive work are justified as part of the efforts to develop guidance on how to prevent disorders.

Several studies have suggested an association between fatigue development and temporal motor variability. More temporal variability in the EMG amplitude of back muscles (van Dieën et al., 2009) and more spatio-temporal variability of the EMG amplitude within the trapezius muscle (Farina et al., 2008) have been shown to be associated with slower development of electromyographic manifestations of fatigue. Recent studies have suggested that motor strategies, including movement patterns, may change due to, or in order to counteract, fatigue in occupational tasks like sawing (Côté et al., 2002) and hammering (Côté et al., 2008).

Since fatigue does imply a decrease in the capacity to generate muscle force, it could be expected to have a negative influence on task performance in many situations relevant to working life. Earlier studies, however, have shown inconsistent effects of fatigue on performance. In repetitive sawing tasks, force and movement frequency were affected (Côté et al., 2008), whereas others did not find reduced performance expressed as movement time and timing errors in similar activities (e.g. Gates and Dingwell, 2008). These conflicting results may not only be explained by the difference in outcome parameters (Hoffman et al., 1992), but also by alterations in movement patterns that allow maintenance of performance in spite of fatigue (Fuller et al., 2011). The fact that movement patterns may be changed in both spatial and temporal structure while performance is maintained has been shown in several studies of occupational relevance (e.g. Dempsey et al., 2010).

Most of the studies investigating kinematics and performance during fatigue used another exercise protocol to induce fatigue than that used for evaluating its effects (e.g. Huysmans et al., 2008). A realistic temporal load pattern, including rest breaks, is rarely applied, neither for inducing fatigue nor for evaluating its effects (Gates and Dingwell, 2008). In the current study we investigated the temporal strategy of movements and performance in a fatiguing short-cycle repetitive reaching task with intermittent rest breaks performed for one hour. We addressed the following research question: does the temporal movement strategy and performance change during one hour of fatiguing repetitive work?

2   METHODS

2.1 Subjects

Eighteen healthy, male subjects (14 right-handed and four left-handed, mean age 24.2 (SD 4.3) years, weight 79.8 (SD 10.4) kg, height 183.5 (SD 3.4) cm, BMI 23.7 (SD 3.0) kg/m2) volunteered to participate in the study. All subjects gave their written informed consent prior to the start of the study and the study was approved by the Regional Ethical Review Board in Uppsala.

338

2.2 Procedure and task

The subjects performed a one-hour repetitive arm reaching task. Six consecutive work blocks were performed, each consisting of seven minutes performing the repetitive task and a three minutes rest break. During the rest break, subjects performed a memory test in which they were presented letters on a computer screen in front of them, and asked at irregular intervals to recall the last presented letter. The repetitive work task involved lifting and lowering of the arm and was performed with the dominant hand while the other arm was resting on the table in a predefined position.

The repetitive work, illustrated in figure 1A, consisted of moving a 300 g manipulandum held in the right hand between two targets at a frequency of 0.5 Hz as guided by a metronome signal. Subjects were asked to hit the targets in synchrony with the metronome signals, while they were free to choose their own timing strategy and work technique within each half-cycle. Each of the targets was equipped with an electrical connector, which was activated whenever the manipulandum touched the target.

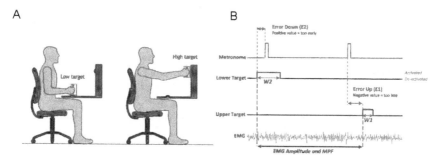

Figure 1. A. Workstation setup. Participants moved the manipulandum upwards from the lower target to the upper target with a predetermined time allowance of 1 s, and vice versa for the downward direction. B. Timing, performance and EMG variables obtained for every work cycle. Waiting time (temporal strategy) in seconds at the high (W1) and low (W2) target was registered as long as the connector was activated. The timing error (performance) at the high target (E1) and low target (E2) was defined as the time difference in seconds between the metronome beat and the touch of the target.

To ensure familiarity with the methods, the task and to offset a learning effect across trials, a training session was performed one day before the experiment. The training was continued until a stable work rhythm was achieved, or for a minimum of 3 minutes; i.e. about 90 work cycles.

2.3 Measurements

Perceived muscle fatigue in the neck and shoulder area was rated before and after every work block, using the Borg CR-10 scale (Borg, 1982). The participant was acquainted with the Borg scale during the training session.

EMG from the right trapezius, pars descendens was measured by a BioPac MP150 System (BIOPAC Systems Inc, Santa Barbara, CA USA) and amplified with BioPac EMG100C modules. Pre-gelled bipolar Ag/AgCl surface electrodes (AMBU Neuroline 720) were placed 2 cm lateral to the mid-point between C7 and the acromion according to Mathiassen et al. (1995), using an inter-electrode distance of 25 mm. A reference electrode was placed on the C7 spinous process. Prior to the electrode placement, the skin area was shaved, scrubbed and cleaned with alcohol and it was checked that the impedance was below 5Ω. EMG signals were continuously sampled during the entire work bout, band-pass filtered (10–500 Hz) and AD converted (16 bits at 2000 Hz). For each half-cycle of work, the average EMG amplitude was expressed in terms of the RMS value of the signal. EMG amplitudes were normalized (%RVE) using the RMS EMG amplitude during a 15s reference contraction with the arms abducted at an angle of 90 degrees (Suurküla and Hägg, 1987). The mean power frequency (MPF) of the signal was calculated using Welch's method (Welch, 1967).

On basis of the signal from the target connectors (figure 1B), the temporal strategy, measured by the waiting time at the high (W1) and low target (W2), was obtained. Furthermore, performance, expressed in terms of timing error at the upper (E1) and lower (E2) target connectors was calculated. Positive timing errors indicated that participants were ahead of the metronome beat while negative indicated a late arrival to the target.

To analyze changes in the EMG amplitude, MPF, temporal strategy and performance, the average were calculated for the first 30 cycles of each work block ("early") and compared to the last 30 cycles of that work block ("late"). In addition to average values, cycle-to-cycle variability across these cycles was assessed in terms of the median absolute deviation (MAD), as described by Shevlyakov and Vilchevski (2002).

2.4 Statistics

Development of perceived fatigue over time was examined using a two-way ANOVA for repeated measures with work block (1-6) and time (pre/post) as independent variables. Changes across time in EMG amplitude, MPF and EMG amplitude variability were analyzed using a three-way ANOVA for repeated measures with direction (up/down), work block (1-6) and time-in-block (early/late) as independent variables. To examine changes in performance and temporal strategy, a similar three-way repeated measures ANOVA was applied with the independent variables position (high/low target), work block (1-6) and time-in-block (early/late). Interaction effects were post-hoc tested using a one-way ANOVA for repeated measures or student t-tests.

340

3. RESULTS

Perceived fatigue significantly increased during the one-hour task as indicated by a main effect of work block (p<0.001). Furthermore, a significant increase (p<0.001) in perceived fatigue was found within work blocks. Similar findings were obtained for EMG amplitude and MPF. EMG amplitude increased significantly within each block (p<0.001). In the first work block, the EMG amplitude in the upward direction increased by 29% whereas the increase in the last work block was 24%. Smaller increases were found in the downward direction (11% and 15%). The EMG amplitude even increased across work blocks over the one-hour period for both the upward and downward movement (p<0.001). A significant effect of time-in-block was also found for the MPF indicating a decrease of the MPF within work blocks for both movement directions. However, no significant change of the MPF was found across work blocks, suggesting that the overall MPF had not changed after one hour. EMG amplitude cycle-to-cycle variability increased significantly during the one-hour task (p=0.002) and within all work blocks (p=0.002).

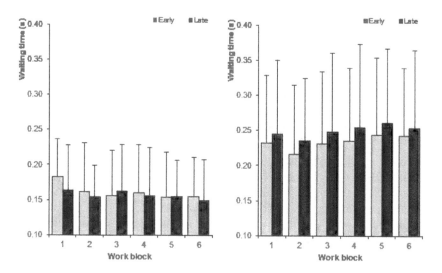

Figure 2. The average waiting times (temporal strategy) at the high (left) and low (right) targets at the start (grey) and end (black) of the each work block. Error bars show the standard deviation between participants

The average waiting time (Figure 2) at the upper target was significantly shorter than at the lower target. The effects of position and work block interacted significantly (p=0.004). Post-hoc testing indicated a significant decrease of waiting time across work blocks at the upper target but no significant change for the lower target. Furthermore, a significant interaction effect between position and time-in-block was found (p=0.003). Post-hoc testing demonstrated a significant increase in waiting time within work blocks at the lower target, whereas waiting time did not significantly change within work blocks at the upper target. The cycle-to-cycle variability of the waiting time did not change significantly within or across work blocks.

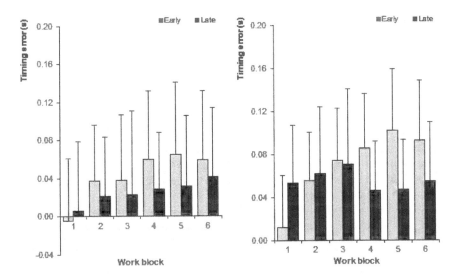

Figure 3. The average timing errors (performance) at the high (left) and low (right) targets at the start (grey) and end (black) of the each work block. Error bars show the standard deviation between participants.

As illustrated in Figure 3, the average timing errors early in the first work block were 0.07s (SD 0.06) and 0.03s (SD 0.07) at the lower and upper target, respectively. This difference was significant for all work blocks (p<0.001). Timing errors increased over the course of all work blocks (p<0.001). Although the average timing errors at the upper target were positive, indicating that participants arrived too early, four out of eighteen participants actually consistently arrived too late at the upper target throughout the one-hour work, while five changed performance from arriving late in the beginning of the first work block to arriving early at the end of the last block. The remaining nine subjects were consistently early on target. A similar inconsistency among subjects was seen for the lower target. Early in the first work block, the cycle-to-cycle variability in timing error was 0.047s at the lower target and 0.041s at the upper. These values did not change significantly within or across work blocks.

A borderline significant three-way interaction effect between position, work block and time-in-block was found. Post-hoc testing indicated a significant decrease in timing errors at the upper target (i.e. E1 in Figure 1B) while changes in timing error within work blocks for the lower target (i.e. E2 in Figure 1B) were inconsistent.

Thus, early in the first work block, the average subject started an average cycle by hitting the upper target 0.005s after the metronome beat (E1 in figure 2), stayed in touch with the upper target for 0.17s (W1) and hit the lower target 0.012s ahead of that metronome beat (E2), then staying in touch with the lower target for 0.23s (W2). Late in the last work block this temporal strategy had changed to an early (0.004s) arrival at the upper target and a shorter stay at the upper connector (0.15s). The arrival at the lower connector was 0.09s early, and the waiting time there was now extended to 0.25s.

4. DISCUSSION

In the current study we investigated changes in temporal motor strategies and performance during series of fatiguing short-cycle repetitive arm movements with interposed rest periods. Even though the task in the current study was simulated and constrained and the level of exertion was higher than in most repetitive occupational work (e.g. industrial assembly work), it did show similarities with the more strenuous types of short cycle assembly work (e.g. Bosch et al. 2007).

All participants developed fatigue during the one-hour work bout, but they differed considerably in fatigue manifestations. One probable reason was that the work load relative to maximal capacity differed between subjects, since they were required to handle the same manipulandum. We found clear signs of increasing fatigue within each work block, both in terms of increased perceived fatigue, and as indicated by a simultaneous increase in EMG amplitude and decline in MPF (e.g. Hägg et al., 2000). Over the entire one-hour task we observed a cumulative increase in perceived fatigue and EMG amplitude but not a consistent decrease in the MPF, which is in line with several previous studies which have found that MPF recovery is fast after exercise (e.g. Elving et al., 2002).

Participants changed their temporal movement strategy along the bout; the waiting time at the upper target was significantly shortened at the end of the 1-hour task. Participants also changed their strategy at the lower target within work blocks; the waiting time was longer at the end of each block. From a biomechanical point of view these are adequate responses to increasing fatigue, since they will reduce the time-averaged shoulder elevation moment produced by the mass of the arm, and thus lower the demand on the upper trapezius muscle. Changes in movement patterns of the upper extremity have been reported after enforced fatigue protocols and Fuller et al. (2011) have suggested that such kinematic changes reflect fatigue adaptation strategies with the purpose of reducing the load on the fatigued body region. In our rather constrained task changes in the temporal movement strategy also occurred, but this was not sufficient to prevent perceived fatigue and development of fatigue manifestations in the upper trapezius muscle. To this end, it should be noted that in our strictly time-constrained task, the latitude for extending waiting times was limited. This might explain that waiting time did increase within blocks, possibly to the extent feasible, but that no further change was possible across the work blocks through the one-hour work period without compromising performance to an unacceptable level. Since repetitive work in an occupational setting is usually performed with lower external loads and allows for more temporal and spatial autonomy, e.g. discretionary micro-breaks between consecutive work cycles (Dempsey et al., 2010), the present findings can only be generalized to working life with due caution.

A significant decrease in performance was found after one hour of repetitive work. As indicated by the positive error values, participants reached on average both targets too early, most pronounced for the lower target. This might be explained by the shorter waiting time at the upper target: participants started their downward movement earlier and as a consequence of that arrived at the lower target too early. This seems to be supported by the on average 0.08s. longer waiting times at the lower target. In the upward movement direction, the gradually longer waiting times at the lower target within each work block did not result in hitting the upper target too late, as indicated by the positive error values. Participants showed a significant decrease in timing error at the upper target within work blocks. This apparent improvement of performance seems to a large extent to be a consequence of hitting the target too early at the beginning of each work block, yet more exact in time late in the block.

The focus of the current study was on changes in temporal strategy and performance and their possible relationships with loading of the trapezius muscle. Obviously, more kinematic measures (e.g. Fuller et al., 2011) would have provided an opportunity to understand in detail the kinematic adaptations underlying the temporal changes observed in the present study.

Although the upper trapezius muscle is generally considered to be a particular important muscle when studying exposure and disorders in the shoulder region, it cannot be ruled out that other shoulder muscles responded differently to the repetitive task and thus would have an influence on or relationship with the temporal strategy. Earlier studies showed that a changed load sharing between synergistic muscles in the shoulders and neck can occur (Palmerud et al., 1995). Since, to our knowledge, no studies have so-far investigated the effects of a changed temporal movement strategy on the relative engagement among and within shoulder muscles in repetitive occupational work, this might be an interesting issue for further research.

The current study showed that the movement strategy while performing a prolonged repetitive task may change, possibly for the purpose of alleviating fatigue, but also that the change may not be sufficient to eliminate fatigue. Whether occupational settings, e.g. in industrial assembly, offer such opportunities to a wider extent, and whether such opportunities would then be utilized by the employees is an interesting issue for further research. The current study also demonstrated that performance may decrease during prolonged repetitive tasks. The performance indicator used in this study was timing error. While relatively small, errors of the present size might still have serious consequences in occupational settings requiring great accuracy and not offering opportunities to repair errors (e.g. standardized short-cycle line assembly). In these work situations adjustments in motor behavior for the purpose of avoiding such errors may not be practicable, and thus fatigue may increase at a greater rate.

344

ACKNOWLEDGMENTS

The authors would like to acknowledge the Centre for Musculoskeletal Research in Sweden and Body at Work, the Dutch Research Center for Physical Activity, Work and Health for their financial support.

REFERENCES

Borg G.A. 1982. Psychophysical bases of perceived exertion. *Medicine & Science in Sports & Exercise* 14: 377-381

Bosch, T., M.P. de Looze and J.H. van Dieën. 2007. Development of fatigue and discomfort in the upper trapezius muscle during light manual work. *Ergonomics* 50: 161-177

Côté, J.N., P.A. Mathieu, M.F. Levin and A.G. Feldman. 2002. Movement reorganization to compensate for fatigue during sawing. *Experimental Brain Research* 146: 394-398

Côté, J.N., A.G. Feldman, P.A. Mathieu and M.F. Levin. 2008. Effects of fatigue on intermuscular coordination during repetitive hammering. *Motor Control* 12: 79-92

Dempsey, P.G., S.E. Mathiassen, J.A. Jackson and N.V. O'Brien. 2010. Influence of three principles of pacing on the temporal organisation of work during cyclic assembly and disassembly tasks. *Ergonomics* 53: 1347-1358

Dieën van, J.H., E.P. Westebring-van der Putten, I. Kingma, M.P. de Looze. 2009 Low-level activity of the trunk extensor muscles causes electromyographic manifestations of fatigue in absence of decreased oxygenation. *Journal of Electromyography and Kinesiology* 19: 398-406

Eurofound. 2010. Changes over time – First findings from the fifth European Working Conditions Survey. European Foundation for the Improvement of Living and Working Conditions: Dublin

Elfving, B., D. Liljequist, A. Dedering and G. Németh. 2002. Recovery of electromyograph median frequency after lumbar muscle fatigue analyzed using an exponential time dependence model. *European Journal of Applied Physiology* 88: 85-93

Farina, D., F. Leclerc, L., and Arendt-Nielsen, et al. 2008. The change in spatial distribution of upper trapezius muscle activity is correlated to contraction duration. *Journal of Electromyography and Kinesiology* 18: 16-25

Fuller, J.R., J. Fung, and J.N. Côté. 2011. Time-dependent adaptations to posture and movement characteristics during the development of repetitive reaching induced fatigue. *Experimental Brain Research* 211: 133-143

Gates, D.H. and J.B. Dingwell. 2008. The effects of neuromuscular fatigue on task performance during repetitive goal-directed movements. *Experimental Brain Research* 187: 573-585

Hägg G.M., A. Luttmann and M. Jäger. 2000. Methodologies for evaluating electromyographic field data in ergonomics. *Journal of Electromyography and Kinesiology* 10: 301-312

Hoffman, M.D., P.M. Gilson, T.M. Westenburg, and W.A. Spencer. 1992. Biathlon shooting performance after exercise of different intensities. *International Journal of Sports Medicine* 13: 270-273

Huysmans, M.A., M.J.M. Hoozemans, and A.J. van der Beek, et al. 2008. Fatigue effects on tracking performance and muscle activity. Journal of *Electromyography and Kinesiology* 18: 410-419

Mathiassen, S.E., J. Winkel and G.M. Hägg. 1995. Normalization of surface EMG amplitude from the upper trapezius muscle in ergonomic studies - a review. *Journal of Electromyography and Kinesiology*, 5: 197-226

Palmerud, G., R. Kadefors, and H. Sporrong et al. 1995. Voluntary redistribution of muscle activity in human shoulder muscles. *Ergonomics* 38: 806-815

Shevlyakov, G. and N. Vilchevski. 2002. Robustness in data analysis. VSP, Utrecht

Suurküla, J. and G.M. Hägg. 1987, Relations between shoulder/neck disorders and EMG zero crossing shifts in female assembly workers using the test contraction method. *Ergonomics* 30: 1553-1564

Takala, E.P. 2002, Static muscular load, an increasing hazard in modern information technology. *Scandinavian Journal of Work, Environment and Health* 28: 211-213

Welch, P.D. 1967. The use of fast Fourier transforms for the estimation of power spectra: A method based on time averaging over short modified periodograms. *IEEE Transactions on Audio and Electro Acoustics* 15: 70-73

CHAPTER 37

Kitting as an Information Source in Manual Assembly

Anna Brolin[1,2], Gunnar Bäckstrand[1,3], Peter Thorvald[1], Dan Högberg[1], Keith Case[1,2]

[1]Virtual Systems Research Centre
University of Skövde
Skövde, Sweden
anna.brolin@his.se

[2]Mechanical and Manufacturing Engineering
Loughborough University
Loughborough, United Kingdom

[3]Research and Development
Swerea IVF
Stockholm, Sweden

ABSTRACT

In manual assembly, a strategy to meet the goal of efficient production is the increased use of kitting as a material supply principle. Even though kitting is already implemented in industry, there are still uncertainties regarding the effects of introducing kits, particularly from a human factors perspective.

This paper presents initial steps in the development of a method to be used for the evaluation of kitting. This from an information source point of view and for studying effects related to productivity and quality. The methodology is projected to act as a foundation for how to carry out a subsequent comprehensive case study. The purpose of the case study is to explore how kitting affects the cognitive workload compared to the ordinary material rack combined with part numbers used in the current manufacturing industry. This is done by measuring productivity; time spent on assembling a product, and quality; number of assembly errors. One step in the methodology development process, which is described in this paper, was to conduct a pilot study, primarily to test the methodology related to the selection of measurement parameters, as well as for getting experiences from running the methodology with real test subjects.

Keywords: manual assembly, kitting, cognitive ergonomics, information use

1   INTRODUCTION

In the automotive industry, well designed and presented information is vital for the assembly personnel to perform effective and accurate assembly operations. Unfortunately this is not the reality in many of the Swedish automotive plants. Due to increased customer demands and global competitiveness the companies have been forced to radically increase the product variation while at the same time becoming more efficient. This in turn has resulted in information overload (Sheridan, 2000; Wilson, 2001; Himma, 2007) that combined with stress, leads to an increased cognitive workload for the assembly workers (Bäckstrand, et al., 2005; Brolin, et al., 2011a).

One primary solution among current Swedish automotive manufacturers is the introduction of philosophies concerning standardisation of operations and eliminating waste throughout the entire production system, such as "The Toyota way" and "LEAN" (Liker, 2004). This has resulted in most of the factories also having developed a need to adapt methods and advanced technology to be able to support their staff in the work towards continuous improvement. However, the technology and methods already exist but due to insufficient knowledge, such as poorly thought out solutions and lack of understanding, the companies have invested in advanced equipment without understanding and investigating the workers' need.

A study was conducted at several Swedish automotive plants with the purpose of exploring which methods and equipment that was used to support the assembly personnel in performing the assembly task (Brolin, et al., 2011b). The observations showed investments such as pick-light, pick-voice, graphic displays, ordinary paper sheets and a material supply principle called kitting, which is currently used widely within the automotive industry. The kitting method was primarily introduced as a logistic tool, mostly due to the expansion of the material racks alongside of the assembly line. The use of kitting can be described as a way of presenting the assembler with a kit of components that together supports one or more assembly operations for a given product (Bozer & McGinnis, 1992; Hanson & Brolin, 2011). However, because of the assemblers' reasoning and experience the kit also functions as a carrier of information that complements or even replaces conventional assembly instructions. The benefit, from a cognitive ergonomics perspective, is that the assembler can focus on the assembly process, i.e. issues related to *how* to assemble, and strongly reduce the attention required for the decision process, i.e. issues related to *what* to assemble. Earlier studies have showed that a decreased cognitive workload can result in increased productivity and quality (Bäckstrand, 2009; Thorvald, 2011). This paper explores a method used to evaluate kitting from an information point of view and the kits possibility to increase productivity and quality.

The purpose of the case study is to explore how kitting affects the cognitive workload compared to the ordinary material rack combined with part numbers used in the current manufacturing industry. This is done by measuring productivity; time spent on assembling a product, and quality; number of assembly errors. One step in the methodology development process was to conduct a pilot study, primarily to test the methodology related to the selection of measurement parameters, as well as for getting experiences from running the methodology with real test subjects.

2   METHOD

The set up consisted of three differently arranged assembly stations where all of the assembly stations used the same assembly product, which contained 37 components (Figure 1). The study was carried out with 18 participants, all engineering staff and students, with three participants assembling simultaneously, one at each station. The two parameters that were explored in this study were productivity and quality.

Figure 1. The assembly product – a LEGO car.

Workstation one emulated a traditional assembly station, presenting material through a material rack including several boxes with attached part numbers indicating a certain component (Figure 2). The material rack also contained components with associated part numbers that was not included in this assembly task, with the purpose to give a more accurate assembly situation as performed in manufacturing plants where product variants are common.

Figure 2. Workstation one – the industry look-a-like.

The assembly instruction for workstation one was illustrated in a traditional way, given on a paper sheet that contained the part numbers in a beforehand decided order for the assembly operation. The instruction included two pictures that showed the result of the assembled product.

Workstation two presented all the relevant components in one box, called an unstructured kit, and used step-by-step pictures as assembly instruction (Figure 3). This set of material presentation suggests that the assembler only has to search for components in one focused area. The instruction was influenced by LEGO-instructions that often are spoken of as clear and easy-to-use instructions.

350

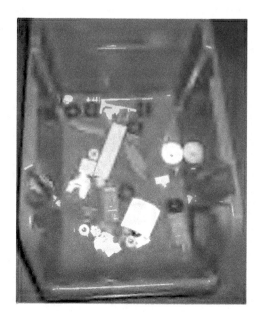

Figure 3. Unstructured kit as used in workstation two.

Workstation three used a similar setup as station two, presenting one box with all the relevant components. However, the box at this station contained separate sections where each component was placed in the same way as the assembly operation, a structured kit (Figure 4). The assembly instructions were the same as in workstation two, but included additional figures attached to the kitting box illustrating the direction of the assembly process.

Figure 4. Structured kit as used in workstation three.

The purpose of the pilot study was to evaluate if the forthcoming case study's experimental design was valid. The case study will follow the same structure; to compare material racks combined with part numbers towards unstructured and structured kits combined with additional step-by-step instructions.

3   RESULTS

The result showed the time it took for each participant to assemble the product (Table 1).

Table 1. Results of the pilot study.

Workstation	Sets (minutes / assembled product)							
	1	2	3	4	5	6	Sum	Average
1. Material rack	*	*	*	*	*	*		
2. Unstructured kit	2,37	3,27	4,51	8,31	3,27	4,32	26,05	4,42
3. Structured kit	2,54	4,27	2,33	2,06	4,00	2,55	17,75	2,96

*Did not assemble within reasonable time (> 9 min)

The pilot study indicates that how material is presented, and thereby information, is of great importance and influences productivity. Further, the few assembly errors indicate that the quality parameter is difficult to use, which shows that productivity should be the only measurement. These contemplations support the experimental design for the upcoming case study and therefore the study will continue.

4   CONCLUSION AND DISCUSSION

One conclusion, based on the results of the pilot study, is that quality cannot be used as a parameter, which was the initial idea. Having just a few assembly errors in the pilot study means that the forthcoming case study needs a tremendous amount of assembled products or a large amount of historic quality data that can be compared with the case study results. Since no previous data exists and there is an uncertainty towards getting a large amount of assembled products, the conclusion is to only use productivity as a measurable parameter. Other metrics that were not identified but had an unclear effect was for example time of day and assembly experience.

Stress is usually a common feature in manual assembly. In order to simulate this aspect during the pilot study, the instructor walked around between the workstations, constantly calling out the current time. The instructor then observed the difference in stress-level among the different workstation. The conclusion was

352

that workstation one, using a material rack, was believed to possess the highest stress level which may be because the assembler is not receiving an overall view of the product and the assembly process due to poorly presented information. However, the stress level at workstation two and three seemed to be the same, situated on a fairly comfortable level.

Another interesting note was the assemblers' reaction after they all had finished the experiment. Each participant was faced towards the assembly object which meant that the participants were not able to see each other or other workstations during the experiment. Afterwards, when facing the other assemblers, the participants that had worked at the workstation with the material rack (station one) were all of the opinion that the other two workstations were perceived as easier. Also interesting is that until then, the participants assembling at station one were of the opinion that themselves were being slow, blaming themselves. Not reflecting on the poor information presented to the assembler.

This was also the case when participants from several Swedish automotive companies performed the pilot study at an industrial workshop. All participated in the study and displayed a great interest, which shows the urgency and importance of improving the productivity as well as the environment for the assembler.

In the forthcoming case study it has been decided that the case study will only study how assemblers handle product variants (i.e. products that differ in their components) and not process variants (i.e. products that are assembled differently but contain the same components). This is due to the argument that the process variants can be dealt with through professional development and training. The product variants on the other hand are harder to learn my heart (Bäckstrand, 2009). Of course it is possible to learn which component to choose for one particular product in a certain assembly situation. However, it is still well known that this can easily and quickly change in the automotive industry due to the arrival of new products and rebalancing of the assembly line, resulting in a quality risk.

ACKNOWLEDGMENTS

The authors would like to thank the participants that executed the pilot study as well as the workshop. This work has also been made possible with the support from Vinnova, Swedish Governmental Agency for Innovation Systems, which is gratefully acknowledged.

REFERENCES

Bozer, Y. A. & McGinnis, L. F. (1992). Kitting versus line stocking: A conceptual framework and a descriptive model. *International Journal of Production Economics*, 28 (1), 1-19.
Brolin, A., Bäckstrand, G., Högberg, D. & Case, K. (2011a). Inadequately designed information and its effect on the cognitive workload. In *Proceedings of the*

International Manufacturing Conference, IMC 28, Institute of Technology, Dublin, Ireland.

Brolin, A., Bäckstrand, G., Högberg, D. & Case, K. (2011b). The use of kitting to ease assemblers' cognitive workload. In *Nordic Ergonomic Society*, Oulu, Finland. pp. 77-82. Nordic Ergonimic Society.

Bäckstrand, G. (2009). *Information Flow and Product Quality in Human Based Assembly*. Loughborough University. Loughborough. Mechanical and Manufacturing Engineering.

Bäckstrand, G., de Vin, L. J., Högberg, D. & Case, K. (2005). Parameters affecting quality in manual assembly of engines. In *Proceedings of the International Manufacturing Conference, IMC 22, Institute of Technology*, Tallaght, Dublin, August. pp. 165-172.

Hanson, R. & Brolin, A. (2011). A comparison of kitting and continuous supply in in-plant materials supply. In *The 4th International Swedish Production Symposium*, Lund, Sweden. pp. 312-321.

Himma, K. E. (2007). The concept of information overload: A preliminary step in understanding the nature of a harmful information-related condition. *Ethics and Information Technology*, 9 (4), 259-272.

Liker, J. K. (2004). *The Toyota way: 14 management principles from the world's greatest manufacturer*. McGraw-Hill Professional.

Sheridan, T. (2000). HCI in Supervisory Control: Twelve Dilemmas. In Elser, P. F., Kluwe, R. H. & Boussoffara, B. (eds.). *Human Error and System Design and Manufacturing*. pp. 1-12: Springer-Verlag London Limited.

Thorvald, P. (2011). *Presenting Information in Manual Assembly*. Doctoral dissertation. Loughborough University. Loughborough. Mechanical and Manufacturing Engineering.

Wilson, T. D. (2001). Information Overload: Implications for Health-care Services. *Health Informatics Journal*, 7 (2), 112-117.

CHAPTER 38

A Biomechanical Evaluation of Dynamic and Asymmetric Lifting Using the AnyBody™ Commercial Software: A Pilot Study

X. Jiang, A. Sengupta

New Jersey Institute of Technology
Newark, USA
xj8@njit.edu, sengupta@njit.edu

ABSTRACT

A six-camera motion capture (mocap) system collected dynamic motion data of lifting 30 lb (13.6 kg) weight at 0°, 30° and 60° asymmetry. The mocap data drove the AnyBody™ model, and the study investigated the effect of the asymmetry. Erector spinae was the most activated muscle for both symmetric and asymmetric lifting. When lifting origin became more asymmetric toward right, erector spinae activity was reduced, but oblique muscles increased their share of activity to counter the external moment. Most muscle tensions peaked at the lift initiation phase except left external oblique and right internal oblique. Left external oblique played a minor role in the right asymmetric lifting task, and the difference of activation for right internal oblique may be due to variance of the motion. Surprisingly the lift asymmetry decreased both compression and shear forces at the L5/S1 joint. This finding contradicted the results obtained from other research studies. The reduction in spine forces is postulated to have resulted from the increased oblique muscles' share in the production of back extensor moment. Since these muscles have longer moment arms, they generated lesser spine force to counteract the external moment. The subject also tended to squat as lifting origin became asymmetric, which effectively reduced the load moment on the spine. This factor might also have contributed to reducing spine forces during asymmetric lifting.

Keywords: biomechanical evaluation, asymmetric lifting, AnyBody™ modeling, spinal forces

1   INTRODUCTION

Asymmetric and dynamic lifting occurs in a great variety of workstations, and it is known to be one of the leading causes of occupational lower back disorders (LBDs). Occupational LBDs is a manifestation from overloading of back extensor muscle and spinal tissues during lifting. Biomechanical modeling has been utilized to investigate lifting task characteristics so that the task demands can be kept within a limit, and internal muscles and joints are not injured.

EMG assisted biomechanical models (McGill and Norman, 1986, Marras and Granata, 1997) were developed under the concept that the muscle tension correlates well with the electrode potential. Another category of model, optimization criterion based, assumes that muscles are recruited in such a way that a criterion function is minimized to reduce a biological cost, such as joint compression force (Schultz and Anderson, 1981, Bean et al., 1988) and muscle fatigue functions (Arjmand and Shirazi-Adl, 2006b, Rasmussen et al., 2001, Chung et al., 1998).

A detailed anatomical model of the lower back is beneficial to both categories of models. Current anatomical models of the lower back can not only consider all major muscle groups relevant in lifting activity, but also the muscle model can differentiate individual muscle fascicles of the individual muscle group (Arjmand and Shirazi-Adl, 2006a, de Zee et al., 2007) with consideration of muscle wrapping against bony structures (McGill and Norman, 1986, Nussbaum and Chaffin, 1996, Arjmand and Shirazi-Adl, 2006a, de Zee et al., 2007).

The AnyBody™ Modeling System is commercially available, optimization criterion based modeling software. It provides by far the most detailed human torso musculoskeletal model. The torso model of AnyBody™ has been utilized effectively to validate internal muscle and joint forces (Grujicic et al., 2010, Wu et al., 2009b, Wu et al., 2009a, Wu et al., 2008), but none of the studies investigated the effect of asymmetric and dynamic aspects of lifting. This study implemented AnyBody™ to analyze internal torso loading in asymmetric and dynamic lifting tasks.

2   METHODS AND MATERIALS

One healthy college student (1.73cm, 75kg) without any history of LBD during the past six months performed asymmetric lifting tasks of 0°, 30° and 60° with 30lb (13.6kg) dumbbell weights, placed evenly in a plastic tray, in OptiTrack™ mocap Laboratory. The lift origin was fixed at knuckle height (99cm off the ground), and at a horizontal distance of 53cm from the center of the tray to the vertical body axis, which was dimensionally identical with Marras and Davis's study (Marras and Davis, 1998), so that the results could be compared. Asymmetric angles were taped on a force plate for feet positioning, including a sagittal symmetric position (0°),

356

30° and 60° to the right of the mid-sagittal plane. The force plate was used to collect ground reaction data during the lifting. The force plate data were not used in this study, but will be used later to check the validity of AnyBody™ model (Figure 1).

Figure 1 Asymmetric lifting task configuration.

Before the experiment, the participant put on the OptiTrack™ medium-size mocap suit. With the help of laboratory assistant, thirty-four reflective markers were attached on the suit based on OptiTrack™ standard thirty-four-marker placement protocol (NaturalPoint Inc., 2011). After standard calibration and skeleton setting up procedure instructed by ARENA™ mocap software (NaturalPoint Inc., 2011), the motion data of lifting were collected through OptiTrack™ six-camera tripod setup (NaturalPoint Inc., 2011) with 100 frames/seconds. A thin metal stand supported the plastic tray with dumbbell weight to prevent marker blocking. During the experiment, the participant performed 0°, 30° and 60° lifting tasks in a randomized order. The participant stood straight with feet along with the tape of pre-defined angle, and lifted from the lift origin to upright position without moving feet.

ARENA™ software automatically filled missing frames less than 20, and smoothed data with cut-off frequency of 6 Hz. Gaps more than twenty frames were filled manually by visual inspection. The ".c3d" files were further trunked to capture the lifting activity only. Approximately between 160 to 220 frames were generated by ARENA™ for individual trials. Figure 2 shows the first frames of 0°, 30° and 60° asymmetric lifting simulated in inverse dynamic study by AnyBody™ model respectively.

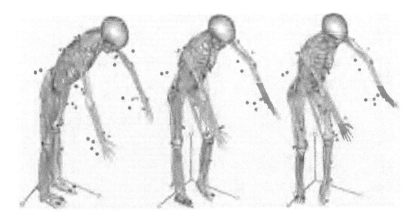

Figure 2 First frames of 0°, 30° and 60° asymmetric lifting initialized in inverse dynamic study by AnyBody™ model.

GaitLowerExtremityProject model in AnyBody's Managed Model Repository1.31 was modified for the experimental task. Because pre-defined marker placement in AnyBody™ is different from reality, parameter and motion optimization algorithm was run before inverse dynamic calculation within AnyBody™ software. On a Sony VAIO® E series laptop computer with 2.2 GHz dual-core CPU and 3GB RAM, inverse dynamic calculation took about 40 second/frame, but parameter and motion optimization lasted for hours depending how accurate the initial marker placement is.

3   RESULTS

3.1   Muscle forces

AnyBody™ models muscle fibers in each muscle fascicles, for example, erector spinae is divided into a total of 29 fascicles on each side (de Zee et al., 2007). To obtain the approximate contribution by each muscle group, the fascicle forces were summed over the normalized duration of lifts (Figure 3).

ES was the most activated muscle for both symmetric and asymmetric lifting. Generally, RES and LES became less activity as the lifting became more asymmetric. Oblique muscles became more active as the lifting became more asymmetric. Majority of the muscles were most active during the lift initiation phase, with exceptions for LEO and RIO. Since at the lift origin the load is farthest from the spine, as well as the upper body is maximally bent, the stronger muscle activity is expected. LEO played a minor role in right asymmetric lifting task, and the difference of activation for RIO may be due to variance of the motion.

However, some observations cannot be properly explained. The zig-zag pattern of oblique activation may be due to the dynamic effect of lifts, resolution of mocap, or error tolerance of AnyBody™ calculation. More data from different subjects

358

need be collected for conclusive results. The more oblique forces for 0° or 30° than 60° at certain instances were also not explainable.

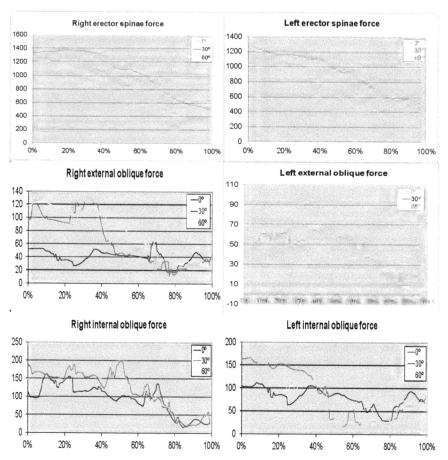

Figure 3 Right erector spinae (RES), left erector spinae (LES), right external oblique (REO), left external oblique (LEO), right internal oblique (RIO), left internal oblique (LIO) force development during the lifting

3.2   L5/S1 joint forces

L5/S1 joint compression, anterior-posterior (A-P) shear and lateral shear forces over the normalized duration of lifts are presented in Figure 4. Compression and A-P shear forces followed the similar pattern, which was identical with ES muscle forces. At the beginning and the end of lifting, the joint loads were steadier than between, probably due to the requirement of movement control. Comparing with 0° lifts, L5/L1 maximum compression force reduced from 3156N to 2963N by 6.1%

and 2888N by 8.5% for 30° and 60° respectively; maximum A-P shear force increased 2.3% to 568N for 30°, but reduced 6.5% to 519N for 60° respectively comparing with 0° from 555N; absolute lateral shear force reduced from 52.6N for 0° to 44.6N by 15.2% and to 23.8N by 54.8% for 30° and 60° respectively. In general, joint forces reduced as lifting origin became more asymmetric.

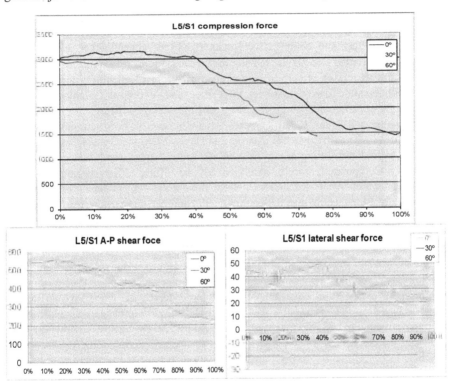

Figure 4 L5/S1 compression, anterior-posterior (A-P) shear and lateral shear forces during the lifting

4   DISCUSSION & CONCLUSION

ES is the main extensor of trunk. When the ES fascicles of one side act together, they produce combined lateral flexion and rotation to the same side (Palastanga et al., 2002). During asymmetric lifts, the support of the external load is shifted from the large ES muscles to smaller, less capable oblique muscles (Marras and Mirka, 1992). Biomechanically, ES has smaller moment arm than oblique muscles referring to lumber joint, so ES is less efficient to support external moment generated by upper body weight and hand loads. When the support of the external moment shifts from ES to oblique muscles, which also means shifting to more efficient muscles, the joint forces should reduce. However, oblique muscles are much weaker than ES, so they are less activated during symmetric lifting to minimize muscle fatigue. Furthermore, from observation (Figure 2), the participant tended to squat more as lifting origin became more asymmetric, which may also be a strategy of our body to reduce joint forces.

According to NIOSH (NIOSH, 1981), the tolerance level for compression loading of the spine is expected to be around 3400 N. At this level of compression, micro fractures of the vertebral endplate begin to occur. The threshold limits for spine lateral and A-P shear are probably less than 900 N (Marras and Davis, 1998). Reducing A-P shear and compressive forces should be considered a priority to prevent LBDs (Marras and Davis, 1998). In this study, joint forces did not exceed the limitation. However, if certain factors such as lifting speed, lifting height and lifting weight become more demanding, joint forces may exceed the tolerance level, and long time working under those circumstances may develop LBDs.

The average maximum L5/S1 compression force derived from ten subjects by Marras et al.'s EMG assisted model (Marras and Davis, 1998) was 3600N, 3900N and 4050N for asymmetric lifting of 0°, 30° and 60° toward right, which was presented graphically. Compression forces increased as the lifting origin became more asymmetric, which was contradicted with this study. A-P shear force was approximately 910N, 850N and 830N for 0°, 30° and 60° asymmetry respectively. Comparing with this study, both A-P shear force deceased as the lifting origin became increasingly asymmetric, but the force predicted by Marras et al. was about 350N higher than this study. Lateral shear force predicted by them ranged from 210N to 350N, which was far higher than the values predicted by AnyBody™ in this study. Generally, they found compression and lateral shear forces increased as the lift origin became more asymmetric, whereas A-P shear force decreased. The EMG assisted model is based on the assumption that the muscle tension correlates well with the electrode potential. This assumption should provide that the muscle contraction is isometric, that is muscle fiber lengths remain unchanged during force production (NIOSH, 1992). However during dynamic situation, when muscle fibers generates force as well as change their lengths, sliding action of muscle fibers underneath the fixed surface electrodes, also induces electrode potential (NIOSH, 1992). Unless the dynamic part of the electrode potential is separated from the gross electrode potential, EMG may not accurately estimate force generation by the muscle fibers.

Commercially available AnyBody™ biomechanical model provides by far the most detailed human anatomical model, which is driven by criterion optimization algorithm. To our knowledge, the model has been used for the first time to evaluate dynamic and asymmetric lifting. ES was the most activated muscle for both symmetric and asymmetric lifting. When lifting origin became more asymmetric toward right, ES activity was reduced, but oblique muscles increased their share of activity to counter the external moment. Most muscle tensions peaked at the lift initiation phase except left external oblique and right internal oblique. Surprisingly the lift asymmetry decreased both compressive and shear forces at the L5/S1 joint. This finding contradicted the results obtained from other research studies. More data from different subjects should be collected for a conclusive results, and force plate data should be used later to check the validity of AnyBody™ model.

REFERENCES

ARJMAND, N. & SHIRAZI-ADL, A. 2006a. Model and in vivo studies on human trunk load partitioning and stability in isometric forward flexions. Journal of Biomechanics, 39, 510-521.

ARJMAND, N. & SHIRAZI-ADL, A. 2006b. Sensitivity of kinematics-based model predictions to optimization criteria in static lifting tasks. Medical Engineering and Physics, 28, 504-514.

BEAN, J. C., CHAFFIN, D. B. & SCHULTZ, A. B. 1988. Biomechanical model calculation of muscle contraction forces: A double linear programming method. Journal of Biomechanics, 21, 59-66.

CHUNG, M. K., SONG, Y. W., HONG, Y. & CHOI, K. I. 1998. A novel optimization model for predicting trunk muscle forces during asymmetric lifting tasks. International Journal of Industrial Ergonomics, 23, 41-50.

DE ZEE, M., HANSEN, L., WONG, C., RASMUSSEN, J. & SIMONSEN, E. B. 2007. A generic detailed rigid-body lumbar spine model. Journal of Biomechanics, 40, 1219-1227.

GRUJICIC, M., PANDURANGAN, B., XIE, X., GRAMOPADHYE, A. K., WAGNER, D. & OZEN, M. 2010. Musculoskeletal computational analysis of the influence of car-seat design/adjustments on long-distance driving fatigue. International Journal of Industrial Ergonomics, 40, 345-355.

NATURALPOINT INC., N. 2011. ARENA™ Tutorial Videos [Online]. Available: http://www.naturalpoint.com/optitrack/products/motion-capture/tutorials/arena/ [Accessed 4/20 2011].

MARRAS, W. S. & DAVIS, K. G. 1998. Spine loading during asymmetric lifting using one versus two hands. Ergonomics, 41, 817-834.

MARRAS, W. S. & GRANATA, K. P. 1997. The development of an EMG-assisted model to assess spine loading during whole-body free-dynamic lifting. Journal of Electromyography and Kinesiology, 7, 259-268.

MARRAS, W. S. & MIRKA, G. A. 1992. A comprehensive evaluation of trunk response to asymmetric trunk motion. Spine, 17, 318-326.

MCGILL, S. M. & NORMAN, R. W. 1986. 1986 Volvo award in biomechanics: Partitioning of the L4-L5 dynamic moment into disc, ligamentous, and muscular components during lifting. Spine, 11, 666-678.

NIOSH 1981. Work practices guide for manual lifting, NIOSH Technical Report No. 81-122.

NIOSH 1992. DHHS (NIOSH) Publication No. 91-100: Selected Topics in Surface Electromyography for Use in the Occupational Setting: Expert Perspective.

NUSSBAUM, M. A. & CHAFFIN, D. B. 1996. Development and evaluation of a scalable and deformable geometric model of the human torso. Clinical Biomechanics, 11, 25-34.

PALASTANGA, N., FIELD, D. & SOAMES, R. 2002. Anatomy and Human Movement.

RASMUSSEN, J., DAMSGAARD, M. & VOIGT, M. 2001. Muscle recruitment by the min/max criterion - A comparative numerical study. Journal of Biomechanics, 34, 409-415.

SCHULTZ, A. B. & ANDERSON, G. B. J. 1981. Analysis of loads on the lumbar spine. Spine, 76-82.

WU, J. Z., AN, K. N., CUTLIP, R. G., KRAJNAK, K., WELCOME, D. & DONG, R. G. 2008. Analysis of musculoskeletal loading in an index finger during tapping. Journal of Biomechanics, 41, 668-676.

WU, J. Z., CHIOU, S. S. & PAN, C. S. 2009a. Analysis of musculoskeletal loadings in lower limbs during stilts walking in occupational activity. Annals of Biomedical Engineering, 37, 1177-1189.

WU, J. Z., DONG, R. G., MCDOWELL, T. W. & WELCOME, D. E. 2009b. Modeling the finger joint moments in a hand at the maximal isometric grip: The effects of friction. Medical Engineering and Physics, 31, 1214-1218.

CHAPTER 39

Real-time Measuring System of Eye-gaze Location and Writing Pressure in Calligraphy

Atsuo MURATA, Kosuke INOUE, Takehito HAYAMI and Makoto MORIWAKA

Graduate School of Natural Science and Technology, Okayama University
Okayama, Japan
murata@iims.sys.okayama-u.ac.jp

ABSTRACT

In this study, a basic system which measures simultaneously a line of eye-gaze and a hand movement was developed in order to extract the skilled elements in calligraphy. This system consists of an eye mark recorder and a three-dimensional position measurement device. The coordinates of a line of eye-gaze and a hand movement were unified as a common coordinate system. Even if the head of a participant is not fixed during the measurement, this system enables us to calculate an eye position by preliminarily calibrating the physical relationship between a camera and a receiver of three-dimensional measurement device. To confirm the effectiveness of the developed system and to examine the accuracy of the system, the participants were required to look at fixation points under a different condition of the head posture. Using this system, calligraphic tasks were imposed on participants. Moreover, the writing pressure measurement system was developed in order to add to the simultaneous measurement system of eye-gaze location and the brush tip movement. We made an attempt to extract skilled elements in calligraphy on the basis of the writing pressure and the eye-hand coordination characteristics.

Keywords: eye-hand coordination, calligraphy, real-time measurement system, skilled element, eye movement, writing pressure measurement.

1   INTRODUCTION

In order to understand the skill in calligraphy and make use of this for the proposal of an effective instruction method, the investigation of hand-eye coordination must be essential. Although there seem to be many approaches that made an attempt to develop an effective calligraphy instruction method, few studies paid attention to a hand-eye coordination skill. In calligraphy, we are much interested in how we gaze at paper and draw characters on paper.

Tchalenko et al., 2003 and Miall and Tchalenko, 2001 investigated eye movement and voluntary control in portrait drawing, and compared eye-hand coordination between skilled painters and novices. They concluded that such an approach would be effective for the understanding of portrait drawing skill. In Proctor and Dutta, 1995 and Singley and Anderson, 1989, it is mentioned from the viewpoint of cognitive science that the coordination of visual input, interpretation and motor system is important to acquire skills. Vickesr, 1995 compared eye-gaze location between expert and novice basketball shooters, and found that mean frequency, duration, location and onset of eye-gaze differed between two groups. In Vickesr, 1995, however, the eye-hand coordination has not been discussed.

An analysis of skill acquisition process is a very important in the field of cognitive science. In order to achieve the traditional skills efficiently, the research focusing on the cognitive processes is necessary and indispensable.

This study pays attention to the relationship between eye and hand movement as one element of the skill in calligraphy. Therefore, a system that can simultaneously measure the brush tip movement and the eye gaze position during calligraphy has been developed and has been verified the potential for its application to calligraphy.

Although Murata et al., 2009 made an attempt to simultaneously measure the brush tip movement and the eye gaze position during calligraphy, this system imposed a constraint that the head movement was not permitted at all on participants. The constraint made it difficult to apply the developed system to the extraction of skilled calligraphy element in eye-hand coordination for longer duration. Therefore, the measurement system that permits participants to move their head to some extent must be developed in order to extract skilled elements in a systematically designed experiment for a large number of participants.

In this study, a system that can simultaneously measure the movement of a brush tip and the eye-gaze position during a calligraphy task has been developed in order to understand the calligraphy skill. Using this measurement system, the skill elements of calligraphy were explored and extracted from the viewpoint of eye-hand coordination. We judged that not only the eye-hand coordination but also the writing pressure during calligraphy is important to understand the difference of skills between novices and experts in calligraphy. Therefore, the writing pressure measurement system was added to the developed system. A system that can simultaneously measure the writing pressure of brush tip and the eye-hand coordination during calligraphy was used to extract the skilled elements in a calligraphic task from the viewpoint of eye-hand coordination and the modulation of writing pressure.

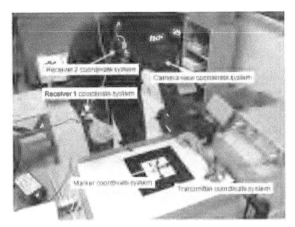

Figure 1　Coordinate systems in the developed system.

2　MEASURING SYSTEM

First, we describe the simultaneous measurement system of eye gaze position and hand movement.

2.1　System Configuration

The system consists of a wearable type eye mark recorder (Nac Image Technology, EMR-9) and a movement tracking system using electromagnetic field (Polhemus, FASTRAK). The wearable type eye mark recorder consists of a head unit equipped with the view camera and the eyeball camera. This system is not influenced by the head sway, and the eye position be captured at any time. FASTRAK calculates relative position of the transmitter and the receiver. Therefore, the system can expressed the movement of the head and the hand in the same coordinate system by coordinate transformation.

The transmitter was installed in the upper left corner on the work space, and the receivers were fixed to the end of brush (Receiver1) and the side of eye mark recorder head unit (Receiver2).

2.2　Unification of coordinate system

This section describes the method to unify each coordinate system to one coordinate system. Each coordinate system on the system configuration is demonstrated in Figure 1.

The eye gaze and the brush tip position should be treated in the common coordinate system. There are expressed by converting each data from a receiver coordinate system to a transmitter coordinate system. Using transformation

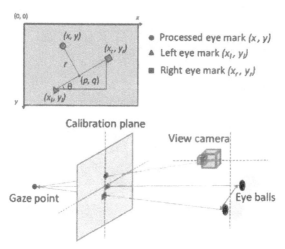

Figure 2 Geometrical relation between eye tracking device and each eye marks on the camera screen coordinate.

matrix ^tT_r including an elements of a rotation $R_{3\ 3}$ and a translation $P_{3\ 1}$ in a simultaneous coordinate system makes this possible as follow.

$$\begin{bmatrix} X_t \\ Y_t \\ Z_t \\ 1 \end{bmatrix} \quad {}^tT_r \begin{bmatrix} X_r \\ Y_r \\ Z_r \\ 1 \end{bmatrix} \qquad (1)$$

$(X_t\ Y_t\ Z_t\ 1)^t$ and $(X_r\ Y_r\ Z_r\ 1)^t$ are vectors of the same point in the transmitter and the receivers coordinate system. Conducting these calculations, each data can be unified and expressed in the transmitter coordinate system if the position of brush tip and eye gaze are expressed in the receiver coordinate system.

2.3   Method to calculate brush tip and eye-gaze positions

The unified coordinate system can easily express the position of brush tip in the receiver coordinate system because the receiver is attached to the end of brush.

It can be obtained by designing so that brush tip may locate at 1 cm in y axis of the receiver coordinate system. The left and right eye mark can be acquired in the camera screen coordinate system of EMR-9. From the geometrical relationship shown in Figure 2, the eye mark can be obtained after binocular disparity processing as follows.

$$x = p + r \sin \theta \qquad (2)$$
$$y = q + r \sin \theta \qquad (3)$$

As the eye mark after binocular disparity processing is a coordinate value of the camera screen coordinate system, it is necessary to unify it to the receiver coordinate system. It is possible to convert it into the receiver2 coordinate system by using the transformation matrixes defined by Eq.(1).

3   METHOD

Two experiments were conducted to confirm the effectiveness of the measurement system. One confirmed the accuracy of method to calculate the brush tip and eye gaze locations. The other measured eye-hand coordination in calligraphy and verified the potential as a means to extract skilled element necessary for improving calligraphy performance.

3.1   Participants

Twenty-two undergraduate or undergraduates (from 21 to 26 years old) participated in the experiment. They were all healthy and had no orthopedic or neurological diseases. Twelve were expert who had learned calligraphy at least for twelve tears. Ten were novices who had never learned calligraphy.

3.2   Experiment-Accuracy evaluation of system-

3.2.1   Design and procedure

The success of measurement of the gaze positions are greatly influenced by the accuracy of the eye tracking device. Three accuracy evaluations were preliminarily carried out. Firstly, the calculation accuracy of coordinate transformation (camera screen to transmitter) was evaluated in mouse clicking operations to the five points (X, Z) = A(100, 50), B(300, 50), C(200, 150), D(100, 250), E(300, 250) on the camera screen. Secondly, the calculation accuracy of eye gaze in fixed head position was evaluated by participants to watch the five points on the work space. Thirdly, the calculation accuracy of eye gaze swing head was evaluated by requiring participants to keep watching the one point C(200, 150) on the work space.

3.2.2   Compensation of the system

The accuracy of coordinate transformation was within ±5mm for all of five points. The accuracy of eye gaze measurement in the fixed head position was within ±25mm for all of five points. The accuracy of eye gaze in the head swing condition was within ±20mm.

3.3   Experiment -Application to calligraphy-

The eye-hand coordination was measured using the developed system to confirm whether this system is useful to extract skilled elements in calligraphic tasks. The task was to write predetermined six kinds of Kanji characters using a brush. Writing Kanji requires us to compose some strokes according to the type

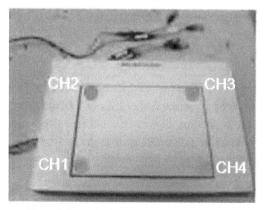

Figure 3 Apparatus for measuring writing pressure.

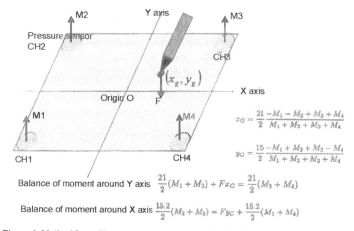

$$x_G = \frac{21}{2}\frac{-M_1 - M_2 + M_3 + M_4}{M_1 + M_2 + M_3 + M_4}$$

$$y_G = \frac{15}{2}\frac{-M_1 + M_2 + M_3 - M_4}{M_1 + M_2 + M_3 + M_4}$$

Balance of moment around Y axis $\quad \frac{21}{2}(M_1 + M_2) + Fx_G = \frac{21}{2}(M_3 + M_4)$

Balance of moment around X axis $\quad \frac{15.2}{2}(M_2 + M_3) = Fy_G + \frac{15.2}{2}(M_1 + M_4)$

Figure 4 Method for writing pressure measurement and calculation of center of gravity.

of Kanji. The writing pressure was also measured during the calligraphy task using the apparatus in Figures 3 and 4.

4   RESULTS

4.1   Preliminary experiment-Accuracy evaluation of system-

The accuracy of coordinate transformation and eye gaze measurement in the fixed head position is shown Figure 5.

4.2   Experiment-Application to calligraphy-

Example of the positions of each brush tip stroke and eye-gaze are plotted

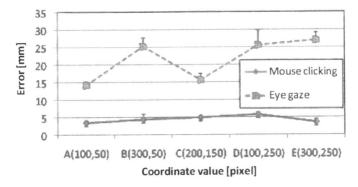

Figure 5 Error as function of pointing type and target point.

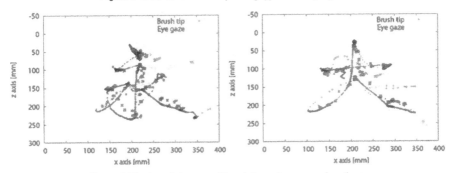

Figure 6 Display of changes of brush tip and eye-gaze locations.

shown in Figure 6. Other examples of each brush tip stroke and the corresponding eye-gaze are shown in Figures 7 and 8.

4.3   Writing pressure during calligraphy

The writing pressure was simultaneously measured with the eye-hand coordination using apparatus shown in Figure 3. In Figure 4, the method for measuring writing pressure and calculation of COG (Center of Gravity) is summarized. The writing pressure differed between novices and experts. It tended that the writing pressure of experts was more variable than that of novices.

5   DISCUSSION

5.1   Preliminary experiment-Accuracy evaluation of system-

The accuracy of measurement system was found be remarkably affected by the accuracy of eye camera (eye mark recorder). A further elaboration of measurement accuracy of the eye camera should be necessary to assure higher accuracy in eye-hand coordination measurement system.

370

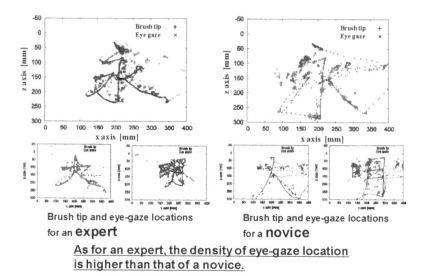

Brush tip and eye-gaze locations for an **expert**

Brush tip and eye-gaze locations for a **novice**

As for an expert, the density of eye-gaze location is higher than that of a novice.

Figure 7 Comparison of density of eye-gaze locations between an expert and a novice.

5th stroke when writing "永"(forever)

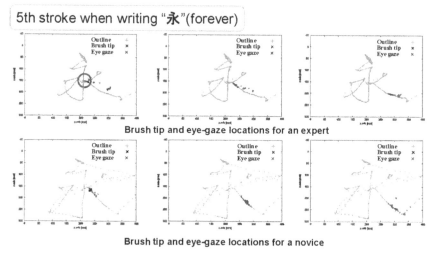

Brush tip and eye-gaze locations for an expert

Brush tip and eye-gaze locations for a novice

As for an expert, the location to be written next is confirmed beforehand. On the other hand, a novice always tends to gaze at a brush tip.

Figure 8 Comparison of relationship between brush tip and eye-gaze locations between an expert and a novice.

5.2   Experiment-Application to calligraphy-

As pointed out by Pelz et al., 2001, in skilled tasks such as portrait drawing and calligraphy, eye-hand coordination must play an important role in skill acquisition. Vickesr, 1995 pointed out that eye movement characteristics of skilled basketball player are different from that of novice players. Unfortunately, Vickers, 1995 did not discuss the hand movement in accordance with the eye movement. On the other hand, Tchalenko et al., 2003 and Miall and Tchalenko, 2001 discussed the eye-hand coordination in drawing. Inferring from these studies, we assumed that effective eye-hand coordination is essential in calligraphy, and developed a systems that can simultaneously measure the locations of eye-gaze and brush tip. Such a system enabled us to explore the characteristics of eye-hand coordination in more detail and systematically.

As shown in Figure 7, the density of eye-gaze location for an expert tended to be higher than that of a novice. This must mean that an expert concentrates more on the Kanji character than a novice during a calligraphy task. A novice tended to gaze at other sites such as an ink container other than the paper to be written on.

It seems that the eye-hand coordination strategy differs between the skilled and the novice participant (See Figure 8). The eye movement characteristics between skilled and novice participants differs not only during the writing task but also before the writing task. The skilled expert tends to look at the wider area of the paper before beginning to write a Kanji character. The novices, on the other hand, tend to look at only the area around 1st stoke. From this finding, we can conclude that skilled experts image the output of a Kanji character before starting to write, and such a skill may lead to skilled Kanji calligraphy.

Moreover, the skilled participant doesn't look at the stroke which he is trying to write, but look at the next stroke. In other words, experts seem to have a more predictive skill in calligraphy. In the baseball game, it seems that a skilled infielder or outfielder has such a predictive skill, and can catch up with a ball faster than a novice player. Different from the skilled participant, the novice consistently looks only at the start point of each stroke. This is indicative of the less predictive characteristics of novices.

5.3   Writing pressure during calligraphy

As well as the eye-hand coordination characteristics, the difference of writing pressure was observed between novices and experts. The writing pressure of experts tended to be more variable than that of novices, while the writing pressure of novices tended to be less variable and monotonous. In calligraphy, the modulation of writing pressure is, in general, essential for creating unique and artistic works. The more skill one gets, the more modulation one's work has. With the predictive characteristics in eye-hand coordination above, the modulation was found to play an important role in calligraphy.

To understand the knowledge of eye-hand coordination task, we developed a system that can measure unified the gaze location and the hand position as a coordinate system simultaneously. The developed system allowed head movement to some extent. A preliminary experiment was conducted to confirm the accuracy of developed system. As for the accuracy of eye gaze, the measurement error of the eye tracking device was more influential than the error caused by the coordinate transformation. Moreover, the calligraphic tasks were carried out to demonstrate the effectiveness of developed system. Using this system, participant's eye gaze position and hand movement are simultaneously measured in the same coordinate system. The developed system might be promising to extract skilled element that promote quick skill acquisition.

Future research should examine the process of skill acquisition by means of the developed system for simultaneously measuring the eye-hand coordination and the corresponding writing pressure. The process or stage of learning should be divided in more detail in future research. An effective method for learning calligraphy should be proposed on the basis of the extracted results necessary for skill acquisition: the predictive eye-hand coordination and the modulation of writing pressure. A method to master such a calligraphy skill in a short time should be proposed by providing learners with both the eye-gaze location and the corresponding writing pressure somehow.

REFERENCES

Carpenter,R.H.S. 1991. Eye Movements, Macmillan Press.

Miall,R.C. and Tchalenko,J. 2001. A painter's eye movements: A study of hand and eye movement during portrait drawing, Leonard 34(1): 35-40.

Murata,A. and Moriwaka,M. 2009. Skill of Eye-Hand Coordination in Calligraphy - Difference of Skill of Hand-Eye Coordination between Expert and Novice-, Proc. of Fifth International Workshop on Computational Intelligence & Applications: 316- 319.

Pelz,J., Hayhoe,M. and Loeber,R. 2001. The coordination of eye, head, and hand movements in a natural task, Experimental Brain Research 139(3): 266- 277.

Proctor,R.W. and Dutta,A. 1995. Skill Acquisition and Human Performance, SAGE Publications.

Singley,M.K. and Anderson,J.R. 1989. The Transfer of Cognitive Skill, Harvard University Press.

Tchalenko,J., Dempere-Marco,L., Hu,X.P. and Yang,G.Z. 2003. Eye movement and voluntary control in portrait drawing, In The Mind's Eye-Cognitive and Applied Aspects of Eye Movement Research, eds. Hyona,J., Radach,R. and Deubel,H., Elsevier: 705-727.

Vickesr,J. 1995. Gaze Control in Basketball Foul Shooting, In Eye Movement Research: Mechanisms, Processes, and Applications, eds. Finley,J., Walker,R. and Kentridge,R., Elsevier: 527-541.

CHAPTER 40

Use of a Monocular See-Through Head-Mounted Display while Walking: A Comparison of Different User Interfaces

Kazuhiro Tanuma, Kenta Kurimoto*,
Makoto Nomura**, Miwa Nakanishi**

*Keio University, Yokohama, Japan,
kaz0414@gmail.com, kenta.kurimoto@gmail.com,
miwa_nakanishi@ae.keio.ac.jp
**Brother Industries, Ltd., Nagoya, Japan
makoto.nomura@brother.co.jp

ABSTRACT

Optical see-through head-mounted displays (OSDs) enable viewing of digital images overlaid on the real world. Because of their hands-free advantage, OSDs are expected to be introduced in the industry as task support tools. In this study, we compared how well users could visually recognize the real world while walking and referring to information using OSDs and conventional media to utilize OSDs safely and efficiently. From the results of the experiment, we verified that OSDs enable an adequate level of awareness and accuracy. In particular, we found that the awareness of central and upper visual fields was greater using OSDs than conventional media. These results suggest that the performance of an OSD as a reference medium while the user is in motion is acceptable, and particularly, the awareness of the front view is higher using OSDs than conventional media.

Keywords: Optical see-through head-mounted display, Users' safety, Awareness of visual field.

1   INTRODUCTION

An optical see-through head-mounted display (OSD), a new type of display that has been recently commercialized, enables users to view digital images overlaid on the real world. It is expected to be applied in various areas because of its high-resolution full-color projection and a small and light body. Because OSDs offer advantages such as hands-free and see-through features, they are being considered particularly for use as task support tools for industrial tasks that require using both hands. In fact, our previous research found that when workers performed wiring tasks by referring to a manual displayed by an OSD, human error decreased remarkably and task efficiency increased by more than 15% compared to using a paper manual (Nakanishi et al., 2006). OSDs have more advantages compared with conventional media, and to utilize them in many situations safely and efficiently, a greater consideration of users' safety is required. We studied the safety of referring to information displayed by OSDs while a wearer was walking, which is a significant problem relating to the environment in which the OSD is used, the range of use, and users' freedom of movement.

In this study, OSDs are compared with conventional media to determine how well users can recognize the real world while walking and referring to information from each medium.

2   METHOD

2.1   Experimental Outline

In this study, assuming that users refer to task-related information whenever necessary while they are walking in industrial work environments, we carried out an experiment in which participants referred to information displayed on an OSD while walking on a treadmill (Figure 1). To examine the users' awareness of their visual field, we fixed light-emitting diodes (LEDs) in a grid-like pattern throughout the participants' field of vision, based on typical human binocular visual perception. We asked them whether they could detect when the LEDs flashed. We prepared four experimental conditions that used various media including the OSD (Figure 2 and Table 1).

Table 1. Experimental conditions

Condition	Media	Background color
OSD (B)	Optical see-through head-mounted	Black
OSD (W)	display (Brother Industries, Ltd.)	White
PDA	iPod (Apple Inc.)	Black
Paper	A4 paper printed on double faces	White

2.2 Participants

Twenty male students and four female students participated in the experiment. Their average age was 22.2 and their average height was 169.6 cm. All of them had normal vision. We obtained the informed consent of the individuals who agreed to participate in this experiment.

2.3 Experimental Environment and Apparatus

Figure 3 shows the overhead view of the experimental environment. The participants, holding media in their hand (or wearing OSDs on their nondominant eye), walked on the

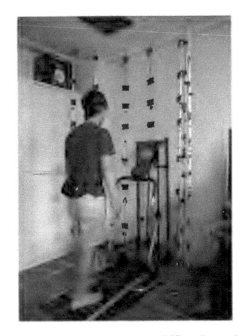

Figure 1 Participant wearing an OSD walks on the treadmill

treadmill (AFW3009, Alinco Inc.). While walking, they watched a 10.1″ tablet PC (ICONIA TABW500, Acer Inc.) fixed in front of them at the center of their visual field. Around the participants, 46 LED panels (NT-16, EK Japan Co., Ltd., Figure 4) were fixed in a grid-like pattern.

Figure 2 Participant wearing an OSD

376

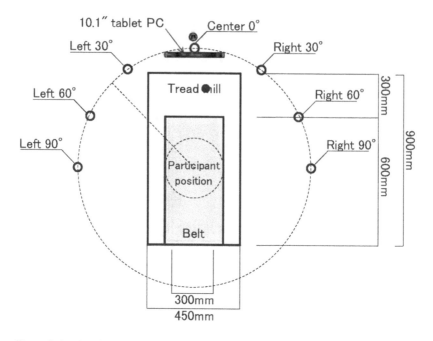

Figure 3 Overhead view of the experimental environment. Four or seven LEDs were fixed on

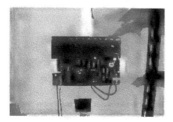

Figure 4 An LED panel

Figure 5 Text content displayed on the tablet PC

2.4　Experimental Task

The participants watched the tablet PC from the center of their visual field while walking on the treadmill at 0.8 km/h. The tablet PC displayed text content for 2 min, and the text changed every 3 s (Figure 5). Also, each medium, including the OSDs, synchronously displayed similar content. The participants compared the text content displayed on the two devices whenever a trigger randomly appeared on the tablet PC. If they perceived that the text content was different, the participants touched the tablet PC. Participants were asked to report whether they detected any one of the 46 LEDs flashing randomly for 1 s during that task. The above process was repeated four times using two types of OSDs and PDAs and printed media as conventional media. In addition, we repeated another process four times as a control in which the participants reported whether they detected flashing LEDs without using any medium while walking on the treadmill.

2.5 Data

The time taken to touch the tablet PC was recorded to calculate the accuracy of reading the contents, and the time and position of the flashing LEDs detected by the participants were recorded to calculate the rates at which they detected the flashing LEDs.

3 RESULTS

3.1 Detection rate of flashing LEDs

Figure 6 shows the detection rate of the flashing LEDs of each medium. At most positions, there were no differences between the rates at which the flashing LEDs were detected by participants using OSDs and other media. However, a chi-square test applied to the detection rate using each medium showed a difference between the detection rate of participants using OSDs and other media at the two leftmost LEDs in the peripheral visual field. These LEDs were located at 90 deg left and 0 deg depression and 90 deg left and 30 deg depression (Figures 7 and 8). We considered that the OSD equipment that the participants wore on their left eye reduced their visual field.

Based on this consideration, we carried out an additional experiment that was a modification of the previous tasks. The triggers that were displayed randomly on the tablet PC were removed, and the participants compared the text content displayed on the two devices each time the content was changed.

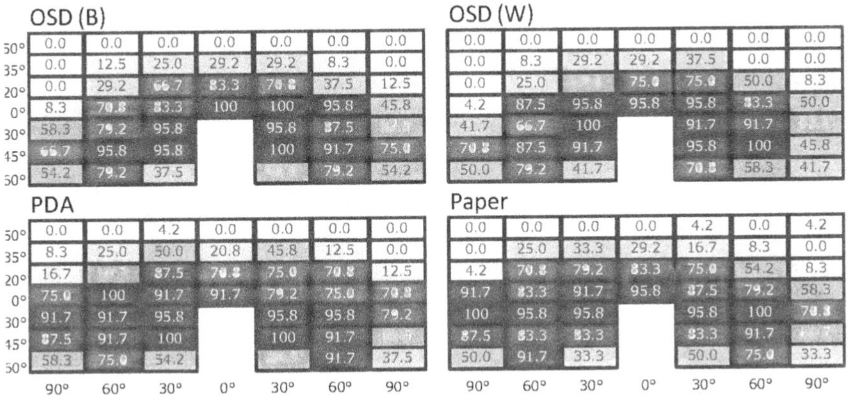

Figure 6 Detection rates of the flashing LEDs (unit: %)

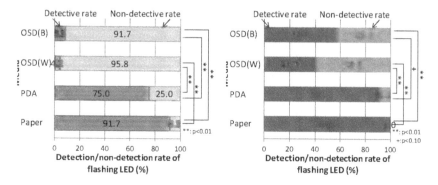

Figure 7 Detection rates of the flashing LEDs at 90 deg left and 0 deg depression

Figure 8 Detection rates of the flashing LEDs at 90 deg left and 30 deg depression

Figure 9 shows the detection rates of the flashing LEDs when the participants used each medium during this experiment. All detection rates during this experiment were lower than those in the previous experiment, and at most of positions, there were no differences between the detection rates of the flashing LEDs for participants using OSDs and other media. As in the previous experiment, we applied a chi-square test to the detection rates of the participants using each medium. We found that the detection rates differed when the participants used OSDs and other media at the same two positions (Figures 10 and 11). In addition, we found a difference at the center of visual field that was located at 0 deg left/right and 20 deg depression (Figure 12). We considered that this difference occurred because the OSDs gave the participants the advantage of viewing information overlaid in the center of their visual field.

Furthermore, aside from the above experiments, we carried out another experiment in which the participants wore Zyl eyeglasses (Figure 13) instead of OSDs and walked on the treadmill while watching the tablet PC from the center of their visual field. In this condition, when the detection rate of any one of 46 LEDs

OSD(B)

	90°	60°	30°	0°	30°	60°	90°
50°	0.0	0.0	0.0	0.0	0.0	0.0	4.2
35°	0.0	0.0	16.7	8.3	0.0	0.0	4.2
20°	0.0	20.8	50.0		33.3	8.3	0.0
0°	8.3	58.3				75.0	16.7
30°	50.0	54.2			8.3		25.0
45°	62.5	75.0				8.3	33.3
60°	41.7	62.5	37.5		50.0	45.8	25.0

OSD(W)

	90°	60°	30°	0°	30°	60°	90°
50°	0.0	4.2	0.0	0.0	0.0	0.0	0.0
35°	4.2	8.3	12.5	4.2	4.2	0.0	0.0
20°	0.0	20.8	45.8	45.8	33.3	4.2	0.0
0°	12.5	50.0			33.3	75.0	16.7
30°	37.5	62.5					25.0
45°	58.3	70.8				75.0	29.2
60°	37.5	45.8	45.8		50.0	50.0	12.5

PDA

	90°	60°	30°	0°	30°	60°	90°
50°	0.0	0.0	0.0	0.0	0.0	4.2	0.0
35°	0.0	0.0	4.2	0.0	4.2	0.0	0.0
20°	0.0	37.5	37.5	16.7	33.3	25.0	4.2
0°	54.2	70.8			66.7	66.7	20.8
30°							33.3
45°							45.8
60°	33.3	50.0	16.7		58.3	66.7	33.3

Paper

	90°	60°	30°	0°	30°	60°	90°
50°	0.0	0.0	0.0	0.0	0.0	4.2	0.0
35°	0.0	0.0	0.0	4.2	0.0	0.0	4.2
20°	0.0	29.2	33.3	16.7	12.5	0.0	4.2
0°	33.3	70.8	66.7	79.2	70.8	54.2	20.8
30°	87.5	87.5	95.8		83.3	95.8	45.8
45°	70.8	75.0	62.5		70.8	87.5	54.2
60°	41.7	41.7	12.5		12.5	37.5	37.5

Figure 9 Detection rates of the flashing LEDs during the additional experiment (unit: %)

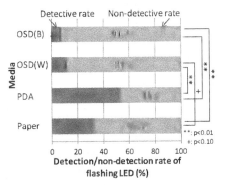

Figure 10   Detection rates of the flashing LEDs at 90 deg left and 0 deg depression during the additional experiment

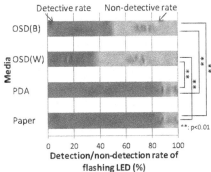

Figure 11   Detection rates of the flashing LEDs at 90 deg left and 30 deg depression during the additional experiment

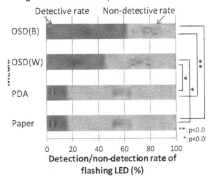

Figure 12   Detection rates of the flashing LEDs at 0 deg left/right and 20 deg depression during the additional experiment

Figure 13 Zyl eyeglasses

flashed randomly was examined as in the previous experiment, we found a difference between the detection rates of the participants using the different devices for the same two leftmost LEDs in the peripheral visual field (Figures 14 and 15). This result indicates that detection rate reduces at some level by using standard eyeglasses.

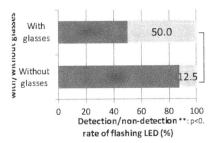

Figure 14 Detection rates of the flashing LEDs at 90 deg left and 0 deg depression with and without glasses

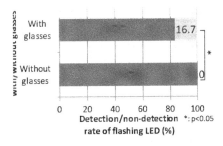

Figure 15 Detection rates of the flashing LEDs at 90 deg left and 30 deg depression with and without glasses

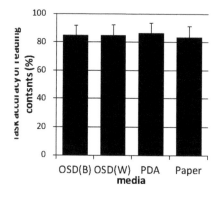

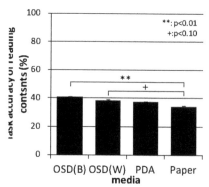

Figure 16 Task accuracy of reading the contents of each medium

Figure 17 Task accuracy of reading the contents of each medium during the additional experiment.

3.2   Task accuracy of reading contents

Figure 16 shows the task accuracy of reading contents obtained using each medium. We found no difference in the accuracy when participants used OSDs and conventional media. This result indicates that OSDs enable users to read contents while walking as accurately as when they use conventional media. However, when we removed the additional experiment in which triggers were displayed on the tablet PC, we found a difference between the task accuracies of reading contents of OSDs and printed papers (Figure 17). From this result, we can infer that the task accuracy of reading contents of OSDs while walking is higher than when reading printed media.

4   CONCLUSIONS

In this study, we assumed that users refer to task-related information whenever necessary while walking in industrial work areas. To determine how the visual field of users of OSDs is affected while they are walking, we carried out an experiment in which the participants referred to information displayed on OSDs while they were walking on a treadmill. In particular, we compared how well the users of OSDs and conventional media can refer to information while walking. We tested the users' ability to visually recognize the real world and accurately refer to information. From the results of the experiment, we verified that OSDs enable an adequate level of awareness and accuracy. Although OSDs worn on the left eye affect the leftmost part of visual field at some level, we found that Zyl eyeglasses also affect the same area. Therefore, further changes in the physical design of the OSDs would be required to overcome the reduction of perception. In contrast, we found that the awareness of the central and upper visual fields was greater using OSDs than conventional media. These results suggest that the performance of an OSD as a reference medium while the wearer is in motion is acceptable. In particular, users' awareness of the front view using an OSD is greater than that using conventional media.

REFERENCES

Akasaka, T., M. Nakanishi, and Y. Okada. "Appropriate Complexity of Image Displayed on Head-Mounted Displays in Augmented Reality." Proceedings of the 12th International Conference on HCI (Human Computer Interaction), Beijing, China, 2007.

Akiyoshi, Y., M. Nakanishi, and Y. Okada. "Enhancement of Cooperation between Workers by Using Scanning Displays." Proceeding of the 2nd International Conference on AHFE (Applied Human Factors and Ergonomics), Las Vegas, NV, 2008.

Miura, T., M. Nakanishi, and Y. Okada. "A Study on Human Interface for Control in Using AR Manual." Proceedings of the 12th International Conference on HCI (Human Computer Interface), Beijing, China, 2007.

Nakanishi, M. and T. Sato. "Digital manual with wearable retinal imaging display for the next innovation in manufacturing." Proceedings of the 3rd International Conference on AHFE (Applied Human Factors and Ergonomics), Miami, FL, 2010.

Nakanishi, M., K. Taguchi, and Y. Okada.. "How can Visual Instruction with See-Through HMD be Effectively Used in Safety-Critical Fields?" Proceedings of APSS (Asia Pacific Symposium on Safety), Osaka, Japan, 2009.

Nakanishi, M., S. Tamamushi, and Y. Okada. 2009. "Study for Establishing Design Guidelines for Manuals Using Augmented Reality Technology -Verification and Expansion of the Basic Model Describing 'Effective Complexity'." *Enterprise Information System*, 21–26.

Nakanishi, M., T. Akasaka, and Y. Okada. "Modeling Approach to "Effective Augmentation" for Designing Manuals with Augmented Reality." Proceedings of the 2nd International Conference on AHFE (Applied Human Factors and Ergonomics), Las Vegas, NV, 2008.

Nakanishi, M., T. Miura, and Y. Okada. "How does the Digital Manual with Controllability and Perfect-Transparency Effect on Workers' Cognitive Processes? -Application of Augmented Reality Technology to Manufacturing Work-." Proceedings of the 4th International Conference on CITSA (Cybernetics and Information Technologies, Systems and Applications), Orlando, FL, 2007.

Nakanishi, M. and Y. Okada. "Development of an Instruction System with Augmented Reality Technology For Supporting Both Skilled and Unskilled Workers." Proceedings of INSCIT (International Conference on Multidisciplinary Information Sciences and Technologies), Merida, Spain, 2006.

Nakanishi, M. and Y. Okada. "Practicability of Using Active Guidance with Retinal Scanning Display in Sequential Operations." Proceedings of IEA (International Ergonomics Association), Triennial Congress in Maastricht, Netherlands, 2006.

Nakanishi, M. and Y. Okada. "Information Sharing by Using Monocular See-Through Head-Mounted Displays for Cooperative Work in Air Traffic Control Systems." Proceedings of the 11th International Conference on HCI (Human Computer Interaction), Las Vegas, NV, 2005.

Nakanishi, M. and Y. Okada. "A Study on Application of a Monocular See-Through Head-Mounted Display for Air Traffic Control System." Proceedings of the 7th International

Conference on WWCS (Work with Computing Systems), Kuala Lumpur, Malaysia, 2004.

Nakanishi, M. and Y. Okada. "A Study on Application of a Monocular See-Through Head Mounted Display for Visual Tracking." Proceedings of IEA (International Ergonomics Association) XVth Triennial Congress, Seoul, Korea, 2003.

Tamamushi, S., M. Nakanishi, and Y. Okada. "Mathematical modeling for determining effective information quantity given by augmented reality manual -overlay on dynamic background-." Proceedings of APSS (Asia Pacific Symposium on Safety), Osaka, Japan, 2009.

Tamamushi, S., M. Nakanishi, and Y. Okada. "Suitable Amount of Information for an Augmented Reality Manual -Superimposing on Dynamic Actual Field of Vision-." Proceedings of the 13th International Conference on HCI (Human Computer Interaction), San Diego, CA, 2009.

Tamamushi, S., M. Nakanishi, and Y. Okada."How to Design of Task-related Information Overlaid by AR Technology -Focusing on Task Characteristics-." Proceedings of the 2nd International Conference on AHFE (Applied Human Factors and Ergonomics), Las Vegas, NV, 2008.

Tanuma, K., T. Sato, M. Nomura, and M. Nakanishi. "Comfortable Design of Task-Related Information Displayed Using Optical See-Through Head-Mounted Display." Proceedings of the 14th International Conference on HCI (Human Computer Interaction), Orlando, FL, 2011.

CHAPTER 41

A Game System for Visually Impaired Utilizing Kinect

Nobumichi Takahashi, Takahiro Shogen, Yoshikazu Ikegami,
Michiko Ohkura

Shibaura Institute of Technology

ABSTRACT

Some game systems for the visually impaired have been developed. However, users must have or wear devices to use these systems. We think that if they do not need to have or wear devices, they will enjoy the system more and be more comfortable.

We developed a game system for the visually impaired utilizing a Kinect sensor and improved the system on the basis of the results of a preliminary experiment with sighted users. We then did an experiment with both sighted and visually impaired users.

INTRODUCTION

We developed some game systems for the visually impaired. However, users must have or wear devices to use these systems. Therefore, their actions are limited and they have to learn how to use these devices. Thus, we think that they will enjoy the system more and be more comfortably if they do not need to have or wear any devices. We therefore developed a new game system for the visually impaired utilizing a Kinect sensor.

This article describes the construction and evaluation of the system.

384

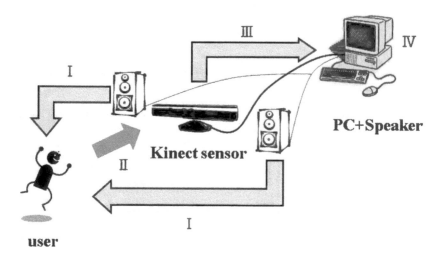

Figure 1 Diagram of system

CONSTRUCTION OF SYSTEM

We did research at a school for the visually impaired to clarify what type of games they want to play. They told us they wanted to play action games. Therefore, we decided to make an action game utilizing gestures.

The content of this game system is that a user gestures correctly to counter enemy sounds. It has four kinds of enemy sounds: cry of bat, roar of monster (left), roar of monster (right), and rolling stone. These sounds correspond with four kinds of gestures: squat down, high knee (left), high knee (right), and jump. Table 1 shows which gestures counter which enemy sounds and gestures. We decided these gestures from the aspect of safety and lower body exercise. This game takes about two minutes to play. When an enemy makes a sound, the user should make the relevant gesture before the enemy makes another sound. If the user does this, he/she can hear the sound that indicates the success and receives a certain number of points. The enemy makes sounds at consistent intervals. When a certain time has passed, the game ends and the user are informed about the total points by sound.

Table 1 Enemy sound and relevant gestures

Name of enemy sound	Relevant gesture
Cry of bat	Squat down
Rolling stone	Jump
Roar of monster (right)	High knee (right)
Roar of monster (left)	High knee (left)

PRELIMINARY EXPERIMENT

We made a prototype and performed a preliminary experiment to evaluate the system by employing eight subjects, all of whom were sighted. We took log data and handed out questionnaires containing the following questions:

- Did you enjoy this game?
- Would you want to play this game again?
- Are the enemy sounds and gestures naturally related?

The evaluation results showed that subjects enjoyed the system. Figure 2 shows the naturalness between enemy sound and relevant gesture. Cry of bat and rolling stone scored highly. However, roar of monster scored lowly. Moreover, subjects gave the following opinions:

- It's hard to distinguish right sounds from left.
- High knee and enemy sound aren't naturally related
- Why don't you use more auditory properties?

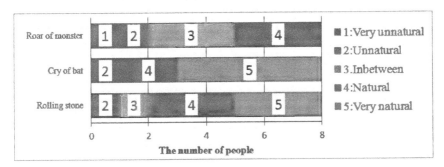

Figure 2 Evaluation of naturalness between enemy sound and relevant gesture

IMPLEMENT OF SYSTEM

On the basis of results of the preliminary experiment, we performed the following improvements:
  • Repositioning speakers for users to more easily distinguish right sounds from left
  • Changing the sounds countered by a high knee
  • Using more auditory properties

First, we repositioned the speakers as shown in Fig. 3. Second, we changed the sounds countered by a high knee. We asked subjects of the preliminary experiment "What kind of sound does a high knee more naturally counter?" From their answers, we changed the sound from "roar of monster" to "cracking of ground". Finally, we increased the number of enemy sounds. There are now two rolling stones: one from the right relevant gesture: "don't move") and one from the left (relevant gesture: "don't move").

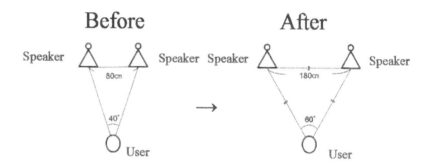

Figure 3 Repositioning of speakers

Table 2 revised enemy sounds and relevant gestures

Name of enemy sound	Hearing direction	Relevant gesture
Cry of bat	Center	Squat down
Rolling stone	Center	jump
Rolling stone	Right	Don't move
Rolling stone	Left	Don't move
Cracking of ground	Right	High knee (right) (
Cracking of ground	Left	High knee (left)

EVALUATION EXPERIMENT

We did evaluation experiment to evaluate the system by employing 24 subjects (12 sighted and 12 visually impaired). We took log data and handed out questionnaires asking the following questions:

· Did you enjoy this game?
· Are the enemy sounds and gestures naturally related?
· Does the gesture interface help to play this game?

Figure 4 shows evaluation results of with the game's enjoyability. No subject found the game unenjoyable. Figure 5 shows evaluation results of the playability of gesture interface. This was also highly evaluated. Figure 6 shows evaluation results of the naturalness of the high knee to the relevant enemy sound. The top graph shows evaluation results of preliminary experiment, and other graphs show evaluation results of the evaluation experiment. The improved game was evaluated to be better than that in the preliminary experiment. Figure 7 shows average and standard variation of game score. There is no difference between visually impaired and sighted users.

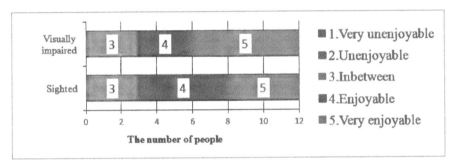

Figure 4 Evaluation of enjoyability

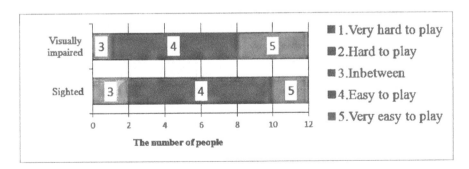

Figure 5 Evaluation of playability of gesture interface

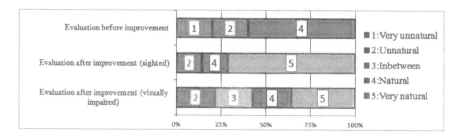

Figure 6 Evaluation of naturalness of high knee to relevant enemy sound

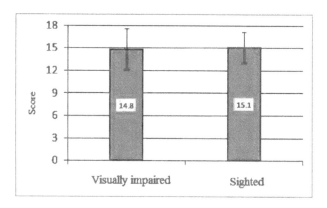

Figure 7 Game score of visually impaired and sighted

CONCLUSION

We developed game system for the visually impaired utilizing a Kinect sensor and evaluated it on the basis of the results of experiments. We found the following:
- This game system is enjoyable for the visually impaired
- Gesture interface helps both sighted and visually impaired users to play the game
- Both the sighted and visually impaired subjects scored the same in the game.

We have problems to solve. We need to improve the game for the visually impaired and evaluate its gesture interface by comparing it with others.

REFERENCES

Y. Ikegami, K. Ito, H. Ishii, and M. Ohkura : Development of a Tracking Sound Game for Exercise Support of Visually Impaired, Human-Computer Interaction, Part II, HCII2011, LNCS 6772, 31/35, Orlando, July.12, 2011.

CHAPTER 42

A Case Study of Participatory Ergonomics in the Quality Assessment of a Drilling "Dog House"

Marco Camilli [1], Giada Forte [1], Giuseppe Massara [1], Luca Spirito [1], Cristian Strinati [2], Simon Mastrangelo [1]

[1] Ergoproject Srl – Rome, Italy
[2] Drillmec Spa – Piacenza, Italy

ABSTRACT

This case study reports on the quality assessment of a drilling cabin (i.e., a dog house) by using a Participatory Ergonomics (PE) approach to the Heuristic Evaluation (HE) compared to an operators' self-reported measurement. In the quality assessment, self-reported rating techniques are broadly used by Human Factors (HF) practitioners with the aim of investigating the subjective components (e.g., satisfaction, comfort and self-efficacy) of the operator work. In the Participatory Heuristic Evaluation (PHE), HF experts evaluate the system together with work-domain experts to extend the detection coverage of potential criticisms in the human-system interaction. In this study, a sample of worldwide drillers reported on the perceived quality of the "dog-house" by a questionnaire, while another sample of drillers evaluated the same type of drilling cabin performing a PHE. Results showed that self-reported questionnaire was associated to higher levels of perceived quality with an undervaluation of potential criticisms. Differently, PHE detected more system's issues thanks to a knowledge exchange between HF and work-domain experts.

Keywords: Drilling rigs, Participatory Ergonomics, Heuristic Evaluation

1   INTRODUCTION

Technological improvements in drilling domain have led towards a recent implementation of new systems and tools (e.g., hydraulic plants and digital displays). While the engineering is required to develop sophisticated, stable and performing systems, the challenge of the safety and quality management consists in the adequate use of methods and techniques aimed at investigating the interaction between the human operator and the system. Traditionally, the Human Factors (HF) and Ergonomics literature is the most comprehensive source of methodologies for investigating the human operator in complex systems. In this domain, the debate on the "qualitative vs. quantitative" approach is still open (see for example Dekker and Woods, 2002 for a discussion in the human-automation context). In the drilling industry, the technical development centered on the operator work is mostly recent and HF contributions are not many, yet. A recent study (Aas and Skramstad, 2010) carried out a first structured assessment of the industrial application of the seven part control center ergonomics design standard ISO 11064, in the Norwegian petroleum industry. The lack of a strong methodological and empirical background in this domain makes the methods selection critical. For example, empirical evidences are not available yet to know which measurement techniques are the most suitable for monitoring driller cognition and performance. Also, the work culture in this field is quite far from HF and Ergonomics knowledge: if an air traffic controller is used to pay attention to the interfaces layout and to his vigilance states, most of drillers assumes extreme physical and cognitive fatigue as obvious and unsolvable. This case study reports on a first attempt to investigate ergonomics and the human factors in a "dog house" on a drilling rig. The research was carried out within a wider R&D projects granted by the Drillmec Spa (a manufacturing company of drilling and workover rigs for onshore and offshore applications) to the Ergoproject Srl (a small company specialized in ergonomics and human factors). The methodological approach to current investigation mixed a quantitative measurement (a self-reported questionnaire) with qualitative and participatory inspections (semi-structured interviews and a participatory heuristics evaluation). This latter method was considered necessary to improve the operators' awareness about ergonomics and to support them in the identification of potential system's criticisms and stress sources. The hypothesis was to find a not strong consistence between findings obtained by the self-reported measurement and by the participatory inspections. The first step of the investigation was a series of semi-structured interviews to the drillers. The interviews were developed on the basis of the "scenario analysis" proposed in the CRIOP (A scenario method for CRisis Intervention and OPerability analysis; see Johnsen & Bjørkli, 2008). Four cognitive factors as partially defined in the Simple Model of Cognition (Hollnagel, 1998) and in the Endsley's Situation Awareness model (1995) were investigated in these surveys. The factors were defined as follow:

Information monitoring – The quality of data and warnings received by the operator.

Comprehension – Integration and interpretation of data from indicators and events from the external environment, into a meaningful whole.

Decision making – Consistency with the standardized procedures and adequate planning to respond to unexpected events (e.g., operative delays).

Action implementation – Execution of the planned actions by respecting the safety rules modality (i.e., without using controls and tools, improperly).

Work environment – A socio-technical factor proposed by the method for CRIOP aimed at investigating the potential issues related to environmental and procedural factors (e.g., lightness and temperature, presence of other operators in the workspace).

The ergonomic issues found by the interviews were evaluated by HF and work-domain experts by carrying out a Participatory Heuristics Evaluation (PHE). The heuristic inspection is a widely accepted technique in HF domain. For example, Nielsen (1994) provided ten useful principles to evaluate the design of the web pages. Also, the ISO 9241-110:2006 reports on seven "dialogue principles" to guide the design of the user-system interaction. Generally, a heuristic analysis requires to a panel of issue-domain experts to evaluate the level of consistency between the system's features and the principles. Differently, PHE adds users (work-domain experts) to the list of expert inspectors under the heuristic evaluation (see also Muller et al., 1998). Finally, this case study represents a first attempt to define a common ergonomics and HF methodology for research and industry, in drilling domain.

2   METHOD

2.1   Participants

Fifteen drillers (all males; mean expertise = 15,1 years, SD = 8,11 years) were involved in this study. The drillers were from oil/gas extraction sites in North Italy (five out fifteen), France (six drillers), Scandinavia (two), Brazil (one) and Colombia (one).

Seven drillers (all males; mean expertise = 16,04 years, SD = 9,84 years) were involved in a participatory heuristic evaluation (PHE) together with three HF experts. The PHE was carried out throughout inspections at two Italian extraction sites and at the system simulator of the manufacturing company.

All drillers involved in the study were trained to the use of the systems according to the requirements provided by the manufacturing company and the industry normative.

2.2   The drilling "dog house"

In this study, it was assessed the quality of the "dog house" implemented in the on shore hydraulic drilling plants manufactured by the Drillmec SPA company (Piacenza, Italy). The driller's workstation was in a prefabricated cabin mounted on the foot of the drilling rig. Inside the cabin, two instruments panels were one on the left and one on the right side of the driller; this latter was in front of a wide windshield that allows monitoring the rig floor from his/her workstation.

Figure 1. A back view of the driller workstation in the simulator (Drillmec® SPA).

As shown in Figure 1, a set of levers, knobs, push buttons and selectors were placed on the panels for activating and controlling several tools (e.g., top-drive and torque wrench), while a series of visual warnings and manometers indicated the systems states and the valves' pressure changes.

2.3   Participatory approach to the quality assessment

Two HF experts carried out three on site investigations with the aim of inspecting the potential issues of the drillers work in the "dog house". This activity was participated also by the R&D manager of the manufacturing company and by a total of seven expert drillers. At the beginning of each survey, a driller was required to simulate the main operative tasks (i.e. drilling, trip-in, trip-out and casing) reporting aloud on his actions, attention allocation, criticisms, doubts and/or general discomfort. After completing this simulation, the driller responded to a semi-structured interview that investigated on specific aspects of the work activity in terms of "information monitoring", "comprehension", "decision making", "action implementation" and "work environment". This participatory and qualitative investigation allowed identifying a series of system issues related to four main quality factors: "system visibility", "error prevention", "layout" and "work environment". On the basis of well-known heuristics in human-system interaction domain (see for example, Nielsen, 1994 and ISO 9241-110:2006), a set of four general principles were defined for guiding the experts' evaluation of the drilling "dog house":

System visibility – The system should allow driller to be continuously informed about what the tools are going on, through visible indicators and an unobstructed view on the rig.

Error prevention – The system should conform to the driller's physical and cognitive skills and constraints, without placing him/her in potentially unsafe or misunderstanding situations.

Layout – The objects' positioning in the cabin should benefit the interfaces' learnability in such a way that the driller can handle the controls keeping the eyes ahead, if necessary.

Work environment – The workspace should be free from discomfort and technical solutions should properly limit external stress factors that could negatively affect driller's safety and performance.

The work-domain experts and the HF experts carried out a participatory heuristic evaluation by assessing the level of consistency between the system features and the "general principles"; the level of consistency was reported by inspectors on a 5-point Likert scale, as follows: "none", "low", "medium", "high", "complete".

2.4   Self-reported quality questionnaire

The system's issues found in the experts' inspection were used for developing a series of statements included in a self-reported quality questionnaire. This measurement was considered necessary to enlarge the drillers' sample both numerically and culturally (i.e., in terms of country of provenience). The items of the questionnaire were revised by a participatory activity that involved several stakeholders: the R&D manager of the manufacturing company, an expert driller and two HF specialists. The questionnaire aims to cover the four main factors defining the quality of the "dog house" (i.e., system visibility, error prevention, layout and work environment).

Table 1. The twelve items of the questionnaire, separately for the quality factors used in the evaluation of the " dog house".

Quality factor	Questionnaire item
System visibility	From the workstation, all displays and indicators / gauges can be adequately monitored "at a glance".
	From the workstation, it is easy to monitor the rig floor (e.g., tubes' container, *top drive* and other staff).
Error prevention	During the operational activities, it happens to accidentally bump into some command.
	While going to press or hold the necessary command, I get confused.

Layout	At the consoles all controls are "at hand".
	The controls on the dashboard are grouped and well separated between them.
Work environment	The noise of the machines is too high.
	In the *dog house*, it is either too hot or too cold.
	The vibrations on the dashboard and on the seat make the work heavy.
	When it is sunny, the brightness in the *dog house* is annoying.
	The environment into the *dog house* is unhealthy because of the dust that comes from the outside.

The questionnaire was developed in a four-languages version (Italian, English, Spanish and French) and it was sent worldwide to the oil/gas extraction sites. After receiving the questionnaire, the drillers could express to what they agree/disagree with the list of statements by using a 5-point Likert scale, as follows: "Strongly disagree", "Disagree", "Neither agree nor disagree", "Agree" and "Strongly agree".

3.   RESULTS

The evaluation scores (range 1-5) for each quality factor (system visibility, error prevention, layout, work environment) were used as dependent variable in a comparison between the "Participatory heuristic evaluation" (PHE) and the "Questionnaire".

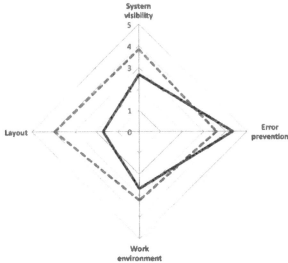

Figure 2. Evaluation scores, separately for "Type of measurement" (participatory heuristic evaluation and questionnaire) and for "Quality factor" (system visibility, error prevention, layout and work environment).

Results showed that the self-reported evaluation by the questionnaire was associated to higher levels of perceived quality of the system for each factor, excluding the "error prevention". In particular, the largest differences between the two methods were associated to the "layout" (MEAN = 3.93 SD = 0.53 by the questionnaire and MEAN = 1.67 SD = 1.15 by the PHE) and to the "system visibility" factor (MEAN = 3.83 SD = 0.52 by the questionnaire and MEAN = 2.67 SD = 1.15 by the PHE). Regards to the "work environment", the two methods reported similar scores (MEAN = 3.21 SD = 0.44 from the questionnaire and MEAN = 2.67 SD = 0.58 from the PHE). Finally, the evaluation of the "error prevention" factor with questionnaire was associated to slightly lower scores respect to the PHE (MEAN = 3.58 SD = 0.54 and MEAN = 4.33 SD = 0.58, respectively).

4.   CONCLUSIONS

In the drilling domain, the recent change from mechanical to hydraulic systems makes available the implementation of new technological solutions that could improve operators' comfort and performance. This technical evolution requires a deeper knowledge of the human operator to center the future system design on his/her skills and constraints. The case study described in this chapter represents a first attempt to identify a suitable HF method to investigate human operator (i.e., the driller) in drilling domain. In particular, a self-reported questionnaire and a participatory heuristic evaluation were mixed and compared between them. Results showed that the operators undervalued potential criticisms and reported a better quality of the workspace by using the self-reported method. Differently, when the operators were involved in a participatory evaluation they showed a better sensitivity to both systems' criticisms and not-comfortable situations. In other words, a participatory approach (including HF and work-domain experts in the evaluation) showed to cover a wider range of potential issues. Of course, these results showed some preliminary evidence that future studies should confirm and validate involving a larger sample of participants (both HF experts and operators). The final aim of this research is to find a valid and reliable methodology to investigate the operator interaction with novel technologies in drilling domain.

ACKNOWLEDGMENTS

This work was supported by an R&D agreement between the Drillmec SPA and the Ergoproject SRL. The authors thank also all drillers involved in this study.

REFERENCES

Aas, A.L. and Skramstad, T. 2010. A case study of ISO 11064 in control centre design in the Norwegian petroleum industry. *Applied Ergonomics*, 42:62-70.

Dekker, S.W.A. and Woods, D.D. 2002. MABA-MABA or Abracadabra? Progress on Human–Automation Co-ordination. Cognition, Technology & Work, 4:240-244.

Endsley, M. 1995. Toward a Theory of Situation Awareness in Dynamic Systems. Human Factors, 37:32-64.

Hollnagel, E. 1998. *Cognitive reliability and error analysis method.* Oxford UK: Elsevier.

ISO 11064-7:2006. Ergonomic design of control centres -- Part 7: Principles for the evaluation of control centres. International Organization for Standardization.

ISO 9241-110:2006. Ergonomics of human-system interaction – Part 110: Dialogue principles. International Organization for Standardization.

Johnsen, S.O. and Bjørkli, C. 2008. CRIOP: A Scenario Method for Crisis Intervention and Operability Analysis (Technical report SINTEF A4312). Accessed January 29, 2011, https://www.sintef.no/Projectweb/CRIOP/

Muller, M.J., Matheson, L., Page, C. and Gallup, R. 1998. Methods & tools: participatory heuristic evaluation. *Interactions*, 5: 3-18.

Nielsen, J. 1994. Heuristic evaluation. In J. Nielsen and R.L. Mack (Eds.), *Usability Inspection Methods*. New York: John Wiley & Sons.

Section IV

Toward an Ergonomic Product

A Calculation Model for Ergonomics Cost-benefit Analyses in Early Product Development Stages

Ann-Christine Falck, Mikael Rosenqvist

Department of Product and Production Development
Chalmers University of Technology
SE-41269, Göteborg, Sweden
annchrif@chalmers.se

ABSTRACT

Increasing international competition between companies has put high focus on cost-cutting actions at all levels in companies and organizations. In product development there are many design requirements to meet and often tough project budgets to keep. Requirements that are considered not profitable will often be neglected, which often affects assembly ergonomics. The objective of this study is to demonstrate the relationship between ergonomics, assembly related quality errors and associated costs and develop a cost-beneficial assessment model. The results showed that ergonomics high risk issues had 5-8 times as many quality errors as low risk issues and the earlier risk issues were found the less were the action costs. A model for cost-benefit analyses was developed based on the obtained quality data.

Keywords: ergonomics, cost-benefit analyses, product development

1   INTRODUCTION

Corrective measures are often made late and reactively when problems have already occurred. Proactive ergonomics risk identification in early product development stages is still unusual although there is today much scientific evidence available that prove both the human and economic benefits (Dul and Neuman,

2009). One of the main reasons may be a lack of knowledge about human and technology interaction and its impact on quality and productivity (Broberg, 1997; Skepper et al., 2000; Langaa Jensen, 2002, Sunwook et al., 2008). Skepper et al. (2000) argued that engineers and designers have poor knowledge of how to apply ergonomics principles. In product development departments there are many requirements for design engineers to meet and often tough project budgets to keep. This places high demands on very clear requirements and cost-beneficial solutions. Wulff et al. (2000) and Sunwook et al. (2008) claimed that requirements that are not easily understood or considered not profitable will be neglected, which often affects ergonomics. In addition, the time pressures in product design make designers reluctant to accept new requirements (Haslegrave and Holmes, 1994; Sunwook et al., 2008). Besides, cost-benefit analyses that demonstrate the profitability of ergonomics interventions are scarce. This is a difficult and complicated task since the straight-out and necessary data and related costs are not easily found in company records and in many cases missing altogether. Bevis and Slade (2003) found that few organizations make follow-ups for as long as necessary to be able to measure the long-term results. Nevertheless, it is urgent to be able to demonstrate the benefits of good ergonomics in economic terms. There are some good examples based on real world data such as by e.g. Hendrick (2003); Beevis and Slade (2003); Yeow and Nath Sen (2006), Goggins et al. (2008). Maudgalya et al. (2008) reviewed eighteen published cases studies with respect to productivity, quality, costs and safety. Regarding workplace safety initiatives, the studies showed an average increase of productivity of 66%, 44% in quality, 82% in in safety records and 71% in cost benefits. In a few reported cases it took only 8 months to obtain payback in terms of monetary investment. Hendrick (2008) presented 23 successful cases of ergonomic interventions in different companies in the U.S. He found that good ergonomics projects typically give a direct cost benefit of 1 to 2, to 1 to 10, with a typical payback period of 6-24 months. He calls attention to the the fact that the language of business is money and accordingly ergonomic project proposals must be expressed in financial terms. Besides Hendrick (2003, 2008) Beevis and Slade (2003), Dul and Neumann (2009), Morse et al. (2009), strongly argue that ergonomists must learn to present ergonomics concerns in business terms.

When there are aggravating assembly conditions such as too high assembly forces, awkward working postures, concealed assembly or tricky assembly in general things are more likely to go wrong and need to be repaired and/or exchanged (Eklund, 1995; Falck et al., 2010). On paced assembly lines there is often too little time for immediate actions. Instead much corrective measures and exchange of parts have to be conducted afterwards in separate repair stations. This circumstance causes late deliveries and results in productivity losses because the work has to be remade. In addition to tasks with high physical demands, Eklund (1995) found that designs that were difficult to assemble caused even more quality deficiencies. Also psychologically demanding tasks contributed to quality deficiencies. Falck (2007) and Falck et al. (2010) made conclusions very similar to Eklund (1995).

Research by Eklund (1997, 1999), Falck (2007) and prolonged experience from the Swedish car industry (Saab Automobile and Volvo Cars) has shown that the ergonomic impact is caused by product design at 60-70%. 30-40% of the problems depend on production conditions in the factory such as work organization, work station design, operator behavior and skill etc. After major design and assembly concepts are determined the possibility to influence these decisions is limited. In the car industry this usually happens several years before production start. Haslegrave and Holmes (1994) and Falck et al. (2010) argued that ergonomics input should be made from the very start of a project near the concept stage to avoid later problems that can be very costly.

2   OBJECTIVES

The objective of this study is to demonstrate the relationship between ergonomics, assembly quality and related costs for correction of assembly errors. A further objective is to develop a calculation model for ergonomics cost-beneficial assessments and comparison of different manual assembly solutions.

3   METHODS

57 manual assembly tasks in a Swedish automotive company were chosen for analysis. 20 assembly tasks were chosen that were assessed to imply high ergonomics risk with harmful impact; 19 tasks were chosen at moderate ergonomics risk level and another 18 tasks at low ergonomics risk level with minor or no harmful impact on operators. The selection of assembly tasks was made in cooperation with ergonomics specialist and responsible engineers in manufacturing engineering. However, some of the assembly tasks were part of the same assembly description, which resulted in 50 assembly tasks (PII) to study. These represented various ergonomics load levels and assembly difficulty as assessed by the ergonomics specialist and manufacturing engineers. The selected tasks were studied during a period of twelve weeks. Quality deficiencies such as failure rate, scrap and costs for correction of assembly related errors were retrospectively collected and analyzed including warranty and repair costs at dealers. The study used data stored in the logging quality databases in the company. However, of the 50 assembly tasks 47 remained because three had to be excluded due to circumstances such as equipment failures and supplier related problems.

3.1   Tracking quality errors and costs

All errors, scrap and related costs of four car variants were tracked. The data was collected from eight different quality tracking systems partly through assistance from responsible quality engineers and team leaders in production.

• **Quality errors and action time online and offline in the plant.**
Because the the study was done retrospectively the established error codes of each assembly task had to be searched out in two quality monitoring systems (Atacq). Each assembly had between 8 and 74 five or six-digit specific error codes, which had to be looked up. These codes then manually had to be transferred to another quality tracking system (Business Object) that stored quality data for longer periods

of time. After input of current codes and necessary main criteria wanted such as car variants, error codes, causing team, repair station, and investigation period, the system presented the numbers and types of errors that had occurred during the chosen period of investigation. The error data was divided into errors that occurred and were taken care of online and errors that had to be corrected offline. This division was a condition for accurate calculation of action times, which differed between action time online that was very short due to high time pressure and action time offline that was much longer due to less time pressure. Reported errors and required action times were converted into an hourly rate that was obtained from the economy department in the factory.

- **Scrap and exchange of parts**

The numbers and costs for parts and components that were replaced (scrapped) due to assembly related problems were obtained from the material coordinators in the plant.

- **Audit and blocked cars**

Each week 27 cars were randomly picked (audit = weekly investigation) from the total amount of built cars for careful analysis of any errors that might have been missed earlier. In the cases where a series of cars were faulty these were taken aside (blocked cars) and carefully examined and fixed before leaving the assembly plant. Quality deviations (errors) found at audit and in blocked car series were included in this study.

- **Warranty and repair costs at dealers**

Problems and errors that are found on the market are continuously reported and logged. Quality teams (VRT and PFU) follow-up concept related assembly issues and product related quality issues of all ready cars that have left the factory. The numbers and costs for errors, which are usually found before the customers receive their cars, are logged. Errors and action costs associated with the selected assembly tasks were obtained from responsible quality engineers and included in the study.

- **Serious quality deficiencies found in the market**

Problems that might cause serious accidents or require frequent repair office visits are taken care of by a special team (FARG). This team deals with problems that can result in recalls of sold cars or in requests to bring cars to repair shops as soon as possible. Usually these problems cost a lot of money for the company. The numbers and action costs related to this study were investigated through assistance from responsible quality engineers.

3.2 Action time and cost-analysis for quality errors in production

The 47 assembly tasks for analysis represented 24 interior and 23 exterior assembly tasks of four car variants. The frequency, number of quality errors and costs were obtained from the different quality tracking databases described in section 3.1.

- **Calculation of action times online and offline in the plant**

The action times for errors online were on average 2.2 minutes (Falck, 2010). Action times offline were obtained by the teams responsible for correction of errors offline. These average action times were then used for calculation of the costs for

corrective measures offline. A labor cost of 360 SEK/hour* was used, which was obtained from the economy department in the plant.
*(Exchange rate at midtime of the study, February, 2019: 1SEK = 0,10 EUR or 1 EUR = 9,89 SEK and 1 SEK = 0,14 USD or 1 USD = 7,26 SEK).

4   RESULTS

- **Failure rate, action times and action costs**

The study covered four different car variants that were built on paced assembly lines during a period of twelve weeks. 47 tasks of 47061 cars were analyzed regarding failure rate associated with manual assembly (Table 1). All errors found in company records and quality monitoring systems were included. For 38 of the assembly tasks it was also possible to calculate both the failure rate and costs for corrective measures related to 26219 of the cars (Table 5,6). In Table 1 the results show that the increased risk of failure rate was 7.8 times for the high load level tasks and 5.3 times for the the moderate load level tasks compared to the low load level tasks. Altogether, the errors related to high and medium load level issues composed 92.9% of all 47 analyzed tasks.

Table 1. Failure rate of all assembly tasks related to ergonomics load levels.

Load level	No. PII (Assembly Task)	Total failure rate	Quality errors - percentage share	Error/PII	Increased risk of failure rate
High	16	5045	55,1%	315	7,8 times
Moderate	18	3455	37,8%	192	5,3 times
Low	13	650	7,1%	50	1
All	47	9150	100%	195	

For 38 of the 47 tasks the failure rate could be divided into errors that were corrected online and errors that must be corrected offline due to more time consuming repair. The average assembly time for the 38 tasks was 1789 TMU:s (just about 64 seconds per car since 1 TMU = $1/28^{th}$ of a second). During twelve weeks this includes 1782420058 TMU:s (about 1060964 minutes or roughly 17683 assembly hours). Table 2. shows the distribution of errors that were taken care of online and offline. 3570 errors (67.1%) were handled online, which is about twice as many as the 1747 errors (32.9%), which were taken care of offline. Scrapped parts and components were altogether 433 pieces and included in the total failure rate of 5750 errors in the assembly plant. 18 errors that were found by from Audit were included in the numbers of errors offline since these were taken care of there. No errors in blocked car series were found in this study. The failure rate and action costs reported by the VRT/PFU and FARG teams are shown in Table 3. Here, the numbers of errors of the moderate load level tasks by far exceed the numbers of the high load level assembly tasks, 23 errors compared to 284. Besides, they are 1.7 times as expensive. The low load tasks only caused two errors but at a repair cost of

Table 2. Load level task errors divided into online and offline measures.

Online			Offline		All			
Load level	Errors online	Error /PII	Error off line	Error /PII	Total no. of errors	Scra pped items	Total no. of errors	% of all errors
High (n=14)	1942	138,7	1195	85,4	3137	176	3313	57,6
Moder ate (n=14)	1106	79	497	35,5	1603	204	1807	31,4
Low (n=10)	522	52,2	55	5,5	577	53	630	11
All (n=38)	3570	93,9	1747	46	5317	433	5750	100

1220 SEK each. The high amount of errors and and costs of the moderate load level tasks were mainly caused by one single task, the front side window assembly that alone resulted in 267 errors and 219423 SEK in repair costs (not shown in Table 3).

For calculation of action times online (Table 4) a standard time of 2.2 minutes was obtained from previous analyses in the plant (Falck, 2010). However, the action times offline are not logged in any system and the only way to know was to ask very experienced operators to estimate the average action time for each type of error that was taken care of. The action times varied considerably and were between 2-180 minutes per error depending on type of error and degree of severity. Therefore, the average action times were used for calculation of the costs for corrective measures offline (Table 5, 6). For that purpose a labor cost of 360 SEK/hour was used, which was obtained from the economy department in the plant. Table 4 shows

Table 3. Failure rate and action costs in the market.

Load level (tasks)	Errors reported by VRT and PFU	Errors reported by FARG	Total failure rate	Action cost/ error, SEK	Action cost/ task SEK	Total action cost, SEK
High (n=14)	23	0	23	466,17	765,86	10722
Moderate (n=14)	284	0	284	799,76	16223,64	227131
Low (n=10)	2	0	2	1220	224	2440
All (n=38)	309	0	309	777,65	6323,50	240293

that the overall offline action times were 20. 24 minutes/error, i.e. 9.2 times longer compared to the action times/error online of 2.2 minutes. The low load level tasks specifically had a much higher average action time offline of 54.49 minutes/error. This was caused by two assembly tasks that required very time-consuming disassembly and replacement of parts. However, the average action times/error on- and offline for all load levels did not differ much. The total action costs of the 38 assembly tasks are shown in Table 5. In the plant the high load level tasks corresponded to 48.7 % of the costs, the moderate load level tasks to 41.1% and the low load level tasks to 10.2%. However, in the market the moderate load level tasks accounted for 94.5% of the total repair cost mainly due to one of 14 tasks (the front side window), which alone caused 267 of 284 errors found in the market (see comments about Table 3). (These errors cost 821.80 SEK each to repair in the market compared to corresponding repair cost in the plant of 78.08 SEK, which was 10.5 times lower). The high and medium load level action costs corresponded to 93.4% of the total action costs in the plant and market for 5427 (Table 2 and 3) of all 6059 errors reported (Table 6). Altogether, high and medium load level tasks caused 89.6% of all errors and 93.4% of the costs. The low load level action costs were 6.6% of the total action costs for 10.4% of all errors. Table 6 shows the increased costs per error the later actions are taken. The lowest action costs/error are online whereas issues taken care of offline are 10.53 times more costly to fix. Errors fixed in the market are 12.2 times as costly as errors fixed at plant level. Consequently, further efforts should be made to find errors and fix them as soon as possible.

Table 4. Failure rate and action times (minutes) in the plant.

Online				Offline			All	
Load level	Errors on line	Mean action time/ error	Total action time	Errors off line	Total action time	Mean action time/ error	Mean action time/ error	
High (n=14)	1942	2,2	4272,4	1195	22728	19,02	8,607	
Moderat e (n=14)	1106	2,2	2433,2	497	9637	19,39	7,529	
Low (n=10)	522	2,2	1148,4	55	2997	54,49	7,184	
All (n=38)	3570	2,2	7854	1747	35362	20,24	8,127	

Table 5. Total action costs of 38 assembly tasks in the plant and in the market.

Load level	Action costs in the plant, SEK	Plant %-age of action costs	Action costs in the market, SEK	Market %-age of action costs	Total action costs, SEK	%-age of all action costs
High (n=14)	177993,93	48,7	10722	4,5	188715,93	31,1
Moderate (n=14)	150462,19	41,1	227131	94,5	377593,19	62,3
Low (n=10)	37420,82	10,2	2440	1,0	39860,82	6,6
All (n=38)	365876,94	100	240293	100	606169,94	100

Table 6. Cost comparisons between different levels of action.

Action level	Total action costs, SEK	Failure rate	Cost/error, SEK	Increased costs
Online	47113	3570	11,52	1
Offline	211968	1747	121,33	10,53 times compared to online
Total plant	365877	5750*	63,63	1
Market	240293	309	777,65	12,2 times compared to total plant
All	606170	6059	100,04	

* including scrapped items reported on plant level.

4.2 A calculation model for cost-benefit analyses

Based on the detailed quality data obtained in this study, a calculation model was developed. The purpose of the model is to support decision making when necessary in the development of assembly solutions. The model enables cost-benefit analyses and comparisons between different assembly solutions and tasks and considers a majority of all costs in the plant and in the market. However, the costs must be known. In this study, for each assembly task the numbers of quality errors online and offline and their respective action times were known. The amount and costs of scrapped items, quality remarks in audit and blocked cars were available. The labor costs could be calculated. The action costs of quality errors on the market or at the customer were also obtained. The costs for lost brand image are very difficult to estimate and such classified (secret) information could not be obtained but should be included if possible. Additionally, the costs for work-related sick leave and rehabilitation should be included for a complete calculation. However, these costs were not calculated since this was not included in the objective of the study. Thus, based on all data obtained, a principle calculation will look like this:

$$C = W(N_{on} \times Ta_{on} + N_{off} \times Ta_{off} + N_{au} \times Ta_{off} + N_{yard} \times T_{ty}) + N_{scrap} \times C_{scrap} + C_{fb} + WRSL + C_{fcomp} + C_{rec} + C_{bw}$$

Number of errors:
N_{on} = number of quality errors online
N_{off} = number of quality errors offline
N_{au} = number of audit quality remarks
N_{scrap} = number of scrapped items
N_{yard} = number of cars with errors in the yard (if there are such cars)
N_{fb} = number of factory blocked cars

Action time (minutes):
Ta_{on} = action time online
Ta_{off} = action time offline
T_{ty} = Transfer time of cars in the yard

Costs:

C = total costs for all manual assembly related errors
W = labor cost/time unit
C_{scrap} = scrap cost/item
WRSL = cost for work related sick leave and rehabilitation
C_{fb} = cost for errors of factory blocked cars (Tracy)
C_{fc} = cost for errors of Factory complete cars (VRT/PFU)
C_{rec} = cost for recall/repair of cars distributed to the customers (FARG)
C_{bw} = cost for lost brand image and customer´s dissatisfaction (badwill).

5.   DISCUSSION/CONCLUSIONS

A practical and simple calculation model is often asked for by engineers but such a model cannot be simple for two reasons: Suitable calculation data is often missing in companies and there are many factors to consider for a complete calculation that prevents simplicity. However, the results in this study show that the majority of assembly errors and related costs (95-97%) were found and solved in production. For practical reasons and in many cases these error data and associated costs might be sufficient enough as decision support in the choice of manufacturing and ergonomics solutions. Table 6 clearly demonstrates that the later errors are fixed the more costly they are. Errors found and fixed in the market were 12,2 times more costly compared to errors found and fixed in the plant. Errors fixed online were 10,3 times less costly than errors fixed offline in the plant. The most profitable is to foresee and prevent assembly errors altogether by making holistic assessments as early as possible in product development stages. The earlier a holistic approach, prediction of failure rate and action costs can be made, the more money can be saved altogether. Besides, both ergonomics and quality issues can be proactively solved at the same time.

ACKNOWLEDGEMENTS

This project is funded by Vinnova AB within the program "Sustainable production: Decision Support for Early Estimation of Cost of Poor Quality". The project was carried out in the Production Area of Advance at Chalmers University of Technology, Göteborg, Sweden, in collaboration with Volvo Car Corporation, Volvo Trucks and Swerea IVF. The support is gratefully acknowledged.

REFERENCES

Beevis, D., Slade, I.M. (2003). Ergonomics-costs and benefits. *Applied Ergonomics*, 34 (5), pp. 413-418 (part 1).Booker, J.D., Raines, M./Swift, K.G. (2001). *Designing Capable and Reliable Products*. Butterworth-Heinemann, Oxford, ISBN: 9780750650762.
Broberg, O. (1997). Integrating ergonomics into the product development process. *International Journal of Industrial Ergonomics* 19: 317-327.

408

Dul, J., Neumann, P. (2009). Ergonomics Contributions to Company Strategies. *Applied Ergonomics 40 (2009)745-752.*

Eklund, J. (1995).Relationships between ergonomics and quality in assembly work. *Applied Ergonomics Vol. 26, No. 1, pp 15-20, 1995.*

Eklund, J. (1997). Ergonomics, quality and continuous improvement – conceptual and empirical relationships in an industrial context. *Ergonomics, 1997, Vol. 40, No. 10, 982-1001*

Eklund, J. (1999). Ergonomics and quality management – humans in interaction with technology, work environment and organization. *International Journal of Occupational Safety and Ergonomics. 5(2), 143-160.*

Falck, A., (2007). *Virtual and Physical Methods for Efficient Ergonomics Risk Assessments – A Development Process for Application in Car Manufacturing.* Thesis for the Degree of Licentiate of Philosophy. Department of Product and Production Development, Chalmers University of Technology, Göteborg, Sweden. Report no. 21. ISSN 1652-9243.

Falck, A., Örtengren, R., Högberg, D: (2010). The Impact of Poor Assembly Ergonomics on Product Quality: A Cost-Benefit Analysis in Car Manufacturing. *Human Factors and Ergonomics in Manufacturing and Service Industries 20 (1) 24-41 (2010).*

Googins, R.W., Spielholz, P., Nothstein, G.L. (2008). Estimating the effectiveness of ergonomics interventions through case studies: Implications for predictive cost- benefit analysis. *Journal of Safety Research 39(2008) 339-344.*

Haslegrave, C.M., Holmes, K.(1994). Integrating ergonomics and engineering in the technical design process. Applied ergonomics 1994 **25**(4) 211-220.

Hendrick, H.W. (2003). Determining the cost-benefits of ergonomics projects and factors that lead to their success. *Applied Ergonomics, Vol. 34, pp. 419-427.*

Hendrick, H.W. (2008). Applying ergonomics to systems: Some documented "lessons learned". *Applied Ergonomics 39 (2008) 418-426.*

Maudgalya,T., Genaidy, A. Shell, R. (2008). Productivity-quality-cost-safety: A sustained approach to competitive advantage – a systematic review of the National Safety Council's case studies in safety and productivity. *Human Factors and Ergonomics in Manufacturing, 18(2), 152-179.*

Langaa Jensen, P. (2002). Human factors and ergonomics in the planning of production. *International Journal of Industrial Ergonomics, 29 (2002) 121-131.*

Morse, M., Kros, J., Scott Nadler, S. (2009). A decision model for the analysis of ergonomic investments. *International Journal of Production Research,* Vol. 47 (21), pp. 6109-6128.

Skepper, N., Straker, L., Pollock, C. (2000). A case study of the use of ergonomics information in a heavy engineering design process. *International Journal of Industrial Ergonomics 26 (2000) 425-435.*

Sunwook, K., Seol, H., Ikuma, L.H., Nussbaum, M.A. (2008). Knowledge and opinions of designers of industrialized wall panels regarding incorporating ergonomics in design. *International Journal of Industrial Ergonomics 38 (2008) 150-157.*

Wulff, A., Westgaard, R., Rasmussen, B. (2000). Documentation in large-scale engineering design: information processing and defensive mechanisms to generate information overload. *International Journal of Industrial Ergonomics,* Vol. 25, pp. 295-310.

Yeow, P., & Nath Sen, R. (2006). Productivity and quality improvements, revenue increment, and rejection cost reduction in the manual component insertion lines through the application of ergonomics. *International Journal of Industrial Ergonomics, 36 (2006) 367– 377.*

CHAPTER 44

Evaluating the Effectiveness of Haptic Feedback on a Steering Wheel for BSW

*Jaemin Chun[1], Gunhyuk Park[2], Seunghwan Oh[1], Jongman Seo[2],
In lee[2], Seungmoon Choi[2], Sung H. Han[1]**

[1]Department of Industrial & Management Engineering, POSTECH, South Korea
[2]Department of Industrial & Management Engineering, POSTECH, South Korea
*shan@postech.edu

ABSTRACT

Driver's inattention to blind spots is one of the main causes of lane change-related car accidents. Efforts have been made to eliminate such blind spots but there still exist some areas around the vehicle that cannot be directly observed by the driver. Also, drivers cannot always focus on the rear-view mirror or side mirrors that help them to monitor other vehicles near blind spots. So, for this study, we used a collision warning system based on a different modality to find out whether it can enhance drivers' awareness of other vehicles in blind spots. A haptic feedback was applied on a steering wheel for BSW (Blind Spot Warning). Haptic feedbacks are known to have several advantages over other modalities. First, they do not require any extra visual load which is already heavily placed upon drivers while driving in general cases. Second, haptic warnings are transmitted to the driver in a private way so other passengers are not disturbed by warnings like flashing lights or loud alarms. Third, haptic feedbacks are free from ambient noises that are generated by various sources around the driver. In this study, the effectiveness of haptic feedback was evaluated by comparing the performance and preference measures of the haptic condition with the non-haptic condition. For the performance measure, the collision prevention rate was analyzed as well as the minimum distance between the vehicles after collision events were carried out. As for the preference measure, the participants were asked to score the perceived usefulness and the overall satisfaction of the provided collision warning. For this experiment, a total number of twenty-four males in two different age groups (30-40 yrs and 50-60 yrs) were recruited to define the age effect. A virtual driving simulator with a software program and devices was developed and collision scenarios were provided to the participants. In the experimental scenario, participants were to perform as drivers and were asked to follow a preceding car which changes lanes from time to time. A haptic warning signal was presented on the steering wheel when a vehicle approached from the driver's blind spot while the participant moved into the other lane to follow the preceding car. Haptic feedbacks were generated in separated locations (left and right) of the steering wheel corresponding to the direction of the possible collision event. The warning was provided to the participants 4 seconds before an expected collision. To isolate the vibration of each haptic feedback and prevent them transferring from one side to the other, the rim of the steering was sawed-off. As shown in the evaluation result, the effect of haptic BSW was valid to the participants. They avoided possible collision situations better with the help of haptic BSW. Also, when the participants successfully avoided the collision event, the mean minimum distance between their vehicle and the other vehicle increased, indicating the decreased possibilities of an accident. In terms of age, the younger group has shown better performance than the older group due to their superior detectability of haptic stimuli and motor skills. As for preference measures, the participants felt haptic BSWs as useful and were highly. The older participants in particular have shown higher level of perceived usefulness and satisfaction with haptic BSWs. From the result of our study, we conclude that the haptic BSW on a steering wheel could be helpful for safe driving.

Keywords: haptic, collision warning, blind spot

1   INTRODUCTION

Blind spots on the sides of a vehicle are closely related to drivers' safety when changing a lane (Svenson et al., 2005). Drivers are frequently unaware of the other vehicle's approach due to blind spots. Several monitoring techniques were introduced to minimize blind spots and help drivers to obtain a better vision of them (Becker et al., 2005, Krips et al., 2003). However, even when blind spots are effectively removed, drivers can still meet problematic situations when they fail to observe the side mirrors at the appropriate time. As the most important and essential task of drivers is to monitor the situation of the road ahead, drivers should be aware of the movement of the preceding car even when they are making a lane change. This means that drivers should constantly shift their visual attention between the road and side mirrors. As a result, from time to time, drivers unintentionally lose their visual attention from side mirrors at critical moments during a lane change. This is especially true for novice drivers who experience greater difficulty when switching their attention from road to side mirrors. Thus, an effective warning is needed to direct driver's attention to side mirrors when necessary.

Several attention-gathering techniques using visual and auditory cues were developed by automobile manufacturers. However, even with visual BSWs such as flashing lights, drivers still have some chances of missing the warning signal because they cannot identify the warning when their attention is focused on the road. Also, a BSW with an auditory alarm can be blocked by ambient noises surrounding the driver (Ryu et al, 2010). Recently, to overcome these limitations of visual and auditory warnings, haptic feedback has been introduced. Haptic feedback is suggested as an effective warning signal in terms of reaction time (Carlander et al., 2007) and in effectively providing the warning signal solely to the driver through a direct contact (Chun et al., 2010). Based on empirical data and expert review, Campbell et al. (2007) evaluated the effectiveness of various haptic warning systems (brake pulse, accelerator counter pulse, seat shaker and etc.) applied on diverse warning situations. However, compared to other types of haptic warning systems, relatively limited evaluation was conducted on the haptic steering wheel. So, in our study, we evaluated the effectiveness of the haptic steering wheel as a BSW through an experiment.

2   METHODS

2.1   Apparatus

A driving simulator with a software program and devices was developed to present a virtual driving environment and the possible collision events to the participants. Also, an adjustable sitting buck of the actual size and layout of a vehicle was arranged. A 50-inch and two 23-inch LCD monitors were used to provide frontal view and side views (Figure.1) and these displays were calibrated to enable a seamless surrounding view for the driver.

Figure 1 Driver's vision in virtual simulator

To generate haptic BSW, three off-the-shelf vibration motors were attached on each side (left and right) of the steering wheel (Figure.2) and the specification of the vibration feedback (frequency:120 Hz; amplitude: 0.014 mm; periodic envelope: 0.2 seconds on-time and 0.1 seconds off-time) was determined through a pilot test. To isolate the vibration feedback from each side of the steering wheel, the rim of the steering wheel was physically disconnected by sawing. Participants were asked to wear ear plugs and headphones (with white noise played) to block the sound cues being generated from the vibration actuators.

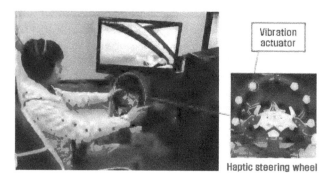

Figure 2 Haptic steering wheel (Yellow circles are locations of vibration actuators)

2.2 Participants

As human sensitivity to haptic stimuli degrades with age, we considered two age groups (30-40yrs and 50-60yrs) and recruited 12 participants for each group. The average age of the younger group was 39.5 while it was 54.0 for the older group.

2.3 Experiment design and scenario

For each age group, two warning conditions were evaluated: non-haptic and haptic. In non-haptic condition, the participants had to detect the collision risk only by observing side mirrors, which is a general case. When the haptic BSW was provided, the participants identified the collision risks with the help of a vibration feedback. To evaluate the effectiveness of warning conditions, several performance and preference measures were used. For performance measures, the collision prevention rate and the minimum distance were collected. The minimum distance was defined as the distance between the participants' vehicle and the other vehicle at a time when the participants successfully avoided the collision event. After the experiment, the participants were asked to give scores to the preference measures including the perceived usefulness and the overall satisfaction. When grading the overall satisfaction of each warning type, we asked the participants to also consider the inconvenience brought about by the proposed warning type.

In the experiment, the participants' task was to follow a car running at a speed of 80km/h and changing lanes occasionally. When they move into the other lane to follow the preceding car, a possible collision scenario plays out; a high speed vehicle from the blind spot suddenly approaches the participants' vehicle. To avoid any possible crash, participants were asked to manipulate the brake pedal and the steering wheel. The possible collision events were provided to each participant for

20 times and a haptic BSW was activated 4 seconds before a collision. The participants' behavior was video-taped to observe any interesting reactions.

3 Results and Discussions

A t-test was conducted to define the statistical significance of warning conditions on performance and preference measures. Regardless of the age group, participants showed a significantly higher collision prevention rate with haptic BSW ($t(477) = 3.09$, $p < 0.0021$ for the younger group and $t(472) = 2.71$, $p < 0.0070$ for the older group) (Figure.3). The collision prevention rate increased from 0.34 to 0.48 for the younger group and from 0.24 to 0.35 for the older group.

While changing lanes to follow the preceding car, the participants had to switch their visual attention frequently between the side mirror and the road and they instantly lost their visual attention on the side mirror during the process. By providing the haptic BSW on the steering wheel, drivers were able to refocus their distracted visual attention to the side mirrors and this resulted in a decreased collision rate. In general, the younger participants were able to perform better with haptic BSW due to their superior cognitive ability of visual and tactile sensation.

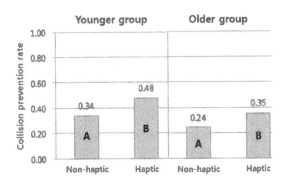

Figure 3 Collision prevention rate

Similarly, when the haptic BSW was provided, the average minimum distance increased significantly for both age groups; $t(458) = -3.26$, $p < 0.0012$ for the younger group and $t(454) = -2.50$, $p < 0.0127$ for the older group (Figure.4). The minimum distance increased from 0.10 m to 0.15 m for the younger group, and 0.06 m to 0.10 m for the older group.

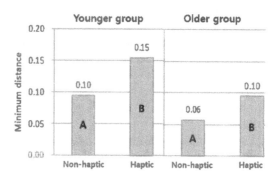

Figure.4 Minimum distance

Not only were the participants of both age groups able to avoid the collision successfully, their minimum distance increased significantly with the haptic steering wheel. This indicates that it is better to have the BSW with 4 seconds of TTC; participants were able to identify the collision risks when there was less than 4 seconds before the crash. If we set the BSW's TTC above 4 seconds, we can reduce the collision risk by increasing the minimum distance. However, care should be taken when determining the warning timing since an earlier warning can increase the rate of false alarm at the same time.

Participants' preference measures were significantly affected by the warning condition. They felt that the haptic steering wheel was useful ($t(22) = -5.45$, p < 0.0001 for the younger group and $t(22) = -5.71$, p < 0.0001 for the older group) and were satisfied with the haptic BSW ($t(22) = -5.23$, p < 0.0001 for the younger group and $t(22) = -5.22$, p < 0.0001 for the older group) even though they felt some discomforts with the unfamiliar warning type. Judging from the score difference between the warning conditions (younger group's usefulness score rose from 41.25 to 79.16 and satisfaction score from 37.50 to 77.92 while those measures increased from 54.58 to 82.92 and from 56.25 to 86.25 respectively in the older group), younger group felt greater advantage with the haptic BSW than the older group (Figure.5).

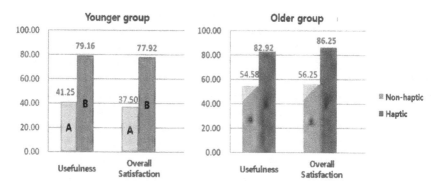

Figure.5 Perceived usefulness and Overall satisfaction

As what we have identified from the video analysis, conservative and safety-oriented driving habits of older drivers (Green, 2009) have resulted in relatively smaller score gap between the warning conditions among the older group than the younger group. Also, the score difference between the preference measures (the perceived usefulness and the overall satisfaction) show that both age groups' level of discomfort experienced by the haptic BSW is not significant. Although the provided haptic BSW was new to participants, according to the participants comments after the experiment, the level of discomfort greatly decreased after experiencing it for several times.

4   CONCLUSIONS

In this study, we empirically evaluated the effectiveness of the haptic steering wheel by comparing the haptic BSW condition with the non-haptic condition. The participants were able to detect the collision risk better and more successfully avoid a possible crash with the haptic BSW. Also, distance between the participants' car and the other car increased when they successfully stopped the vehicle after receiving the haptic BSW at possible collision event. The effectiveness of the proposed haptic BSW was valid for participants of all age groups tested in this study. Participants considered the haptic BSW useful and were satisfied with the way warnings were provided. Also, they quickly became familiar with the haptic stimuli even though they newly experienced it. As shown in our experimental results, we conclude that the haptic BSW on a steering wheel can be used effectively. The simulator and experimental program can also be used to evaluate other haptic feedbacks such as the car seat, seatbelt and so on. In the following works, softer and smoother haptic stimuli with an equally effective specification should be identified so that it can deliver effective warnings as well as satisfy the emotional aspect of drivers. Also, further study is needed to find a more efficient warning timing that can enhance the warning performance without increasing the false alarm rate.

ACKNOWLEDGMENTS

This work was supported in part by a NRL Program 2011-0018641 and a Pioneer Program 2011-0027995 both from NRF.

REFERENCES

Svenson, A.L., Gawron, V.J., Brown, T., 2005. Safety evaluation of lane change collision avoidance system using the national advanced driving simulator. National Highway Safety Administration (NHTSA), Paper No. 05-0249.

416

Becker, L.P., Debski, A., Degenhardt, D., Hillenkamp, M., Hoffmann, I., 2005. Development of a Camera-Based Blind Spot Information System. Advanced Microsystems for Automotive Applications, Part 2, 71-84.

Krips, M., Teuner, A., Velten, J., Kummert, A., 2003. Camera based vehicle detection and tracking using shadows and adaptive template matching. ICECS'03 Proceedings of the 2nd WSEAS International Conference on Electronics, Control and Signal Processing.

Ryu, J., Chun, J., Park, G., Choi, S., Han, S., 2010. Vibrotactile feedback for information delivery in the vehicle. IEEE Transactions on Haptics, vol. 3, no. 2, 138-149.

Carlander, O., Eriksson, L., Oskarsson, P., 2007. Handling uni- and multimodal threat cueing with simultaneous radio calls in a combat vehicle setting. HCII 2007, LNCS 4555, 293–302.

Chun, J., Oh, S., Han, S., Park, G., Seo, J., Choi, S., Han, K., Park, W., 2010. Evaluating the effectiveness of haptic feedback on a steering wheel for FCW. In Proceedings of the 9th Pan-Pacific Conference on Ergonomics (PPCOE), 2010.

Campbell, J.L., Richard, C.M., Brown, J.L. & Marvin. 2007. Crash warning system interfaces: Human factors insights and lessons learned. NHTSA technical report.

Green, M., 2009. Driver Reaction Time [Online] Available. http://www.visualexpert.com/Resources/reactiontime.html

CHAPTER 45

Identifying Affective Satisfaction Elements of a Smartphone Application

Joohwan Park, Sung H. Han, Jaehyun Park, Hyunkyung Kim

POSTECH
Pohang, Korea
Shan@postech.ac.kr

ABSTRACT

Affect is an integrated image toward a product. Affective satisfaction elements influence a user's affect toward a product. The purpose of this study is to derive the user affective satisfaction elements of smartphone applications.

As the smartphone becomes one of the most important products in the consumer electronics industry, researchers keep trying to find the affective satisfaction elements of the smartphone. This study collected vocabularies expressing affect to identify the affective satisfaction elements toward smartphone applications. Focus group interviews (FGI) and literature surveys were used to collect affective vocabularies from smartphone users. As a result, a total of 283 affective vocabularies were obtained. These affective vocabularies were classified and analyzed with K-means clustering and the open card sorting method. Affective satisfaction elements toward smartphone applications were identified based on affective vocabularies, considering three affect perception stages. These results will be useful for measuring the affective satisfaction of smartphone applications.

Keywords: Smartphone, Application, Affect, Affective satisfaction

1   INTRODUCTION

When a user interacts with a product, he or she can experience many affective states. Every usage of a product gives some feelings and sensations to the user.

These feelings and sensations are usually integrated into some images of the product, and these images are called "affect" (Han et al., 2001).

Recently, people increasingly have an interest in the exterior design or aesthetic features of a product. The affective satisfaction has become a key factor of product design so that the products may have competitive advantages in the market. In addition, affective satisfaction is regarded as one of the major constituents to the user experience by many researchers (Kim et al., 2010; Law et al., 2009).

The affective satisfaction elements toward mobile phone usage have widely been studied in the field of affective engineering (Im et al., 2007; Park et al., 2011). However, smartphones, becoming popular in the mobile phone market recently, are considered to have many different features from traditional mobile phones (feature phones). One of the major differences is application usage.

Isomursu et al. (2007) have discussed methodologies to collect user's emotions on mobile application usage. However, affective satisfaction, different from emotion, in smartphone application usage is still not discussed in the field of affective engineering.

In this study, affective vocabularies related to application usage were collected through focus group interviews (FGIs) and literature survey to identify the elements of affective satisfaction toward the smartphone application.

The collected affective vocabularies were analyzed with the K-means clustering method. Three groups were developed in terms of affect perception stages. Then, open card sorting analysis was performed to finalize affective satisfaction elements toward a smartphone application (see Figure 1).

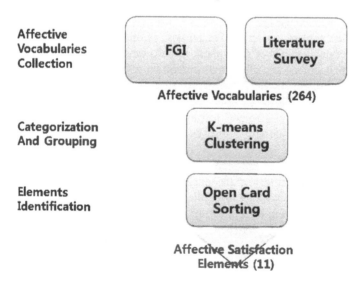

Figure 1. Affective Satisfaction Elements Identification Process

2   COLLECTION OF AFFECTIVE VOCABULARIES

The FGIs and literature survey were performed to collect the affective vocabularies related to smartphone applications.

At first, FGIs were conducted to extract affects that users had felt consciously or subconsciously during their smartphone usage. To extract affect and those elements from users, we collected affective vocabularies from users' speech in the FGIs. A total of 70 subjects participated in the FGIs. They were undergraduate and graduate students who had their own smartphones. The FGI subjects were 23.0 years old on the average (standard deviation of 3.33). The subjects were divided into four homogenous user segments by gender and smartphone OS (iOS and Android). Each session of FGI was held with seven to eight subjects. FGI subjects shared their own application usage experiences during a two-hour session. A total of 423 affective vocabularies were collected from users' speech.

FGI was conducted with specific smartphone OS user groups. To supplement additional affective vocabularies from various smartphone users, a literature survey was conducted. Web pages (user forums, review sites, blogs, advertisements, newspaper articles, etc.) were collected to analyze descriptions and reviews of smartphone applications. A total of 264 affective vocabularies were extracted from the web pages.

The vocabularies collected through FGIs and the literature survey were combined into 283 affective vocabularies. Duplicate vocabularies were merged into one vocabulary, and some improper vocabularies were deleted. Deletion criteria were: (1) vocabulary not describing affective factors but showing a user's emotional status; (2) slang; and (3) vocabulary coming from affective factors of smartphone hardware, not the application. For example, we deleted vocabulary such as "feels angry", because it only shows the emotional status of the user, not the affective factors of the smartphone.

3   IDENTIFICATION OF AFFECTIVE SATISFACTION ELEMENTS

The K-means clustering and open card sorting analyses were conducted to categorize the affective vocabularies and identify the affective satisfaction elements from them.

3.1   Categorization of affective vocabularies

The K-means clustering analysis method was used to classify the collected affective vocabularies into three categories by affect perception stages. The three affect perception stages were defined as: (1) primitive affect stage; (2) descriptive affect stage; and (3) evaluative affect stage (Han et al., 2001; See Figure 2.)

420

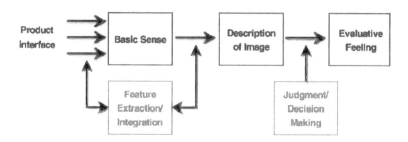

Figure 2. Transition of image/impression of a product (adopted from Han et al., 2001)

Three criteria were defined to categorize the vocabularies by affect perception stages. HCI experts evaluated each vocabulary by using three criteria: (1) how much interpretation or inference was immanent in the vocabulary; (2) how much evaluation was immanent in vocabulary; and (3) how abstract the vocabulary was. K-means clustering analysis was conducted with these evaluation results. As a result, 27 primitive affective vocabularies, 151 descriptive affective vocabularies, and 105 evaluative affective vocabularies were categorized.

3.2   Identification of affective satisfaction elements

Open card sorting and hierarchical clustering analysis were conducted to extract affective satisfaction elements toward a smartphone application. The affective vocabularies were generalized and merged into affective satisfaction elements in this stage. Thirty-two elements were identified from all the affective vocabularies.

The 32 derived elements were evaluated by HCI experts and refined. In this stage, primitive satisfaction elements were eliminated. It is because they are in too early a perception stage of sensation, and they do not have enough affective satisfaction evaluation but only have sensation narration. Some elements sharing a common nature were merged into one element, and elements that can be explained by other elements are eliminated. As a result, a total 11 elements were derived as affective satisfaction elements (see Figure 3). Six elements were classified as being in the descriptive affective element group and five elements as evaluative affective element group.

Smartphone interface

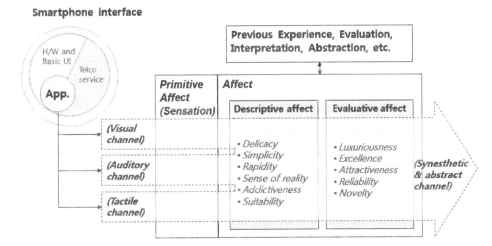

Figure 3. Affective Perception Stages and Affective Satisfaction Elements

4   DISCUSSION

Most of the collected vocabularies about smartphone applications were related to the visual channel. Several vocabularies related to auditory and tactile modality were also collected. However, there were no affective vocabularies related to the other two modalities (olfactory and taste modality). This could be because of the lack of olfactory and taste feedback on the usual smartphone application experiences. Therefore, the affect toward the smartphone application is strongly influenced by visual modality (Desmet and Hekkert, 2007).

Affective satisfaction elements in the primitive affect perception stage (color, volume, shape, brightness, etc.; abolished during the refining process) were influenced and bounded by the interaction channel. The descriptive affective satisfaction elements describe, depict, and explain the images of smartphone application. Lastly, evaluative affective elements (excellence, attractiveness, novelty, etc.) are not bound to one sensation channel and provide more information about users' preferences. Primitive affect is clearly discriminable from other affect. However, the boundary between descriptive affect and evaluative affect is somewhat ambiguous and unclear.

5   CONCLUSION

A total of 11 elements of affective satisfaction elements toward a smartphone application were identified, considering three affect perception stages. These affective satisfaction elements are expected to contribute as a reference in future application development and evaluation processes.

Because of the absence of olfactory and taste modality feedback, vocabularies related to those modalities were not collected and analyzed. In future studies, the affective satisfaction elements of olfactory and taste modality need to be developed. Methods to evaluate the affective satisfaction of smartphone applications also need to be developed.

ACKNOWLEDGEMENTS

This work was supported by the Mid-career Researcher Program through a National Research Foundation of Korea (NRF) grant funded by the Ministry of Education, Science and Technology (MEST) (No. 2011-0000115).

REFERENCES

Desmet, P., and P. Hekkert. 2007. Framework for product experience. *International Journal of Design* 1: 57-66.

Han, S. H., M. H. Yun, and J. Kwahk, et al. 2001. Usability of consumer electronic products. *International Journal of Industrial Ergonomics*, 28: 143-151.

Im, H., S. H. Han, and W. Park, et al., "Analysis of Human interface elements and evaluation of user satisfaction of mobile phones." Fall conference of the ergonomics society of Korea, Pusan, Korea. 2007.

Isomursu, M., M. Tahti, and S. Vainamo, et al. 2007. Experimental evaluation of five methods for collecting emotions in field settings with mobile applications. *International Journal of Human-Computer Studies*, 65: 404-418.

Kim, H. K., S. H. Han, and J. Park, et al. "Theoretical perspectives on user experience concept." *spring conference of Korea institute of industrial engineers*, Jeju, Korea, 2010.

Law, E. L.-C., V. Roto, and M. Hassenzahl, et al. "Understanding, scoping and defining user experience: a survey approach", *Proceedings of the 27th international conference on Human factors in computing systems CHI 09*, Boston. MA, 2009.

Park, J., S.H. Han, and H.K. Kim, et al. 2011. Developing elements of user experience for mobile phones and services: survey, interview, and observation approaches, *Human Factors and Ergonomics in Manufacturing & Service Industries*, Accepted.

CHAPTER 46

Identifying Elements of User Value for Smartphones through a Longitudinal Observation

Jaehyun Park, Sung H. Han

Department of Industrial and Management Engineering,
Pohang University of Science and Technology (POSTECH), Republic of Korea
shan@postech.edu

ABSTRACT

Even though researchers in the fields of sociology, psychology, and marketing have widely studied the concept of value itself for decades, the user value of products and services have recently begun to receive attention from researchers in the fields of human-computer interaction (HCI) and user experience (UX). Individuals can pursue several value elements at the same time. However, only parts of those value elements could be satisfied by a certain product or service. For example, a smartphone, no matter how many smart functions it has, would fail to meet one's need for salvation. The purpose of this study is to reveal the value elements that would be met by smartphone use. To achieve this, value elements were collected and merged to cover the spectrum of universal value. Then, the revised Day Reconstruction Method (DRM), one of the most frequently used methods for a longitudinal observation, was applied to collect the experiences of smartphone users. After analyzing how smartphone experiences can help to satisfy values, fifteen smartphone user value elements such as convenience and pleasure were extracted. The results of this study can help designers to evaluate their alternatives by using it as a checklist. Further study might develop concrete and reliable measurements to evaluate one's value elements.

Keywords: user value, longitudinal observation, smartphone

1   INTRODUCTION

Value, we pursue, can be defined as desirable states of existence or modes of behavior (Rokeach, 1968). Scores of years ago, psychologists investigated value elements (Kahle, 1983; Rokeach, 1968). Happiness, equality, pleasure, excitement, and beauty are candidates for value elements that any individual can pursue. Adopting this framework from psychologists, researchers in the field of marketing and business administration became concerned about how values might affect customer behavior (Gutman, 1982; Vinson et al., 1977). They focused on customers' purchasing decisions.

As studies on user experience (UX) were widely conducted in the field of human-computer interaction (HCI), the value from product or service experiences began to receive attention. Value, termed user value by UX researchers, has been regarded as a key constituent of product or service experiences (Boztepe, 2007; Park et al., 2011). Elements of user value are met when a user interacts with a certain product or service. The elements may vary with the products and services targeted. For example, elements of smartphone user value (e.g., convenience and pleasure) are different from those of driver value (e.g., safety and comfort).

The purpose of this study is to determine the elements of smartphone user value. A set of elements was developed so as to cover an entire spectrum of product/service user value. Then, elements of smartphone user value were selected from the set, based on the results of a longitudinal experiment collecting users' diverse experiences.

2   ELEMENTS OF USER VALUE

Above all, we comprehensively investigated related studies in the fields of psychology, sociology, marketing, design and HCI. Candidates for value elements were collected from the literature and merged. Then, a list including 55 value elements was drawn up (Table 1). This list was regarded as a superset of smartphone user value elements in this study.

Table 1 A list of value elements developed

Value elements				
Accomplishment	Courage	Happiness	Money	Salvation
Altruism	Courtesy	Honesty	National security	Self-actualization
Ambition	Creativity	Identity	Obedience	Self-control
Appearance	Curiosity	Independence	Peace	Self-respect
Beauty	Equality	Individuality	Physiological need	Sensitivity
Cheer	Excitement	Inner harmony	Pleasure	Social recognition
Cleanness	Family security	Intelligence	Privacy	Status
Comfort	Forgiveness	Justice	Relaxation	Tenderness
Competence	Freedom	Kinship	Reminiscence	Tradition
Confidence	Friendship	Logic	Responsibility	Trust
Convenience	Generosity	Love	Sacredness	Wisdom

The list was assumed to comprehensively cover the spectrum of universal value. Note that the list included almost every value element proposed by prominent value researchers, such as Rokeach (1968) and Schwartz and Bilsky (1987). However, some elements seemed to be similar to others because we allowed non-exclusiveness between elements to some degree. In addition, value judgments are excluded, if possible, from the process of drawing value elements. Value judgments, such as ranking value elements by importance, seem to be beyond the scope of engineering.

3   METHODS

After drawing up the list of value elements, a longitudinal observation targeting smartphone users was conducted to choose the elements of smartphone user value from the list. A longitudinal approach was used because the passing of time was regarded as one of the most important factors in analyzing user experience. The revised DRM (Day Reconstruction Method) was used in this study for sampling experiences of smartphone users.

Each participant was asked to keep diaries, consisting of several episodes, three times a day over a week (Figure 1). One episode described one smartphone-related event, including information on the title, time, place, description, and interpretation of the event. Participants were also asked to provide reasons why they thought about what they did, because inducing participants to be aware of the experience and its value is important (Reynolds and Gutman, 1988).

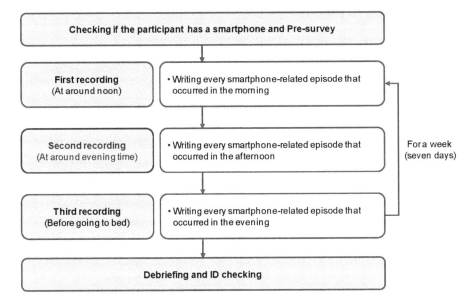

Figure 1 A procedure for conducting revised DRM

Twenty four participants, 24.4 (±3.4) years old on average, were recruited according to the time at which they had purchased their smartphone: 1) users who had bought smartphones less than a month before, 2) users who had bought smartphones more than a month before, and 3) users who planned to change smartphones within a month. Each group consisted of eight participants. Note that a month would probably be enough to learn how to use a smartphone and to impart a subjective meaning to it (Karapanos et al., 2009; Kim and Han, 2010).

4   RESULTS

A total of 705 episodes were collected. In order to extract smartphone user value elements, each episode was mapped onto value elements. A one-to-many relation was accepted at this time. A total of 402 linkages between episodes and value elements were mapped. Three criteria were applied in this process. First, whether the smartphone had met a participant's subjective and important values was investigated. Second, whether the smartphone had worked properly and normally was investigated. Third, to make sure the mapping was performed without a jump in logic, episodes were conservatively judged.

Finally, 15 value elements explaining the top 95% of linkages and being regarded as smartphone user value elements were selected: convenience, pleasure, money, friendship, beauty, curiosity, relaxation, comfort, tenderness, privacy, confidence, kinship, happiness, reminiscence, and excitement (Table 2).

Table 2 Smartphone user value elements and their episode examples

Elements	Episode examples
Convenience	A navigation application allows me to reach my destination by myself
Pleasure	I enjoy music and watch soap operas and movies through my smartphone
Money	I call or text my friends for free with VoIP or messaging applications via Wi-Fi
Friendship	SNS applications (e.g., Facebook) allow me to form friendships
Beauty	This smartphone case fulfills my aesthetic requirements
Curiosity	The mobile web always informs me regarding what I am curious about
Relaxation	When I am mentally fatigued, smartphone games make me relaxed
Comfort	I lay down on a bed, stretch, and feel comfortable having a phone on hand
Tenderness	I often call and ask about my parents
Privacy	An application helps me to check if my registration number was stolen or not
Confidence	When I watch a baseball game on my smartphone, my friends envy me
Kinship	I became a member of the community of iPhone users after I purchased it
Happiness	I feel happy when I manage a small web community through my smartphone
Reminiscence	Pictures stored in my smartphone album remind me of the past
Excitement	I and my friends have a good time with the Jenga application

5 DISCUSSION

The list reflects the usage patterns of Korean smartphone users. The UX of a certain product or service may depend on cultural or social backgrounds. In practice, Our Mobile Planet (2011) shows that only 12% of Korean smartphone users visited websites after viewing mobile advertisements, while 32% of Chinese smartphone users did so. Value elements that Chinese users feel to be met by the usage of smartphones may be different from those on the list provided in this study.

A longitudinal observation, as was used in this study, has pros and cons. It can cover various situations and the passage of time, which influences UX (Karapanos et al., 2009; Mäkelä and Fulton Suri, 2001). However, it cannot examine every minute of smartphone experiences in detail, unlike the in-depth interview and think-aloud methods. To compensate for this defect, this study increased the frequency of experience sampling as compared to that of the original DRM.

6 CONCLUSION

This study focused on extracting the user value elements of a smartphone. User value elements may vary with products or services; user value elements of a smartphone can be different from those of an automobile. User value elements of a certain product or service can be a subset of universal value elements. In this study, a total of fifty-five elements that an individual can pursue were developed. Then, user value elements of a smartphone were selected based on the results of a longitudinal observation.

The results of this study can be used as a checklist in order to investigate how newly-designed smartphones satisfy user value. In addition, this study has significance in terms of the process of extracting the user value elements of a smartphone. Similarly, in the future, user value elements of an automobile or home appliance can be drawn.

ACKNOWLEDGMENT

This work was supported by Mid-career Researcher Program through the National Research Foundation of Korea (NRF) grant funded by the Ministry of Education, Science and Technology (MEST) (No. 2011-0000115)

REFERENCES

Boztepe, S. 2007. User value: competing theories and models. *International Journal of Design* 1: 55-63.
Gutman, J., 1982. A means-end chain model based on consumer categorization processes. *Journal of Marketing* 46: 60-72.

428

Kahle, L.R. 1983. *Attitudes and Social Adaptation: A Person–Situation Interaction Approach*. London: Pergamon Press.

Karapanos, E., J., Zimmerman, J., Forlizzi, et al. 2009. "User experience over time: An initial framework." Proceedings of the 27th International Conference on Human Factors in Computing Systems (CHI 2009), Boston, MA 729–738.

Kim, H.K., S.H., Han. 2010. "A longitudinal study on the importance of user experience elements in mobile service." Proceedings of the 2010 Fall Conference of the Korean Institute of Industrial Engineers, CD format, Seoul, Korea.

Mäkelä, A., J., Fulton Suri. 2001. "Supporting users' creativity: Design to induce pleasurable experiences." Proceedings of International Conference on Affective Human Factors Design, Singapore, 387–394.

Our Mobile Planet. 2011. "Measuring global smartphone impact." Accessed December 7, 2011, from http://www.ourmobileplanet.com/

Park, J., S.H., Han, H.K., Kim, et al. 2011. Developing elements of user experience for mobile phones and services: survey, interview, and observation approaches. *Human Factors and Ergonomics in Manufacturing & Service Industries*, In press.

Reynolds, T.J., J., Gutman. 1988. Laddering theory, method, analysis, and interpretation. *Journal of Advertising Research* 28: 11-31.

Rokeach, M. 1968. *Beliefs, Attitudes and Values: A Theory of Organization and Change*, San Francisco: Jossey-Bass, Inc.

Schwartz, S.H., W., Bilsky. 1987. Toward a psychological structure of human values. *Journal of Personality and Social Psychology* 53: 550-562.

Vinson, D.E., J.E., Scott, L.M., Lamont. 1977. The role of personal values in marketing and consumer behavior. *Journal of Marketing* 41: 44-50.

CHAPTER 47

Development of an Affective Satisfaction Model for Smartphones

Hyun K. Kim, Sung H. Han, Jaehyun Park, Joohwan Park

Pohang University of Science and Technology(POSTECH)
Pohang, South Korea
shan@postech.ac.kr

ABSTRACT

Affect has come into the spotlight because customers have come to pay more attention to product design. This research has developed an affective satisfaction model to verify affect elements and to determine the importance of each of the elements. Participants in an evaluation experiment rated their affective satisfaction using a 0–100 scale (0 = least satisfied; 100 = most satisfied) with respect to the appearance of their smartphones and their three most frequently used applications. Six affect elements related to smartphone appearance and 11 aspects of smartphone applications were evaluated. The data from the experiment were analyzed using the multiple regression method. The affective satisfaction model developed in this study can provide background information for the future development of smartphones and applications

Keywords: Affect, Affective engineering, Smartphone, Application

1   INTRODUCTION

Affect has attracted a great deal of attention in product design, and the terms, affective engineering, affective architecture, and affective design are widely used. Affect represents the images or feelings that users have while interacting with products or services (Han et al., 2001). Although it is an ambiguous concept, it is considered an important factor that has an influence on customers' purchases. Affect was originally a significant concept in the design research field, but has also become a significant issue in the field of human-computer interaction (HCI). Consumer electronics, furniture, and web pages have previously been the objects of affect-related study. Nowadays, mobile phones are attracting such research attention because they have become a necessity in daily life (Hong et al., 2008, Yun et al., 2003).

Research on affective satisfaction with smartphones requires two considerable changes to be made in relation to previous research. First, previous research on affective satisfaction related to mobile phones considered only appearance, whereas studies on affective satisfaction in terms of smartphones have to take applications into account. There is no concrete and universal distinction between smartphones and feature phones, but many researchers agree that the major difference is the applications on smartphones. Second, earlier research only focused on determining affect elements or identifying product design features that led to users' affective satisfaction. Researchers used affect elements of a mobile phone subjectively because no verification procedures had been developed. In current research, the affective elements that compose affective satisfaction need to be verified.

An affective satisfaction model is a model that can verify affect elements and identify the importance of each of the elements. In this research, multiple regression analysis was used to develop an affective satisfaction model of smartphones' appearances and applications. The study identified affect elements related to smartphones and the importance of each element. The verification procedure developed in this study will be applicable to other affect-related studies that involve different types of products.

2   METHODS

Affective satisfaction with smartphones is divided into two parts: satisfaction with appearance and satisfaction with applications. In terms of the former, there have been many studies on affective satisfaction with feature phones' appearance. Affect related to smartphones' appearance is not different from that related to the appearance of feature phones. In this study, six affect elements connected to appearance were identified (Park et al., 2011), which were color, delicacy, simplicity, texture, luxuriousness, and attractiveness. In the case of applications, a focus group interview and a literature survey were conducted because there is a paucity of previous research. Eleven affect elements related to applications were identified, which were addictiveness, attractiveness, delicacy, excellence, luxuriousness, novelty, rapidity, reliability, sense of reality, simplicity, and suitability.

An experiment was conducted through the Internet, recruiting many unspecified persons who had their own smartphones. The participants entered their age, sex, and the model of their smartphone, along with the names of the three most frequently used applications. Then, they evaluated their affective satisfaction in terms of appearance and the three types of applications identified using a 0–100 scale (0 =

least satisfied; 100 = most satisfied).Smartphone users have 50 kinds of applications on average, but only three or four applications are used very actively(Falaki et al., 2010).Therefore, although the study only collected affective satisfaction with three frequently used applications, this could represent the affective satisfaction with smartphone applications as a whole.

After identifying various types of applications, the study classified them into seven categories according to their main functions: 1) conversation/social network service (SNS), which represents applications used for exchanging messages or calling; 2) entertainment, which includes applications that provide amusing and interesting content; 3) information offering, which involves applications that provide useful and essential information; 4) commercial related, representing applications created by companies for their commercial operation; 5) convenience related, which covers applications that make users' lives easier; 6) setting/utility, which includes applications for changing the state of the smartphone system; and 7) application store, which allows users to buy new applications.

It was acknowledged that participants who were not motivated and qualified for the assessment might take part in the web-based survey. Therefore, a verification procedure was developed to obtain more reliable experiment results (Cho et al., 2011). People who filled in an invalid score, used a particular score more than 75% of the time, used the same score for more than 10 continuous questions, or completed the survey too quickly were regarded as unmotivated participants. Additionally, the users' Internet protocol (IP) address and a social security number were checked to prevent multiple participations.

3 RESULTS

3.1 Participants

A total of 273 participants took a part in the online survey over seven days. Only 244 participants were selected after the verification procedure. Eighty-five percent of the participants were male and 15% were female. They had diverse brands of smartphones, specifically Samsung, Apple, HTC, LG, Motorola, Blackberry, SKY, Sony Ericsson, and so on. A total of 211 different applications, particularly Kakao talk, Facebook, and Web browsers were identified as frequently used applications, while communication/SNS, information offering, and entertainment were the representative application categories of the frequently used applications.

3.2 The affective satisfaction model for smartphones

3.2.1 The affective satisfaction model for smartphone's appearance

The multiple regression analysis was used to analyze the affective satisfaction with smartphones' appearance (Table 1). The dependent variable was affective satisfaction and the independent variables were color, texture, delicacy, simplicity, luxuriousness, and attractiveness. All possible regression method was used and Mallows' Cp criterion was considered as the model selection criterion. The adjusted R^2 value of the model was 0.82, and there was no multicollinearity problem because all variance inflation factors (VIF) were lower than 10. Mallow's Cp statistic indicated that the regression model had the least bias when six elements were included in the model. As expected, the six elements had a significant effect on affective satisfaction.

Table 1. Multiple regression model for smartphones' appearances

Target	Adj R^2	Affective satisfaction model
Appearances	0.82	9.55 + 0.12*×(Color) + + 0.11*×(Luxuriousness) + 0.19*×(Simplicity) + 0.13*×(Delicacy) + 0.13*×(Texture) + 0.22*×(Attractiveness)

※Note : *p<0.05

3.2.2 The affective satisfaction model for smartphone applications

The multiple regression analyses were conducted to analyze significant affect elements according to each application category. Affective satisfaction with total applications was analyzed and then affective satisfaction with representative categories of the most frequently used applications (conversation/SNS, information offering, and entertainment) was investigated. Table 2 shows the regression model for all smartphone applications and each application category. The dependent variable was affective satisfaction and the independent variables were delicacy, simplicity, rapidity, suitability, excellence, reliability, addictiveness, novelty, luxuriousness, attractiveness, and sense of reality. All possible regression models were used and the least biased model was selected using Mallow's Cp statistic. The adjusted R^2 values of all models were more than 0.70 and all VIF values were less than 10.

Table 2. Multiple regression models for smartphones' applications

Target	Adj R^2	Affective satisfaction model
Total application	0.77	7.26 + 0.10*×(Delicacy) + 0.05*×(Simplicity) + 0.10*×(Rapidity) + 0.02 ×(Sense of reality) + 0.06*×(Addictiveness) + 0.22*×(Suitability) + 0.04*×(Luxuriousness) + 0.17*×(Excellence) + 0.07*×(Attractiveness) + 0.04*×(Reliability) + 0.04*×(Novelty)
Conversation /SNS application	0.78	10.22 + 0.12*×(Delicacy) + 0.11*×(Rapidity) + 0.09*×(Addictiveness) + 0.27*×(Suitability) + 0.05*×(Luxuriousness) + 0.12*×(Excellence) + 0.05*×(Attractiveness) + 0.05*×(Reliability) + 0.04×(Novelty)
Information offering application	0.79	11.88 + 0.11*×(Delicacy) + 0.15*×(Rapidity) + 0.09×(Suitability) + 0.21*×(Excellence) + 0.09*×(Attractiveness) + 0.13*×(Reliability) + 0.09*×(Novelty)
Entertainment application	0.77	3.93 + 0.14*×(Delicacy) + 0.06×(Sense of reality) + 0.28*×(Suitability) + 0.29*×(Excellence) + 0.18*×(Attractiveness)

※Note : *p<0.05

4   DISCUSSION

This study has verified affect elements by analyzing an affective satisfaction model for smartphones' appearance and applications. In the case of the affective satisfaction with smartphone appearance, the six affect elements used in the study were all significant. In previous studies, the researchers identified different affect elements related to mobile phones' appearance and the number of elements ranged from 6 to 13 (Table 3). However, if phone designers consider only the six elements mentioned in this study, they can still fulfill the greater part of users' affective satisfaction.

In terms of affective satisfaction with smartphone applications, 10 affect elements were significant; sense of reality was the one exception. Interestingly, the affective satisfaction model of smartphone applications exhibited lower adjusted R^2 values than that of appearance, which means that people show more diverse and subjective affect when using applications. Affect arises from humans' five senses, which are sight, hearing, touch, smell, and taste and it is especially easily influenced by visual imagery (Desmet and Hekkert, 2007). However, applications have software characteristics like formlessness and intangibility, which are less related to human sight (Bateson, 1979; Zeithaml et al., 1985). This may lead to diverse, subjective characteristics of affect.

434

Table 3. The previous affect studies with mobile phones

Study	Affect elements
Han et al. (2004)	Color, Texture, Harmoniousness, Luxuriousness, Granularity, Simplicity, Rigidity, Salience, Attractiveness, Overall satisfaction
Hong (2005)	Simplicity, Attractiveness, Delicacy, Color, Luxuriousness, Overall satisfaction
Khalid & Helander (2004)	Portable, Sturdy, Enjoyable, Dignified, Cheerful, Natural, Delightful, Stimulating, Comfortable, Dazzling, Mature, Fashionable, Friendly, Cute, Futuristic
Cuang et al. (2001)	Traditional, Heavy, Hard, Nostalgic, Large, Masculine, Obedient, Hand-made, Coarse, Plagiaristic, Rational
Chen & Chuang (2008)	Originality, Utility, Completeness, Pleasure, Satisfaction of form, Simplicity

Affect elements related to each application category exhibited different features. Eight affect elements in the communication/SNS, six in the information offering, and four in the entertainment category were significant. Delicacy, attractiveness, and excellence were identified as significant elements in all of the application categories. Novelty is no longer an important element in the conversation/SNS category because the conversation/SNS function is one of the basic functions of mobile phones. In the case of the regression coefficient of reliability, the value in the information offering category (β=0.13) was approximately twice that for the total application value (β =0.04), which means that participants consider trustworthy information to be an important factor. Moreover, the regression coefficient for attractiveness (β=0.18) in the entertainment category was approximately twice that for the total application value (β=0.07), which means that users consider prefer pleasing, engaging, interesting, and attractive applications in the entertainment category. In this way, the characteristics of each application category were analyzed and application developers should focus on the specific affect elements according to the application's category.

5   CONCLUSION

This study surveyed affective satisfaction with smartphones through the Internet and analyzed affective satisfaction models using the multiple regression analysis. The results of this study allowed important affect elements of smartphones' appearance and applications to be identified and compared.

The affective satisfaction model developed in the study can be utilized in developing smartphones and applications, as well as evaluating affective satisfaction related to smartphones. Affect elements that have significant and higher coefficient values than the others should be considered in the development and evaluation process. Moreover, the verification procedure that was suggested in this study can be applied to other affect research involving various products. As a follow-up study, design factors of smartphones and applications related to each affect element can be analyzed and this would offer design concepts directly to developers.

ACKNOWLEDGMENTS

This work was supported by Mid-career Researcher Program through the National Research Foundation of Korea (NRF) grant funded by the Ministry of Education, Science and Technology (MEST) (No. 2011-0000115)

REFERENCES

Bateson, J.E.G. 1979. Why we need service marketing, in: Ferrell, O.C., Brown, S.W., Lamb Jr., C.W. (Eds.), Conceptual and theoretical developments in marketing. American Marketing: 131-146.

Chen, C.C., and Chuang, M.C. 2008. Integrating the Kano model into a robust design approach to enhance customer satisfaction with product design. International journal of Production Economics: 667-681.

Cho, Y., Park, J., Han, S. H., and Kang, S. 2011. Development of a web-based survey system for evaluating affective satisfaction. International Journal of Industrial Ergonomics 41(3): 247-254.

Cuang, M.C., Chang, C.C., and Hsu, S.H. 2001. Perceptual factors underlying user preferences toward product form of mobile phones. International Journal of Industrial Ergonomics 27: 247-258.

Desmet, P., and Hekkert, P. 2007. Framework for product experience. International Journal of Design 1 (1): 57-66.

Falaki, H., Mahajan, R., Kandula, S., Lymberopoulos, D., Govindan, R.,and Estrin, D. 2010. Diversity in smartphone usage. In MobiSys '10: Proceedings of the 8th international conference on Mobile systems, applications and services, New York, USA.

Han, S. H., Kim, K. J., Yun, M. H., Hong, S. W., and Kim, J. 2004. Identifying mobile phone design features critical to user satisfaction, Human Factors and Ergonomics in Manufacturing, 14(1), 15-29.

Han, S.H., Yun, M.H., Kim, K., and Kwahk, J. 2000. Evaluation of product usability: development and validation of usability dimensions and design elements based on empirical models. International Journal of Industrial Ergonomics 26: 477-488.

Hong, S.W. 2005. A methodology for modelling and analyzing user's affective satisfaction toward consumer electronic products, Unpublished Ph.D. Dissertation, POSTECH, Pohang, South Korea.

Hong, S.W., Han, S.H., and Kim, K. 2008. Optimal balancing of multiple affective satisfaction dimensions: A case study on mobile phones. International Journal of Industrial Ergonomics 38(3-4): 272-279.

Khalid, H. M., and Helander, M. G. 2004. A framework for affective customer needs in product design. Theoretical Issues in Ergonomics Science 5(1): 27–42.

Park, J., Han, S.H., Kim, H.K., Cho, Y., and Park, W. 2011. Developing elements of user experience for mobile phones and services: survey, interview, and observation approaches. Human Factors and Ergonomics in Manufacturing & Service Industries, Accepted.

Yun, M.H., Han, S.H., Hong, S.W., and Kim, J. 2003. Incorporating user satisfaction into the look-and-feel of mobile phone design. Ergonomics 13-14: 1423-1440

Zeithaml, V.A., Parasuraman, A., and Berry, L.L. 1985. Problems and strategies in services marketing, Journal of Marketing 49: 33-46.

CHAPTER 48

Determination of Ringtone Volumes of Mobile Phones: Applying Signal Detection Theory

Heekyung Moon, Sung H. Han*

Pohang University of Science and Technology
Pohang, South Korea
gomsak@postech.ac.kr, shan@postech.ac.kr*

ABSTRACT

Regardless of the ambient noise level, mobile phones always ring at a preset volume, causing inconvenience for users. This study aimed to find the appropriate ringtone volumes for different frequencies and for various noise levels. This study implemented the key concepts of signal detection theory to achieve the objective of this study. An experiment was conducted with 30 participants, using the forced choice tracking method. In the experiment, the signal was a pure tone and the noise was white noise. There were two factors: the ambient noise level (70 and 80dB) and the frequency of a signal (500, 1000, and 4000Hz). The results showed that the ringtone volume should be 10~15dB louder when the ambient noise level increases from 70dB to 80dB. In addition, the subjects were most sensitive to the pure tone with a frequency of 500Hz.

Keywords: ringtone volume, signal detection theory, frequency, noise

1   INTRODUCTION

Despite the recent technological breakthroughs in the mobile phone industry, people still experience inconvenience when adjusting the ringtone volume of mobile phones. Regardless of the ambient noise level, mobile phones always ring at a preset volume. As a result, people either fail to hear their ringtones in noisy places or are frightened by the loud phone ringing in quiet places.

Many researchers have been developing technologies to adjust the ringtone volume based on the ambient noise level (ETRI, 1999; Samsung Electronics Co., 2006; Motorola Inc., 2008). However, they did not specify the exact volume level needed for each noise level. Meanwhile, Yoo and Park (2000) suggested the appropriate ringtone volume (dBA) by surveying the subjective preferences in relation to a user's position, sex, ringtone type, environment, etc.

Until now, it is difficult to find research studies that considered a human's cognitive performance when adjusting the ringtone volume. Therefore, this study aimed to find the appropriate ringtone volumes for different frequencies and for various noise levels based on signal detection theory.

2   METHOD

2.1   Subject

Thirty subjects participated in this study. They consisted of 15 males and 15 females, and their mean age was 22.2 ($\pm$2.3). We conducted pure tone audiometry and distributed several questionnaires about hearing to select subjects without any hearing problems. Every subject who was allowed to participate had hearing loss less than 25dB, corresponding to the ISO standard definition of normal. Also, they had no medical history and living habits related to hearing problems.

2.2   Experimental design

A within-subjects design with two factors was used in the experiment. The two factors were the ambient noise level (70 and 80dB) and the frequency of a signal (500, 1000, and 4000Hz). The presentation order of the six experimental conditions was determined using the Latin square balancing technique to minimize the carryover effects, such as learning and fatigue. One session was comprised of just one condition and after each session, a five-minute break was given. Large sounds could easily cause a fatigue, plenty of rest was encouraged.

A pilot study was conducted to select the ambient noise level and frequency level of a signal. People could easily detect a small ringtone when the ambient noise level was less than 60dB, for example, in an office in the daytime or inside a house. Environments with noises greater than 90dB seemed to be an unusual situation for users, so two noise levels, 70 and 80dB, were selected. And frequency levels were composed of the major frequencies of commercial ringtones, such as basic melodies, classics, pop songs, etc.

A ringtone is masked by a noise differently depending on their frequencies. To reduce the difference in masking, this study chose a white noise as an ambient noise. Because the frequency of the signal should be uniform during the experiment, the signal was given as pure tone.

2.3 Apparatus

A commercial audiometer (model: DB-15000, 500~4000Hz, 0~100dB) was used for pure tone audiometry. White noise and a pure tone were created by Audacity Portable v1.3, and the volume of each sound was adjusted by Mp3Gain v1.2.5. Loudness level and the frequency of sounds were measured next to the subject's ear by using a sound-level meter (model: SC-160). An iPad was used to provide a secondary task for the subjects.

2.4 Procedure

In this study, the forced choice tracking method (Gescheider, 1997) was utilized to obtain the intensity (dB) of a signal corresponding to a 90% chance of signal detectability. This means that the hit rate and sensitivity (d') are 0.9 and 1.81, respectively, according to signal detection theory.

The subjects were given two observation intervals: one composed of only white noise and another composed of white noise plus a pure tone. Two intervals were given randomly for three seconds each. After listening to the two intervals, the subjects chose one interval which they thought contained a pure tone. If the answer is not correct, the intensity of the signal was increased by 3.0dB. After nine correct responses (not necessarily consecutive), the intensity was decreased by 3.0dB. At the same time, they calculated four fundamental arithmetic operations on the iPad.

At a 70dB noise level, the first signal was a 49dB pure tone and at an 80dB noise level, the first signal was a 58dB pure tone. If twenty successive answers were correct and the intensity range of the answers was within 2dB, the session ended. An appropriate intensity was obtained by averaging the last twenty answers.

3 RESULT

3.1 Appropriate volume level of a ringtone

The appropriate volume level of a ringtone, which could be detected 90% of the time, was obtained via the intensity of the pure tone according to the ambient noise level and the pure tone frequency (Table 1).

Table 1 Intensity of the pure tone according to the noise level and the pure tone frequency

Factor level	500Hz	1000Hz	4000Hz
70dB	45.8 (±3.69)	51.7 (±3.64)	49.5 (±2.37)
80dB	56.7 (±3.67)	62.9 (±2.72)	65.0 (±1.95)

3.2 Analysis of variance (ANOVA) and post-hoc analysis

According to the two-way ANOVA, both main effects (ambient noise level and pure tone frequency) and the interaction between the two factors significantly affected the intensity of the pure tone (0.05 α level). A Student-Newman-Kelus (SNK) test and a simple effect test were applied to significant effects.

According to a SNK test, the intensity of the pure tone was greater when the noise level was 80dB than 70dB. Also, the intensity of the pure tone was less when the frequency was 500Hz than 1000Hz or 4000Hz. However, there is no difference between the intensity of 1000Hz and 4000Hz pure tones.

The result of a simple effect test is depicted in Figure 1. At a 70dB noise level, the intensity of a pure tone with a 500, 4000, or 1000Hz frequency increased (Figure 1, Left). The intensity of pure tone was lowest when the frequency was 500Hz regardless of the noise level. However, the intensity between 1000Hz and 4000Hz pure tone was not different at an 80dB noise level. Regardless of the frequency, the intensity of the pure tone was higher when the noise level was 80dB than 70dB (Figure 1, Right).

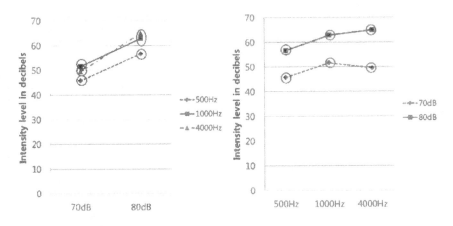

Figure 1: Results of a simple effect test. (Left) Intensity in decibels according to the ambient noise level. (Right) Intensity in decibels according to the pure tone frequency. Conditions within the same ellipse are not significantly different at the 0.05 α level.

4 DISCUSSIONS

4.1 Comparison with real ringtones

To compare the obtained intensity of a pure tone with the intensity of a real ringtone. This study measured the intensity of ringtones by using widely used mobile phones (models: SHW-M110S and AIP4-32 삼성? 애플?). The average of the maximum intensities was above 70 dB when 50cm away from the mobile phone. This level is about 5dB higher than the obtained intensities at an 80dB noise level.

Often, people set the ringtone as the maximum intensity level so as not to miss the ringing. However, this level is unnecessarily high, even when the surrounding environment is loud. Furthermore, in a quiet environment (less than 60dB), people would experience the great inconvenience of hearing the noisy ringing.

4.2   Relation with the Equal Loudness Contours

According to the equal loudness contours, human hearing is the most sensitive to sounds of 500 and 4000Hz, rather than 1000Hz, when the intensity range is from 40dB to 70dB (Sanders and McCormick, 1993).

This characteristic was identified in a simple effect test (Figure 1). The intensity of a pure tone differed significantly between 1000Hz and 4000Hz. In this experiment, the subjects were more sensitive to the 500Hz pure tone, rather than the 4000Hz tone, regardless of the noise level. However, the intensity of the 4000Hz pure tone was significantly different from that of the 1000Hz pure tone only at a 70dB noise level. In other words, people can easily recognize a ring tone of a lower frequency in a noisy environment.

5   CONCLUSIONS

This study suggested the appropriate intensity of ringtones of various frequencies at various ambient noise levels. This study obtained the intensity of pure tone corresponding to 90% detectability using the forced choice tracking method based on signal detection theory.

The subjects could recognize pure tones that are about 15~20dB below the noise level. For all frequencies used in this study, the appropriate intensity of pure tone increased as the noise level increased. In addition, they are the most sensitive to the 500Hz pure tone, so the main frequency of ringtone should be considered as well as the ambient noise level when setting up the intensity of the ringtone.

The result could be used as a standard value for adjusting a ringtone in accordance with ambient noise. In addition, this can also be used to design other auditory signals, especially a warning alarm. Further research is expected to consider other factors, such as the user's situation and the relative location of the mobile phone.

ACKNOWLEDGMENTS

This work was supported by the Mid-career Researcher Program through a National Research Foundation of Korea (NRF) grant funded by the Ministry of Education, Science and Technology (MEST) (No. 2011-0000115).

REFERENCES

ETRI. 1999. Apparatus and method for volume control of the ring signal and/or input speech following the background noise pressure level in digital telephone. Appl. No.1019990054900 (19991203), Korean Intellectual Property Office.

Gescheider, G. A. 1997. *Psychophysics: the fundamentals*, 3rd edition., New Jersey: Lawrence Erlbaum Associates.

Motorola, Inc. 2008. Methods and Devices for Adaptive Ringtone Generation. Appl. No. 11617807 (2006.12.29), the US Patent Office

Samsung Electronics, Co. 2006. Method for Controlling bell on noise in mobile communication terminal. Appl. No. 1020060015059 (2006.02.16), Korean Intellectual Property Office.

Sanders, M.S. and McCormick, E.J. 1993. *Human Factors in Engineering and Design*, 7th edition, New York: McGRAW-HILL, Inc.

Yoo, S. and Park, S. 2000. Survey on the Mean Opinion Scores on the Sound Level of Mobile Phone Ring Tones *The Journal of the Acoustical Society of Korea* 19 (6): 23-27.

CHAPTER 49

Design and Evaluation of Methods to Aid One-handed Input in Mobile Touch Screen Devices

Seunghwan Oh[1], Sung H. Han[1], Jongman Seo[2], D. Park1*

[1]Department of Industrial & Management Engineering, POSTECH, South Korea
[2]Department of Computer Science and Engineering, POSTECH, South Korea
shan@postech.ac.kr

ABSTRACT

This paper developed new techniques to press touch keys that are located at difficult areas to press and evaluated the usability of the new techniques in mobile devices. Two techniques involve pointing to the touch key in the upper left-hand corner of screen when the user uses only one-handed. When the user drags the display with his or her finger, the first technique rearranges the touch key positions for easy selection. The second technique involves the creation of an assisted cursor in the upper left area of the screen when the user touches a point in the lower right-hand area of the screen. The user then selects the touch key with the assisted cursor. A within-subject design was used in an experiment, in which one within-subject variable (three levels: the traditional manner and the two proposed manners) was manipulated. A total of 24 subjects in their 20s participated in the experiment. The experimental task was to press the touch key as quickly and accurately as possible when the target key appeared on the screen using the different techniques. We collected three objective measures: task completion time, number of errors, and number of re-grips. The results showed that the proposed method resulted in a better performance in terms of number of re-grips, but a worse performance in terms of task completion time. The results of this study can be used to design user interfaces for touch screen mobile devices with large screens.

Keywords: Touch screen, Mobile devices, Technique to aid selection, Usability

1   INTRODUCTION

Touch screen technology enables reduction of the screen size of a device so that input and output occur simultaneously on a single screen (Kwon et al., 2009; Hobart, 2005; Colle and Hiszem, 2004). Thus, many devices such as mobile devices, kiosks, and navigation devices currently use touch screen technology (Lee, 2010; Shneiderman, 1991). However, this technology exhibits low accuracy of input compared with existing ones that use physical buttons because it is difficult to recognize the actual pressing on the screen due to the lack of visual, tactile, and auditory feedback (Lee, 2010; Ostroff and Shneiderman, 1988). As a result, there is an increased need for touch screen-related research, such as improving accuracy of input. Some studies have been conducted to investigate usability issues in touch screens, such as touch key sizes, spacing, and location (Park et al., 2008; Park and Han, 2010; Vogel and Baudisch, 2007). Park et al. (2008) and Park and Han (2010) investigated the effect of touch key size and location on usability and subjective satisfaction with a screen divided into 25 areas (5 $\times$ 5). A touch key located near the center of the screen resulted in high performance and satisfaction in terms of task completion time and subjective satisfaction, while a touch key located near the left side of the screen exhibits high performance in terms of fewer errors when participants perform a target selection task. Vogel and Baudisch (2007) developed a new pointing technique they call "shift" that helps to select small touch keys on the screen, and evaluated new approaches in terms of performance (including error rate and task completion time) and subjective satisfaction. When the user touches the point on the screen, the shift technique creates a small screen that shows a copy of the occluded screen area and places this in a non-occluded location. In the results of using the "shift" technique, finger input and pen input result in similar levels of performance and subjective satisfaction.

Mobile devices tend to be bigger and bigger. Apple and Samsung Electronics, which manufacture mobile devices, have released new, larger models. This trend has resulted in a new usability problem such that it is difficult to select the touch key at the corners of the screen when the user uses only one hand. Although many studies have been conducted on touch screens, the usability problem outlined here has not been covered in previous research. A study by Karlson et al. (2006) revealed that users tend to use only one hand when they use a mobile device while standing or walking. A pre-survey was conducted on a sample of smart phone users to identify problems with smart phones. From the pre-test for users of touch screen devices, 65% of participants used one hand to operate the devices. In addition, 56% of participants thought that it was difficult to select touch keys located far away from the user's hand. Thus, we proposed two techniques to aid in pointing a touch key in the upper left-hand corner of the screen by using one hand, and evaluate these techniques in terms of usability.

444

2   METHOD

2.1   EXPERIMENTAL DESIGN

This experiment used a one-factor (touch key selection technique) within-subject design. The touch key selection technique had three levels such as one traditional manner and two proposed techniques.

When the user drags the display diagonally using his or her finger, the first technique rearranges the touch key position for easy selection. Specifically, the touch keys in the upper left-hand corner of the screen move to the center. When the screen is quartered, the second technique creates an assisted cursor in the upper left area of screen when the user touches a point in the lower right-hand area and holds it for more than 0.5 seconds. Thus, the user selects the touch key using the assisted cursor (Fig. 1).

In this experiment, the dependent variables were three objective performance measures. The objective measures were task completion time, number of errors, and number of re-grips. Among the objective measures, the number of re-grips is the number of changing the posture of hand when a participant performs the selection task.

Figure 1 Examples of the touch key selection techniques (from left: traditional manner, rearranging the touch key position, creating assisted cursor)

2.2   PARTICIPANTS

A total of 24 participants aged 18–28 years old (average: 22.9 (± 3.2)) participated in this experiment. The participants were recruited via advertisements placed on the POSTECH university homepage. All participants used the same touch screen devices (Model name: SHW-M110S) and frequently used such devices in their daily lives. They had normal visions and normal hands to perform the task of selecting the touch key on the screen.

2.3   APPARATUS

Smart phones (model name: SHW-M110S) based on the Android operating system were used in the experiment. Prototype programs used for the experiment were developed using Eclipse, which is a development tool for the Android operating system.

2.3   EXPERIMENTAL TASK

The experimental task was to press the touch keys as quickly and accurately as possible when the target key in red appeared on the screen (Fig. 1). The layout of the test screen included three pages and each page had 16 touch keys to make the task similar to real mobile device use. The presentation order of the three touch key selection techniques was determined by a balancing technique to reduce the learning effect and fatigue effect. The participants carried out the experiment after they practiced the three touch key selection techniques sufficiently. For each treatment condition, the target selection task was repeated 144 times (48 different touch key locations repeated three times for each location).

3   RESULTS

The analysis of variance (ANOVA) was conducted at a significance level of 0.05. The results showed that the main effects of task completion time and number of re-grips were significant at the 0.05 level. For the statistically significant main effects, the Student-Newman-Keuls (SNK) test was conducted (α=0.05). The ANOVA results are shown in Table 1.

Table 1 Summary of the ANOVA results (p-value)

	Task completion time	Number of errors	Number of re-grips
Touch key selection technique	<0.001*	0.162	<0.001*

(*: significant at α= 0.05)

3.1   TASK COMPLETION TIME

The traditional touch key selection technique showed a lower task completion time than the two touch key selection techniques. As a result of the SNK test for the task completion time, there was a significant difference among the three touch key selection techniques. The result of the SNK test for the task completion time is shown in Figure 2.

446

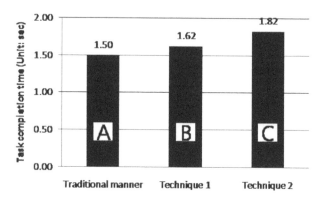

Figure 2. Results of the SNK test for task completion time (technique 1: rearranging the touch key position technique; technique 2: creating assists cursor technique; same alphabet means that there was no significant difference)

3.2 NUMBER OF RE-GRIPS

The traditional touch key selection technique showed a higher number of re-grips than the two touch key selection techniques. As a result of the SNK test for the number of re-grips, there was a significant difference between the traditional one and the two techniques, but there was no significant difference between the two techniques. The results of the SNK test for the number of re-grips is shown in Figure 2.

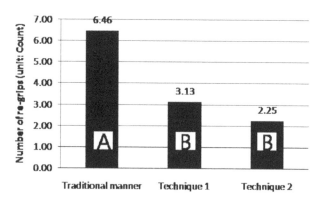

Figure 3. Result of SNK test for the number of re-grips (technique 1: rearranging the touch key position technique; technique 2: creating assists cursor technique; same alphabet means that there was no significant difference)

4   DISCUSSION

The results show that the two proposed techniques allow a higher performance than the traditional one in terms of the number of re-grips, but a lower performance than the traditional one in terms of task completion time. The two proposed techniques require more time than the traditional one to activate the techniques, such as moving to the touch key and creating the assisted cursor. In spite of this result, however, the two proposed techniques exhibit a small number of re-grips. Thus, the two proposed techniques can help to select the touch key in the upper left-hand corner of the screen without re-grips. Users may use touch screen devices with one hand when it is difficult to use the other hand or they are walking on the street (Karlson et al., 2007). The proposed techniques in this study can help users to grip mobile devices stably when they use mobile phones.

5   CONCLUSIONS

This study developed two new techniques making selections easier for the corners keys of touch screen mobile devices and evaluated the usability on actual mobile devices. It was found that the proposed techniques could help users to grip mobile devices with one hand. Future research would be pressing a touch key that is located too near to the user's hand.

ACKNOWLEDGMENTS

This work was supported by Mid-career Researcher Program through the National Research Foundation of Korea (NRF) grant funded by the Ministry of Education, Science and Technology (MEST) (No. 2011-0000115)

REFERENCES

Colle, H.A., Hiszem, K.J. 2004. Standing at a kiosk: effects of key size and spacing on touch screen numeric keypad performance and user preference. *Ergonomics*, 47, 1406-1423.
Hobart, J. 2005. Designing Mobile Applications, *Classic System Solutions*, Inc
Karlson, A.K., Bederson, B.B. and Contreras-Vidal, J.L. 2006. Understanding Single-handed Mobile Device Interaction, Technical Report, University of Maryland, *HCIL*, 2
Kwon S., Lee D., Chung M.K. 2009, Effect of key size and activation area on the performance of a regional error correction method in a touch-screen QWERTY keyboard, *International Journal of Industrial Ergonomic*, 39, 888-893
Lee Y. L. 2010. Comparison of the conventional point-based and a proposed finger probe-based touch screen interaction techniques in a target selection task, *International Journal of Industrial Ergonomic*, 40, 655-662
Ostroff, D., Shneiderman, B. 1988. Selection Devices for Users of an Electronic Encyclopedia: An Empirical Comparison of Four Possibilities, *Information Processing and Management*

Park, Y.S., Han, S.H., Park, J., Cho, Y. 2008. Touch key design for target selection on a mobile phone. *Proceedings of the 10th Mobile HCI conference. Amsterdam, Netherland.*

Park, Y.S., Han, S.H. 2010. Touch key design for one-handed thumb interaction with a mobile phone: Effects of touch key size and touch key location, *International Journal of Industrial Ergonomics*, 40, 68-76

Shneiderman, B., 1991. Touch screens now offer compelling uses. *IEEE Software*, 8 (2), 93-94.

Vogel, D., Baudisch, P. 2007. Shift: a technique for operating pen-based interfaces using touch, *Proceeding CHI '07 Proceedings of the SIGCHI conference on Human factors in computing systems*

CHAPTER 50

Evaluation Method for Product End-Of-Life Selection Strategy

Ghazalli Zakri[1], Murata Atsuo[2]

[1]:Faculty of Mechanical Engineering,
Universiti Malaysia Pahang,
26600 Pekan,
Pahang Darul Makmur,
MALAYSIA
[2]:Department of Intelligent Mechanical System,
Division of Industrial Innovation Science,
Graduate School of Natural Science and Technology,
Okayama University,
Tsushima Naka, Okayama City,
800-8530, JAPAN.
Email address[1]: zakri@ump.edu.my,
Email address[2]: murata@iims.sys.okayama-u.ac.jp

ABSTRACT

Environmental issues are increasingly important due to rapid development of products, and shortage of the landfill spaces. Thus, there is a strong demand for developing the evaluation system that can provide a quick estimation of EOL products. This paper presents an integrated evaluation system which considers the EOL assessments at both product and part levels. This paper serves two objectives. The first objective is to evaluate the product end-of-life (EOL) by integrating an analytical hierarchy process (AHP) with case-based reasoning (CBR). The second objective is to evaluate EOL of parts and components. The integration of an economic, environmental, and disassembly costs model is employed to determine the EOL of parts and components. An example is presented and discussed. The predicted results of the complete method are compared with those of the previous study. The results showed that the integration of the AHP–CBR method has reached a good correspondence with the established methods.

Keywords: case-based reasoning, analytic hierarchy process, end-of-life, economic cost, environmental cost, end-of-life profit

1 INTRODUCTION

Nowadays, the environmental issues are at the forefront of public concerns and many governmental decisions making due to the following factors: (a) wide circulation of short lifetime of consumer goods; (b) reduction of natural resources; (c) shortages of landfill area; and (d) rapid product development. Some products obsolete even though they are still in excellent condition. These problems force the legislators in many countries to enact various bills in order to protect the government. In particular, legislations such as EuP (Eco-design requirements for energy Using Products), WEEE (Waste Electrical and Electronic Equipment), RoHS (Restriction of the use of certain Hazardous Substance in electric and electronic equipment), and ELV (Directives for End-of-Life Vehicle), impose on the manufacturers responsibility to collect, reuse, remanufacture, recycle, and even dispose of products in an environmentally conscious way (The European Parliament and of the Council of European Union, 2000, 2003a, 2003b, Kongar and Gupta, 2005, Herrmann *et al.*, 2006, Anityasari, 2008, Mat Saman, 2009, Ma *et al.*, 2011).

Manufacturers have to seek other essential resources to provide for the input of the materials for manufacturing their products (Anityasari, 2008). This gives a strong demand for secondary resources such as remanufactured parts and recycled materials. This goal can be best achieved by promoting EOL selection strategies. Environmentally conscious manufacturing and product recovery (ECMPRO) is gaining in popularity, especially when new products are developed. Environmental awareness and regulations have pushed manufacturers and designers to recycle, remanufacture, and reuse the components at their end-of-life (EOL). As the consumer demand and government legislation require that manufacturers reduce the quantity of manufacturing waste, the best practice is to treat these products with end-of-life (EOL) processes (Mat Saman, 2009). These processes aim to minimize the amount of waste sent to landfills by recovering materials and parts of old or obsolete products by recycling and remanufacturing them (Kaebernick *et al.*, 2002, McGovern and Gupta, 2007, ElSayed *et al.*, 2010). Therefore, the EOL selection strategy is extremely crucial in enhancing product recovery (McGovern and Gupta, 2007).

Hitherto, most of the studies assume recycling as the main method for EOL strategy (Shih *et al.*, 2006, Shih and Lee, 2007). However, remanufacturing is a more-efficient EOL strategy than recycling (Inderfurth, 2005, Ijomah *et al.*, 2007, Ijomah and Childe, 2007, Ijomah and Chiodo, 2010, Ilgin and Gupta, 2010). In summary, the above-mentioned factors (from (a) to (d)) provide a strong demand for developing the evaluation system, which can provide manufacturers a relatively rough but quick estimation of EOL products. Realizing the importance of the strong demand, this study proposed the integrated evaluation system for EOL selection strategy specifically for remanufacturing.

This paper presents the integrated evaluation system for selecting the EOL strategies. The first objective is to evaluate the EOL selection strategy at the product level. The second objective is to evaluate the EOL strategies of parts and subassemblies from economic, environmental and disassembly aspects at the level of parts and subassemblies.

2   OVERVIEW OF THE INTEGRATION METHOD

We developed the integrated method of analytic hierarchy process (AHP) and case-based reasoning (CBR) to evaluate the EOL options at the product level. The evaluation process begins with gathering information about an EOL product (see Figure 1). The information consists of the identified EOL characteristics of the parts and components. The weights for these characteristics are given on the basis of the AHP.

The nearest neighbourhood algorithm (NN) is applied to find the smallest difference between the input case and the stored case. The stored case with a maximum percentage that more than 85% is considered as the closest case to the input case. If the EOL option of the retrieved case is remanufacturing, the input case will adopt the EOL option of the retrieved case. Next, this product undergoes the evaluation of the EOL option at parts and subassemblies level. It involves the integration of the economic, environmental and disassembly costs (EOL profits). We considered the maximum value of any EOL profits as the most suitable EOL option for the parts or subassemblies.

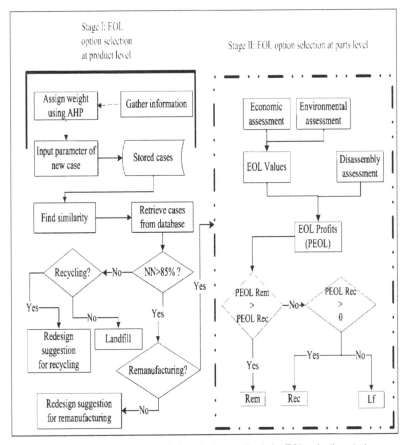

Figure 1. Overview of the integrated evaluation methods for EOL selection strategy

2.1   Product level EOL selection strategy

We employed the integration of AHP and CBR for finding the product EOL strategy. Fifty cases of the successful product EOL from Rose (2000) are adopted as the stored cases. The indices of these cases are defined based on End–of–Life Design Advisor (ELDA) (Rose 2000). In ELDA, candidates for EOL strategies are reuse, repair, remanufacturing, disassembly with material recovery, shredding with material recovery and disposal. In AHP–CBR method, the EOL selection strategy comprises the following steps:

- *Identifying the attributes for case indices.* The attributes such as market demand, technological changes, legislation, and product design. The descriptions of these attributes are given in Table 1.

- *Identifying the case indices for case retrieval.* The case indices such as wear-out life, technology cycle, level of integration, design cycle, number of parts, and reason for redesign. The descriptions of these indices are given in Table 2.

- *Determining the weights by using an Analytic Hierarchy Process.* The purpose of weight is to define the importance of each case index relative to others in retrieving the most appropriate case. An AHP evaluation has been employed for calculating the weight in CBR. The AHP structure is shown in Figure 2. The process of measuring weights is further discussed in Sub-section 2.2

- *Retrieving the similar cases from the case base.* The purpose of this process is to measure the similarity between the retrieved case and the input case. The similarity is basis of the comparison and calculation of a weighted sum of each parameter between the retrieved cases and the input case. The Euclidean distance is usually used to compare similarity between the retrieved cases and the input case (Vong *et al.* 2002). The procedure of retrieving similar cases is discussed in Sub-section 2.3.

Table 1. The attributes that influence the case indices (Rose 2000).

Attributes	Description
Market demand (MD)	Any demand or pressures from the consumers about how the product should feature. For instance, the consumer wants a durable product with a leading edge technology.
Technological change (TCh)	The technological changes influencing consumer to replace their aging product with a new models. For instance, the cellular phone.
Product design (PD)	Any design based on the consumer's voice. It has a major influence on the wear–out life, reason for redesign and technology cycle.
Legislation (L)	Any legislation that concerns to the environment and sustainability e.g. WEEE, ROHS,EPR, EPA and so on.

Table 2 Case indices for product level EOL strategy selection (Rose 2000)

Case Indices	Definition
Wear–out life (WOL)	Duration from which the product is purchased until it no longer meets the original functions. For example, the wear–out life of automobile is approximately 10 – 20 years
Technology cycle (TC)	The period during which the product will be on the leading edge of technology before the new technology makes the original product obsolete. For example the technology cycle of computer is approximately 6 months to 1 year
Level of integration (LOI)	Interrelation between modules and functions. If there are many unique functions, for each module, the level of integration is high
Number of parts (NOP)	Number of assemblies in the product relevant for EOL treatment
Reason for redesign (RFR)	Depends on customer demand, competitor behaviour and scientific progress. It is divided into original design, evolutionary design (either function improvement or aesthetic change), feature change (either function improvement or aesthetic change)
Design cycle (DC)	The frequency that a design team designs a product e.g. the design cycle of an automobile is about 2 to 4 years.

2.2 Determination of weight using Analytic Hierarchy Process

The purpose of weight is to define the importance of each case index relative to others in retrieving the most appropriate case. An AHP evaluation has been employed for calculating the weight in CBR. The attributes are structured in a hierarchical form such as market demands, technological changes, product design and legislation. The individual attributes are the case indices defined in Table 2. The structure of AHP is illustrated in Figure 2.

The main problem in CBR is how to set the weight of each attribute relative to others in retrieving the most similar cases. This difficulty can be overcome by pair–wise comparison in AHP. The weights associated with each attribute are calculated by geometric mean GM_i. The geometric mean and the normalized weight, w_i can be expressed as follows:

$$GM_i = \left[\prod_{j=1}^{n} a_{ij} \right]^{1/n} \tag{1}$$

$$w_i = \frac{GM_i}{\sum_{i=1}^{n} GM_i} \tag{2}$$

,where, a_{ij} is the element in the pair–wise comparison matrix. The $a_{ij} = 1$ when i=j and $a_{ij} = 1/a_{ji}$ when $i \neq j$, n= 1,2,3,......,i. The GM_i is normalized in order to obtain, w_i of each attribute.

454

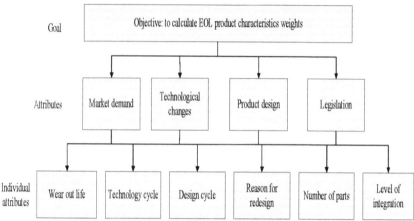

Figure 2. The structures of AHP in AHP–CBR method.

2.3 Retrieval of similar cases

The purpose of retrieving similar cases is to measure the similarity between the retrieved case and the input case. The Euclidean distance is employed to compare similarity between the retrieved cases and the input case (Vong *et al.* 2002). The Euclidean distance can be represented as follows:

$$Total\ Similarity(P_i, Q_i) = \frac{\sum_{i=1}^{n} w_i \times Sim(P_i, Q_i)}{\sum_{i=1}^{n} w_i} \tag{3}$$

$$Sim(P_i, Q_i) = 1 - dist(P_i, Q_i) \tag{4}$$

$$dist(P_i, Q_i) = \frac{\sqrt{(P_i - Q_i)}}{Max(P_i, Q_i) - Min(P_i, Q_i)} \tag{5}$$

However, this is valid for the numerical values. For the non-numerical values, the following equations are applied:

$$dist(P_i, Q_i) = \begin{cases} 0 & if\ P = Q \\ 1 & if\ P \neq Q \end{cases} \tag{6}$$

$$Max(P_i, Q_i) - Min(P_i, Q_i) = 1 \tag{7}$$

, where P_i and Q_i denote the input and retrieved cases of the ith index. w_i is the weight of the ith index. $\sum_{i=1}^{n} w_i$ is the summation of w_i and equal to 1. $Sim(P_i, Q_i)$ and $dist(P_i, Q_i)$ represent the function of similarity of the index and the distance between the input case(P_i) and the retrieved case (Q_i), individually. $Max(P_i, Q_i)$ and $Min(P_i, Q_i)$ denote the maximum value and the minimum value between input and retrieval cases.

The stored cases are screened to determine the cases that have a similarity with the new case of more than 85%. If the similarity value is less than 85%, or the EOL option is at the product level, it is assumed that the new case should be considered for recycling or landfilling of the selective components. Additionally, if the similarity value is more than 85%, or the EOL option is not remanufacturing, it is assumed that the new case should not be remanufactured. The new case should be considered for the redesign of a component so that it can be remanufactured.

2.4 Part level EOL selection strategy

The economic, environmental, and disassembly costs have been employed for selecting the EOL options. The model of EOL selection strategy at part level is illustrated in Figure 1.

The EOL values are defined for evaluating the EOL options at parts and components level. These values are remanufacturing value, recycling value and landfill value (Lee et al. 2001). Additionally, these values represents the assessments of selecting the EOL options from economic and environmental perspectives. These values can be represented as follows:

$$V_{Rem} = (C_{pi} + C_{pEnv}) - (C_{Rem} + C_{EnvRem}) - (C_{Misc} + C_{pEnvMisc}) \quad (8)$$
$$V_{Rec} = (C_{Rec} + C_{EnvRec}) - (C_{Misc} + C_{pEnvMisc}) \quad (9)$$
$$V_{Lf} = -(C_{Lf} + C_{EnvLf}) - (C_{Misc} + C_{pEnvMisc}) \quad (10)$$
$$C_{Misc} = C_{trans} + C_{Process} + C_{Handling} + C_{Storage} \quad (11)$$
$$C_{EnvMisc} = C_{Envtrans} \quad (12)$$

, where C_{pi}, C_{Rem}, C_{Rec}, C_{Lf} represent the lifecycle costs to produce, remanufacture, recycle,and landfilling part i, respectively. C_{Misc} is the miscellaneous cost which include the costs of transportation (C_{trans}), handling ($C_{Handling}$), reprocessing ($C_{Process}$), storing ($C_{Storing}$) the EOL part i. V_{Rem}, V_{Rec}, and V_{Lf}, denote the remanufacturing, recycling, and landfilling values (EOL values), individually.

The EOL profits are included for evaluating the end-of-life parts and subassemblies of the EOL product. The difference between EOL value and disassembly cost are used to find the most suitable and profitable EOL options.. The EOL profit can be represented as follows:

$$P_{EOL_k} = \begin{cases} V_{Rem} - C_{Dis} & if \quad k = remanufacturing \\ V_{Rec} - C_{Dis} & if \quad k = recycling \\ V_{Lf} - C_{Dis} & if \quad k = recycling \end{cases} \quad (13)$$

$$C_{Dis} = t_{dis} \times \dot{C}_{Lb} + C_{tool} + C_{ohd} \quad (14)$$

, where C_{Dis} is the cost to disassemble part i, t_{dis} is the labor time to disassemble part i from an EOL product and $\dot{C}_{Lb}$ is the labour rate per hour($/hrs). C_{tool} and C_{ohd} denote the cost of using tool(s) during disassembling part i and is the overhead cost of disassembled part i, individually. P_{EOL_k} represents the EOL profits of remanufacturing, recycling or landfilling (k) options, respectively.

If the V_{Rem}, is higher than disassembly cost, and the EOL profits of remanufacturing is higher than that of the recycling and landfill, the remanufacturing will be consider as an EOL option. If V_{Rem} is lower than disassembly cost, the part or subassemblies under consideration should be further evaluated for recycling or landfilling.

3 RESULTS OF THE APPLICATION ON THE ELECTRONIC PRODUCTS

Figure 3 and Figure 4 show the weight distributions of each attributes and individual attributes during AHP evaluation, respectively. Figure 5 shows the comparison of EOL selection strategy between the AHP-CBR and traditional

456

approach. Figure 6 and Figure 7 show the comparison of remanufacturing values with disassembly cost, and EOL option between the EOL profits and traditional method, respectively.

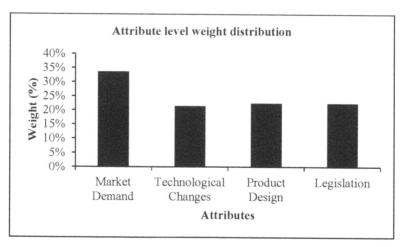

Figure 3. The weights distributions at attribute level.

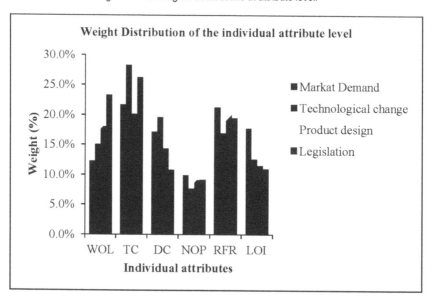

Figure 4. The weights distributions at individual attribute level.

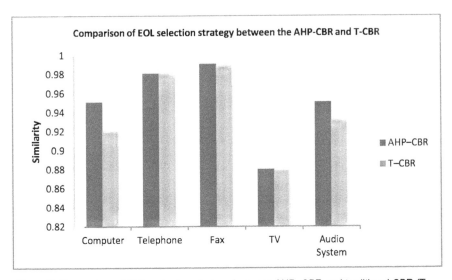

Figure 5. Comparison of EOL selection strategy between AHP–CBR and traditional CBR (T-CBR) methods

Table 3. comparison of EOL profits, NPV and EOL options of seven selected parts of a desk phone

No	Part name	EOL profits			EOL options	
		Rem	Rec	Lf	EOL	Trad.
3	4 screws	-0.271	0.001	-0.245	Rec	Rec
8	Screws	-0.701	0.001	-0.241	Rec	N/A
14	Shift button	-0.07	-0.001	-0.224	Rec	Rec
17	Feature button	-0.02	-0.002	-0.224	Rec	Rec
18	Release button	-0.045	-0.001	-0.224	Rec	Rec
22	Volume control	-0.03	-0.001	-0.224	Rec	Rec

Rem: Remanufacturing, Rec: Recycling, Lf: Landfill, EOL: EOL profit method, Trad: Traditional method

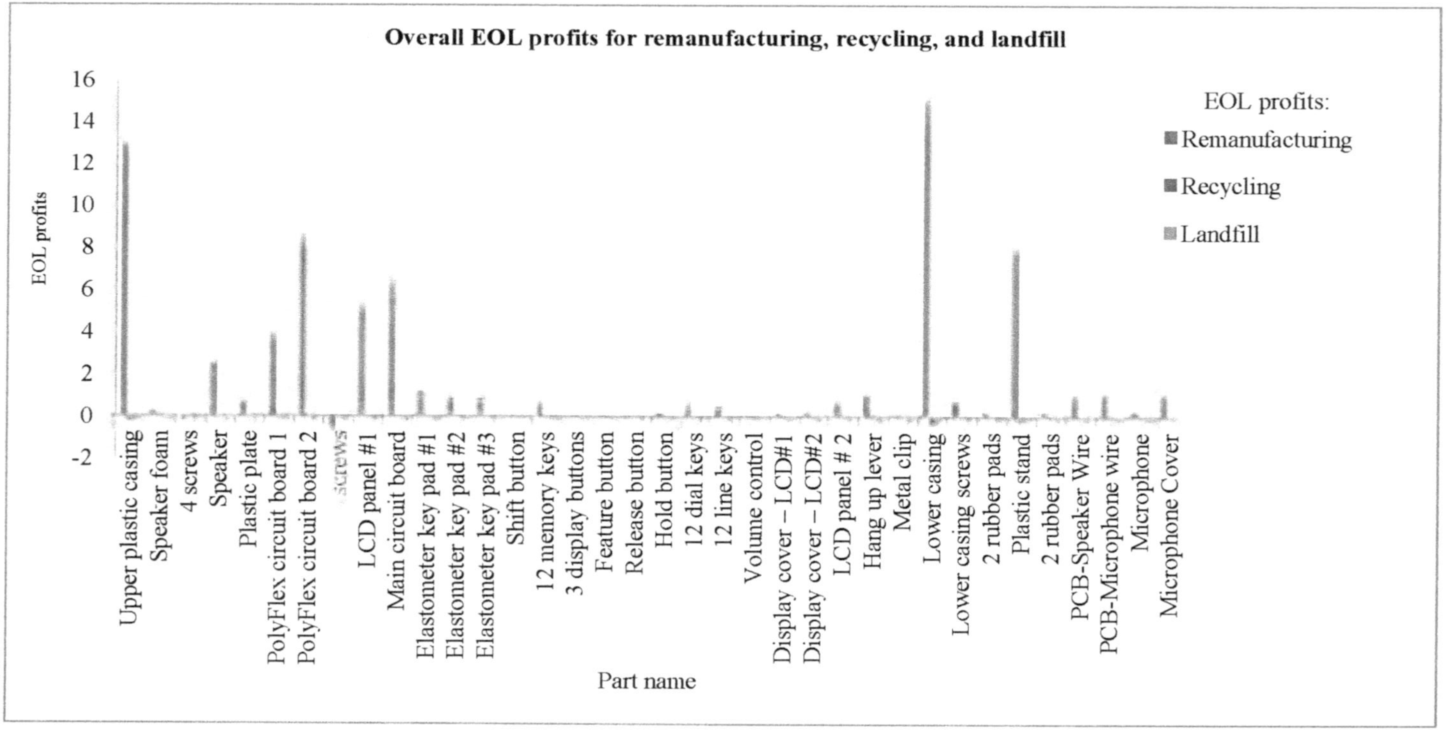

Figure 6. Comparison of remanufacturing values and disassembly cost

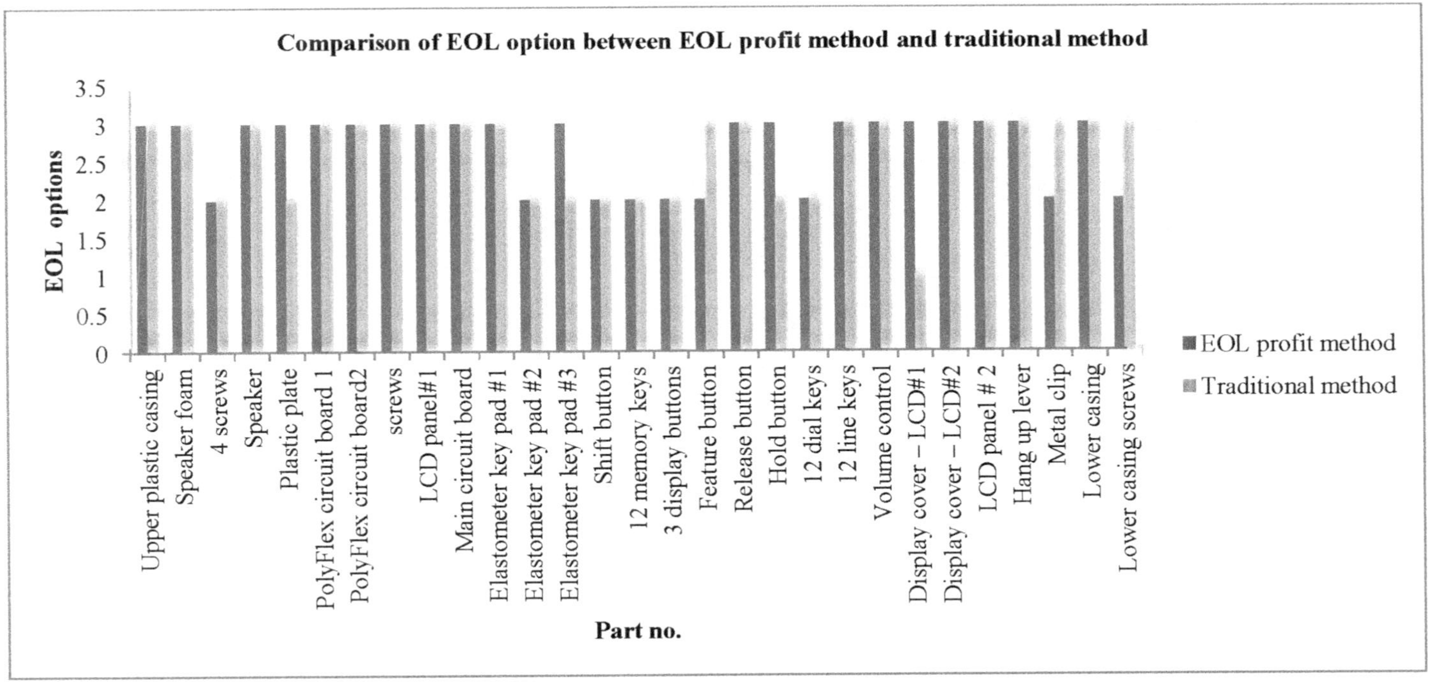

Figure 7. Comparison of EOL option between EOL profits method and traditional method

4   DISCUSSION OF RESULTS

As shown in Figure 3, the attribute of market demand (MD) is the most important, followed, in this order, by the attributes of legislation (L), product design (PD), and technological changes (TCh). Our findings demonstrate that the AHP gives priority weights to the features using pairwise comparison at each level. The possible explanation is that the AHP prioritized and ranked the attributes in the form of ratio scale numbers.

Figure 4 shows that the individual attributes of technology cycle (TC) and reason for redesign (RFR) are very important in selecting the EOL strategy of a product; then the individual attributes of wear out life (WOL), design cycle (DC), level of integration (LOI) and number of parts (NOP) follow. The attribute of MD is very influential in TC, DC, RFR, and LOI. On the other hand, none of the attributes are crucial in NOP. It can be concluded that how to design a product is more important than aiming at reducing the number of parts of the EOL product during selecting the EOL strategy.

Figure 5 shows that the AHP-CBR outperformed the traditional CBR (T-CBR) in finding the similar case. The computer, fax, and audio system give higher similarity values than the T-CBR. Additionally, the telephone and TV show no significant difference of similarity value between our approach and T-CBR. Our finding shows that the AHP-CBR approach is very close to T-CBR approach. One possible explanation is that AHP-CBR approach utilized AHP to provide weights to CBR structurally and systematically.

Another objective in this study is to find the most profitable EOL option at parts level. Figure 6 shows the overall EOL profits of remanufacturing, recycling and landfill. The EOL profit for landfill is always in deficit. As a whole, the EOL profit of remanufacturing outperformed others. This result indicates that it is more profitable to remanufacture these parts than to recycle or landfill from economic, environmental, and disassembly perspectives. Also, our finding demonstrates that it is the most profitable to recycle the parts such as screws, volume control, shift, feature, and release buttons (see **Table 3**). One possible explanation is that the cost to disassemble these parts for remanufacturing is more expensive than recycling them. Additionally, the EOL profit is in deficit if these parts are remanufactured. Figure 7 shows a strong agreement between the EOL profits method and traditional method in EOL strategy selection. Parts such as elastometer keypad#3, hold button, and display cover-LCD#1 are more profitable to remanufacture than recycling or dumping them (see Figure 7). On the other hand, the feature button, metal clip, and lower casing screws are more profitable to recycling than remanufacturing.

5   CONCLUSIONS

We presented an integrated approach for selecting the EOL strategy. The integration of AHP and CBR is used to determine the remanufacturing strategy at the product level. The cost model, which consist of the economic, environmental, and disassembly perspective, is used to evaluate the EOL profits in finding the best EOL option. The developed method was compared with previous studies and showed a good correspondence with the past literatures.

The characteristic of the survey is based on individual participants of a scatter population. It is possible that the survey should be extended to groups of engineers and executives in the manufacturing industry. Additionally, the judgment of weights could be biased due to the experience gap among the participants. This method can be improved by consideration of employing fuzzy analysis to reduce the bias caused by weighting factor and gap of experience.

REFERENCES

Anityasari, M., 2008. Reuse of industrial products – a technical and economic model for decision support. PhD. The University of New South Wales.

Elsayed, A., Kongar, E. and Gupta, S.M., 2010. A genetic algorithm approach to end-of-life disassembly sequencing for robotic disassembly. Proceedings of the 2010 Northeast Decision Sciences Institute Conference. Alexandria, Virginia, 402-408.

Herrmann, C., Frad, A., Luger, T., Krause, F.-L. and Ragan, Z., 2006. Integrating end of- life evaluation in conceptual design. Proceedings of the 2006 IEEE International Symposium on Electronics and the Environment, 2006. Scottsdale,AZ,United States: IEEE, 245-250.

Ijomah, W., Mcmahon, C., Hammond, G. and Newman, S., 2007. Development of design for remanufacturing guidelines to support sustainable manufacturing. Robotics and Computer-Integrated Manufacturing, 23 (6), 712-719.

Ijomah, W.L. and Childe, S.J., 2007. A model of the operations concerned in remanufacture. International Journal of Production Research, 45 (24), 5857-5880.

Ijomah, W.L. and Chiodo, J.D., 2010. Application of active disassembly to extend profitable remanufacturing in small electrical and electronic products. International Journal of Sustainable Engineering, 3 (4), 246-257.

Ilgin, M.A. and Gupta, S.M., 2010. Environmentally conscious manufacturing and product recovery (ecmpro): A review of the state of the art. Journal of Environmental Management, 91 (3), 563-591.

Inderfurth, K., 2005. Impact of uncertainties on recovery behavior in a remanufacturing environment: A numerical analysis. International Journal of Physical Distribution & Logistics Management, 35 (5), 318-336.

Kaebernick, H., Anityasari, M. and Kara, S., 2002. A technical and economic model for end-of-life (eol) options of industrial products. Int. J. Environment and Sustainable Development, 1 (2), 171-183.

Kongar, E. and Gupta, S.M., 2005. Disassembly sequencing using genetic algorithm. The International Journal of Advanced Manufacturing Technology, 30 (5-6), 497-506.

Ma, Y.-S., Jun, H.-B., Kim, H.-W. and Lee, D.-H., 2011. Disassembly process planning algorithms for end-of-life product recovery and environmentally conscious disposal. International Journal of Production Research, 49 (23), 7007-7027.

Mat Saman, M.Z., 2009. Design for end of life value: Methodology and implementation: VDM Verlag.

Mcgovern, S. and Gupta, S.M., 2007. A balancing method and genetic algorithm for disassembly line balancing. European Journal of Operational Research, 179 (3), 692-708.

Shih, L., Chang, Y. and Lin, Y., 2006. Intelligent evaluation approach for electronic product recycling via case-based reasoning. Advanced Engineering Informatics, 20 (2), 137-145.

Shih, L.H. and Lee, S.C., 2007. Optimizing disassembly and recycling process for eol lcd-type products: A heuristic method. IEEE Transactions on Electronics Packaging Manufacturing, 30 (3), 213-220.

The European Parliament and of the Council of European Union, T.E.P.a.O.T.C.O.E.U., 2000. Directive 2000/53/ec of the european parliament and of the council of 18 september2000 on end-of-life vehicles. Brussels: Official Journal of the European Communities.

The European Parliament and of the Council of European Union, T.E.P.a.O.T.C.O.E.U., 2003a. Directive 2002/95/ec of the european parliament and of the council of 27 january 2003 on the restriction of the use of certain hazardous substances in electrical and electronic equipment. Brussels: Official Journal of the European Communities.

The European Parliament and of the Council of European Union, T.E.P.a.O.T.C.O.E.U., 2003b. Directive 2002/96/ec of the european parliament and of the council of 27 january 2003 on waste electrical and electronic equipment (weee). Brussels: Official Journal of the European Communities, 1-15.

CHAPTER 51

Studying Customer Experience for Chinese Fast-Food Service

Na Chen, Pei-Luen Patrick Rau, Ming-Qiang Wang

Tsinghua University
Beijing, China
chenn06@mails.tsinghua.edu.cn, rpl@tsinghua.edu.cn, wangmq06@gmail.com

ABSTRACT

Chinese residents' dietary has been undergoing great changes in recent 20 years. Fast food industry is a typical service industry and has been developing rapidly in China. For Chinese customers, food service is the most important domain of customer experience. Food healthiness is major character of food service. The study aimed to investigate whether current Chinese fast food satisfies customers' nutrition and healthiness requirements and at the same time provides a good customers' experience. The results of the study indicated that (1) Chinese Beef Rice and Beef Noodle (two most popular Chinese fast foods) cannot satisfy customers' nutrition and healthiness requirements, but Beef Rice performed a little better than Beef Noodle. (2) The food nutrition has negative relationship with customer experience of food-related domain. Chinese customers do not know how to select healthy fast food.

Keywords: Chinese fast food, customer requirement for nutrition, customer experience

1   OBJECTIVE AND SIGNIFICANCE

Fast food industry is one feature of urbanization in modern business, and also is a typical service industry. Since from 1987, China fast food industry has been developing rapidly (Ping 2002). In 2009, the annual business revenue of Catering service was 268.6 billion RMB, while the number was just 5.4 billion in 1978. And fast food chain catering enterprises took 45.5 billion of the 268.6 billion (National Bureau of Statistics of China 2010).

With the rapid development, customers pay more attention on experience of fast-food restaurants. Bernd H. Schmitt defined customer experience as a process of strategic management of customers on their total experience and perception of the products or the whole company (Schmitt, 2003). According to the researchers' previous study, there are four domains of customer experience for fast food chains, i.e. Food-Related Domain, Staff-Related Domain, Interior Design-Related Domain, and Convenience-Related Domain. Food-Related Domain is the dominantly most important service domain for Chinese customers (Choe, Rau, Chen, & Jeon 2011).

For further investigation of Chinese customer experience, the researchers designed a series of studies focusing on specific areas. Food healthiness is the most primary area of fast-food industry. And in Food-Related Domain, food healthiness is increasingly drawing customers' attentions. China is a country with a splendid cooking culture and her people are extremely proud of Chinese traditional food, which is thought as "healthy" and "nutritious", such as noodle, baozi, and dumplings. However, China has undergone dramatic changes since the reform and opening, and Chinese citizens' dietary structures and habits vary somewhat with the growing of income (Feng & Shi, 2006; Zhai et al., 2006).

However, there is no persuasive study about this primary area. For better understand of customer experience of fast-food industry, a study focusing on fast food healthiness and nutrition is necessary. Hence, the research question of this study is to investigate whether current Chinese fast food satisfies customers' nutrition and healthiness requirements and at the same time provides a high customers' experience.

The study valued the nutrient contents and some related parameters of two popular Chinese fast foods, Beef Rice and Beef Noodle, so as to investigate Chinese customers' experience and requirements from the aspect of restaurants.

2 METHODOLOGY

Two experiments were conducted: Beef Noodle experiment and Beef Rice experiment. The methodology associated with the experiments is discussed as following.

2.1 Selection of restaurants and meals

Based on the investigation on the website of Dazhong Dianping (www.dianping.com), the study selected 20 most popular nationwide Chinese fast-food chains in Beijing (the capital city of China). The website was the biggest and most powerful consumer guide website in China.

Then, according to the analysis of the menus of these chains, Beef Rice and Beef Noodle were the two most common Chinese fast-food meals. Beef Rice appeared in the menus of 12 restaurants and Beef Noodle appeared in the menus of 7 restaurants.

At the same time, rice and noodle are the most common main food of South and North China. Hence, Beef Rice and Beef Noodle can be considered as typical Chinese fast-food meals.

Based on the locations and menus, 6 fast-food restaurants with Beef Rice and 6 fast-food restaurants with Beef Noodle were selected from the 20 most popular Chinese fast-food chains. All the restaurants were located in University Area of Beijing, where the business was developed and the population is large, to avoid the influence of area differences.

Beef Noodle experiment was conducted in Yonghe King (www.yonghe.com.cn), Master Kong Chef's Table (www.masterkongchef.com), Kyo-Nichi, Malan (www.malan.com.cn), Mr. Lee Beef Noodle (www.mrlee.com.cn), Yipingsanxiao (www.yippin.cn). Beef Rice experiment was conducted in Yoshinoya (www.bjyoshinoya.com.cn), Kungfu (www.zkungfu.com), Hehegu (www.hhg.com.cn), Yonghe King, Banmuyuan (www.banmuyuan.com.cn), Laojiaroubing (www.laojia.com.cn). These branches are selected from nationwide fast-food chains. These chains have strict service and food standards. Hence, the branches are typical for Chinese fast food.

2.2   Selection of measured nutrients and measuring tools

Proteins, fats, carbohydrates, vitamins, water and inorganic salts are the six nutrients necessary for the human body.

The major ingredient of rice and noodle are carbohydrates. Beef, rice and noodle contain a lot of fat and protein. Hence, the study measured the contents of carbohydrate, fat and protein.

Inorganic salts include many chemical elements, such as Na^+, Ca^{2+}, and Zn^{2+}. NaCl (salt) is the most important condiment for Chinese foods. Some studies indicated the association between salt intake and high blood pressure (Law, Frost, & Wald, 1991; Liu, et al., 1979). The greater East Asian, especially for China, usually has a higher salt intake (Zhou, et al., 2003). Hence, the study measured the content of Na^+.

There are many kinds of Vitamin. According to the researchers' experience, Beef Rice and Beef Noodle usually contain little vegetable, so as that the vitamin contents are low in the fast-food meals. Customers do not depend on fast foods to intake water. Hence, the study did not measure the contents of vitamins and water.

In all, the nutrients to be measured include the contents of calorie, fat, carbohydrate, Na^+, and protein. The weights of main food, beef and vegetable were also measured. Besides, prices were recorded as reference.

2.3   Measurement procedure

Two experimenters conducted the experiment. They bought two meals (Beef Rice or Beef Noodle) in one restaurant at the same time. One was used for field measurement. The other one was to be sent to a professional agency for the measurement of the contents of calorie, fat, carbohydrate, and salinity.

3   RESULTS

Chinese Nutrition Society, a national professional and academic group, provides recommended dietary intakes for Chinese in *Chinese Dietary Guide (2007)* and *Chinese Dietary Reference Intakes (DRIs)*. The study took the recommends as references of the meals.

The selected fast-food branches were all located in University Area of Beijing. Most of the customers were college students, college staff, and businessmen. It was reasonable to consider that they were light or moderate work labors and older than 18 years old. Hence, the study selected the recommended values for light and moderate work labors, aging from 18 years old.

Beef Rice and Beef Noodle are usually considered as lunches and suppers by Chinese customers. Chinese Nutrition Society recommends that the intake nutrition from lunches or suppers should take 30%~40% of the whole days'. Hence, the study converted the recommended intake nutrition for one meal.

3.1   The weights of main food, beef and vegetables

The recommended weights of main food, beef and vegetables for one meal are 75~ 60 grams, 15~30 grams, and 90~200 grams. However, for both Beef Rice and Beef Noodle, all the restaurant branches provided much more main food and beef than the recommends. Two braches even did not provide vegetables and the other branched provided much lower vegetables than the recommends.

5 Beef Noodle branches and 2 Beef Rice branches provided more than twice of main food than recommends. 3 Beef Noodle branches and 2 Beef Rice branches provided less than a half of vegetables than recommends. Hence, although both Beef Rice and Beef Noodle are not nutritional balanced, Beef Rice branches performed a little better than Beef Noodle branches.

3.2   The contents of calorie, fat, carbohydrate, and salinity

Table 1 shows the nutrient contents of Beef Rice and Noodle. The recommended energy provided by one meal are 720~1020 kcal for the male and 630~920 kcal for the female. Except for Malan, no branches can provide enough energy for both the male and the female. The energy provided by fat and carbohydrate should take 20%~30% and 50%~60% of the total energy.

Beef Rice of 4 branches contained higher percentages of fat. The other 2 Beef Rice branches satisfied the recommended range. The Beef Noodle of 4 branches contained lower percentages of fat. The other 2 Beef Noodle branches are higher than the recommended range. 3 Beef Rice branches provided less protein then the recommends. 4 Beef Noodle branches provided more protein than the recommends. Only 1 Beef Rice branch and 1 Beef Noodle branch satisfy the protein content. The content of Na^+ of 2 Beef Rice branches satisfied the recommends. The other Beef Rice and Beef Noodle branches provided higher level of Na^+. Especially, the contents of Na^+ of all Beef Noodle branches were more than twice of the recommends. For nutrient contents, Beef Rice branches performed a little better than Beef Noodle branches.

Table 1 Nutrient (fat and carbohydrate) contents of Beef Rice and Noodle and recommended values

Restaurant		Energy			Content / g	
		Total energy / kcal	Fat / %	carbohydrate / %	Protein	Na$^+$
Beef Rice	Yoshinoya	644.36	35.36	50.25	23.19	1.05
	Kungfu	661.13	34.65	49.45	26.29	0.77
	Hehegu	642.37	49.25	39.61	17.88	0.67
	Yonghe King	597.49	25.43	61.58	19.40	0.94
	Banmuyuan	519.04	47.76	27.63	31.93	1.15
	Laojiaroubing	856.16	24.56	62.77	27.13	1.84
Beef Noodle	Yonghe King	450.03	15.24	60.36	27.44	2.75
	Master Kong	659.32	8.89	61.8	48.3	2.56
	Kyo-Nichi	618.76	26.43	46.99	41.12	2.46
	Malan	1540.89	59.23	30.37	40.07	3.72
	Mr. Lee	604.62	15.28	52.83	48.20	4.20
	Yipingsanxiao	533.23	15.80	64.99	25.61	3.96
Recommended values / meal		Male 720~1020 Female 630~920	20~30	50~60	Male 22.5~32 Female 19.5~28	0.66~0.88

3.3 Customer experience and nutrition evaluation

Table 2 shows the customers' experience of food-related domain from previous research (Choe et al., 2012). For evaluation, mark the nutrient contents. Give 3 points if the nutrient's content is in the recommended range. Give 1 point if the content is no more than 50% of the lower limit of the recommended range or no more than 200% of the upper limit of the recommended range. Then, sum up the scores of the branches, as shown in Table 2. Regression results show no significant relationship among evaluation, experience and prices.

Laojiaroubing got the highest evaluation score but the lowest experience score. The branch performed the best as for satisfying customers' nutrition requirements but it provided the worst customer experience. Master Kong Chef's Table got the lowest evaluation score but the highest experience sore.

Table 2 Customer experience and nutrition evaluation of Beef Rice and Noodle branches

Restaurant		Evaluation	Experience	Price
Beef Rice	Yoshinoya	8	4.2	13.5
	Kungfu	10	4.6	15.0
	Hehegu	9	4.5	13.5
	Yonghe King	12	5.0	16.5
	Banmuyuan	6	4.5	18.0
	Laojiaroubing	13	3.5	10.0
Beef Noodle	Yonghe King	7	4.2	18.0
	Master Kong	4	5.7	14.0
	Kyo-Nichi	6	4.5	24.0
	Malan	7	4.4	15.0
	Mr. Lee	6	5.0	14.0
	Yipingsanxiao	7	5.3	13.0

4 DISCUSSION

4.1 Both Beef Rice and Beef Noodle cannot satisfy customers' requirements for nutrition

From the results above, it can be seen that both Beef Rice and Beef Noodle supplied by all restaurants cannot satisfy customers' requirements for nutrition. The meals are unhealthy and unbalanced in nutrition. For Beef Rice, the calorie and the contents of protein are much lower than the standards, and the contents of fat and salinity are much higher. For Beef Noodle, the calorie and the contents of fat are lower than the standards, and the contents of protein and salinity are higher. However, although most Beef Noodle and Beef Rice meals are not nutritional balanced, Beef Rice performed a little better than Beef Noodle.

For all restaurants, including Beef Rice and Beef Noodle, the contents of main food and beef are much higher than the recommends. Customers may just take part of the meal, but it will lead to large waste. Except for Laojiaroubing and Malan, the contents of vegetables are much lower. The weights of main food and vegetables of different restaurants varies a lot, the maximum is almost twice of the minimum.

4.2 Chinese customers do not know how to selece healthy fast food

The results also indicate that the customers' experience show significant negative relationship with whether the food can satisfy customers' requirements for nutrition. But price has no significant relationship with it. It is an interesting finding. Although food-related domain is the dominantly most important domain of customer experience for Chinese customers, it seems that Chinese customers do not know their requirements for food nutrition and they do not know the nutrient contents of some common meals. For example, Laojiaroubing provides the lowest price (just a half the highest) and gets the best nutrition evaluation score (more than twice of the lowest), but its customer experience is the worst. Customers think food-related domain is very important, but they do not what actually kinds of food they need. This is a study direction in the future.

4.3 Beef Rice is more healthy than Beef Noodle

Previous studies indicated that Chinese customers' intakes of animal food and oil increased, so as that the intake of fat increased (Zhai, et al., 2006; Feng & Shi, 2006). The view is consistent with the nutrient contents of Beef Rice, but it is contrast to Beef Noodle. Chinese Beef Rice usually is taken with gravy. The gravy contains a lot of oil and fat. Hence, the fat content of Beef Rice is higher than the standard.

Precious studies also indicated that the residents' intakes of vegetables decreased largely. The conclusion is consistent with our study that most meals do not provide enough vegetables for customers.

Feng and Shi's study indicated that the residents' intakes of protein decreased. The conclusion is contrast to Beef Noodle. The protein content of noodle is 8.6 grams per 100 grams of noodle, while the protein of rice is just 2.6 grams per 100 grams of rice. And rice and noodle take large percentage of weights of fast-food meals. Hence, the protein content of Beef Noodle is contrast with previous study.

5 CONCLUSIONS

Chinese residents' dietary have been undergoing great changes in recent 20 years, since the great changes of Chinese society and economy. The study aimed to investigate whether current Chinese fast food satisfies customers' nutrition and healthiness requirements and at the same time provides a good customers' experience. The results of the study indicated that Chinese Beef Rice and Beef Noodle (two most popular Chinese fast foods) cannot satisfy customers' nutrition and healthiness requirements, but Beef Rice performed a little better than Beef Noodle. And the results also indicated that the food nutrition showed significant negative relationship with customers' experience.

To the best of our knowledge, our work is the primary study to analyze customer experience and requirements of Chinese fast foods. It is also the first to investigate the differences of customers' requirements and experiences. The study provides a primary conclusion that Chinese customers do not know how to select healthy fast food. And the current fast food cannot satisfy their requirements.

Further studies will provide a great interest to better understand Chinese fast food industry and extend to other related areas more focusing on customer experience. It will also be beneficial to further research into predicting customers experience and requirements for Chinese fast foods.

REFERENCES

Schmitt.B.H (2003). *Customer Experience Management: A Revolutionary Approach to Connecting with Your Customers.* New Jersey: John Wiley & Sons, Inc..

Chinese Nutrition Society 2000. *Chinese Dietary Reference Intakes.* Beijing: China Light Industry Press.

Chinese Nutrition Society 2007. *Chinese Dietary Guide.* Beijing: Tibet People's Press.

Choe, P., Rau, PL., Chen, N., & Jeon, G. 2011. *The Study of Customer Service Experience for Franchised Restaurant in Beijing, Tokyo, and Seoul.* Industrial Engineering and Management, in press.

Feng, Z., & Shi, D. 2006. *Chinese food consumption and nourishment in the latest 20 years.* Resources Science, 28(1), 2-8.

Law, M., Frost, C., & Wald, N. 1991. *By how much does dietary salt reduction lower blood pressure? III--Analysis of data from trials of salt reduction.* British Medical Journal, 302(6780), 819.

Liu, K., Cooper, R., McKeever, J., Makeever, P., Byington, R., Soltero, I., et al. 1979. *Assessment of the association between habitual salt intake and high blood pressure: methodological problems.* American journal of epidemiology, 110(2), 219.

National Bureau of Statistics of China 2010. *2010 China Statistical Yearbook.* Beijing: China Statistics Press.

Zhai, F., Wang, H., Wang, Z., He, Y., Du, S., Yu, W., et al. 2006. *Changes of Dietary and Nutrition Status of Chinese Residents and Recommendations for Policies.* Food and Nutrition in China, (5), 4-6.

Zhou, B., Stamler, J., Dennis, B., Moag-Stahlberg, A., Okuda, N., Robertson, C., et al. 2003. *Nutrient intakes of middle-aged men and women in China, Japan, United Kingdom, and United States in the late 1990s: the INTERMAP study.* Journal of human hypertension, 17(9), 623-630.

CHAPTER 52

Effects of Preference on Crape Structure of Saijo Japanese Paper

Hongguang HU *, Yuka TAKAI**, Noritaka SAIKI***,

Takeshi TSUJINAKA***, Mitsuyoshi OCHI***,

Akihiko GOTO**, Hiroyuki HAMADA*

*Kyoto Institute of Technology
Kyoto, Japan
ko@cej.co.jp
hhamada@kit.ac.jp

**Osaka Sangyo University
Osaka, Japan
takai@ise.osaka-sandai.ac.jp
gotoh@ ise.osaka-sandai.ac.jp

***Saijo City
Ehime, Japan
saiki988@saijo-city.jp
tsujinaka1314@saijo-city.jp
ochi930@saijo-city.jp

ABSTRACT

A Japanese paper that is made in Saijo city is called "Saijo Japanese paper", which is one of traditional crafts in Japan. One of the main characteristics of the Saijo Japanese paper is a crape structure. This crape structure is called "Shibo". The Shibo structure is made by craftsmen's hand. The Shibo is fabricated by bending from the edge of Saijo Japanese paper toward front. Structure of Shibo affect by bending angle. Recent years, the demand of Saijo Japanese paper is stagnant. Saijo Japanese paper requires creating new products. However, it is not clear whether the preference of what kind of the Saijo Japanese paper is high. On the other hand, structure of Shibo on Saijo Japanese paper can change easily. We can prepare many

472

types Shibo structure. In this study, the Shibo structures of the Saijo Japanese paper with various making condition were quantified. Further, preference of the Saijo Japanese paper with various Shibo structure were clarified.

In this study, 6 types Shibo structure Saijyo Japanese paper was prepared. One was made by non-expert. The other was made by expert (career of 18 years) with different technique. Shibo structures of 6 Saijyo Japanese paper were measured by using non-contact three-dimensional measuring device (NH-3SP; Mitaka Kohki Co.,Ltd.). The cross section was measured with 100μm Measuring pitch and 50 mm measurement range. The questionnaire of Saijyo-Japanese paper was performed for Japanese and Chinese. Structure measurement results confirmed that sample of usual fabrication by expert had the height of the highest Shibo, and the width of the narrowest Shibo. Subjective evaluation results provided that tendencies of the evaluation to the Saijo Japanese paper of Japanese and Chinese people do not differ greatly.

Keywords: Japanese paper, structure, subjective evaluation

1   INTRODUCTION

From time immemorial, Japanese paper has been fabricated in Saijo City, Ehime Prefecture. This paper called *"Saijo Japanese paper"*. One of the main characteristics of the Saijo Japanese paper is a crape structure. This crape structure is called *"Shibo"*. Fig.1 shows photo of Saijo Japanese paper. The Shibo structure is made by craftsmen's hand. The Shibo is fabricated by bending from the edge of Saijo Japanese paper toward front. Structure of Shibo affect by bending angle. The Saijo Japanese paper is categorized as high-class Japanese paper from characteristic appearance. The Saijo Japanese paper is used as a gift money envelope for wedding or pocket paper for the tea ceremony. However, due to demand downturn of the Japanese paper of these days, examination of the new practical use method of the Saijo Japanese paper is needed.

Therefore, purpose of this study is clarifying structure of the Saijo Japanese paper liked by people. 6 types Shibo structure Saijyo Japanese paper was prepared. One was made by non-expert. The other was made by expert with different technique. Shibo structures of 6 Saijyo Japanese paper were measured by using non-contact three-dimensional measuring device. The questionnaire for preference of Saijyo-Japanese paper was performed in Japan and China.

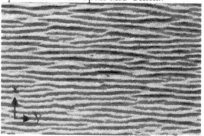

Fig. 1 Saijo-Japanese paper

2.1 Fabrication

In this study, 6 types Shibo structure Saijyo Japanese paper were prepared. Sample name and fabrication conditions were shown in Table 1. Sample A, C, D, E and F were made by expert (career of 18 years) with different technique. Sample B was made by non-expert. Fig. 2 shows photo of making Shibo process. Working table for making Shibo has 13.5 inclinations. At first, Saijo Japanese paper is set on the working table, and has wrinkles lengthened with the brush. Next, Saijo Japanese paper is bent from the edge toward front. Usually, a raising angle of Japanese paper is parallel to the working table. Sample C-F were changed the raising angle intentionally. Fig. 3 illustrates schematic image of making Sample C. The raising angle of Sample C was parallel to the ground. The raising angle of the sample D was 15 degrees from the ground. The raising angle had the relation: Sample A < Sample E < Sample F < Sample C < Sample D.

Table 1 Sample name and fabrication conditions

Sample name	Fabrication conditions
Sample A	Usual fabrication by expert
Sample B	Fabrication by non-expert
Sample C	Fabrication by expert with parallel raising
Sample D	Fabrication by expert with 15 degrees raising
Sample E	Fabrication by expert with ϕ 3 mm stick using
Sample F	Fabrication by expert with ϕ 10 mm stick using

Fig. 2 Photo of making "Shibo" process

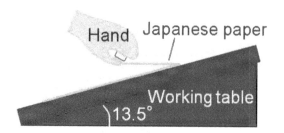

Fig. 3 Schematic image of making "Shibo"

2.2 Structure Mesurment

Shibo structures of 6 Saijyo Japanese paper were measured by using non-contact three-dimensional measuring device (NH-3SP; Mitaka Kohki Co.,Ltd.). The longitudinal direction of Shibo was decided to be y direction. The xz cross section was measured with 100μm measuring pitch and 50 mm measurement range.

2.3 Objective Evaluation

The questionnaire survey about liking and image of Saijo Japanese paper was conducted. Japanese subjects were 50 persons of 17 to 69 years old who live in Saijo City. Chinese subjects were 66 persons of 18 to 69 years old who work for major IT companies.

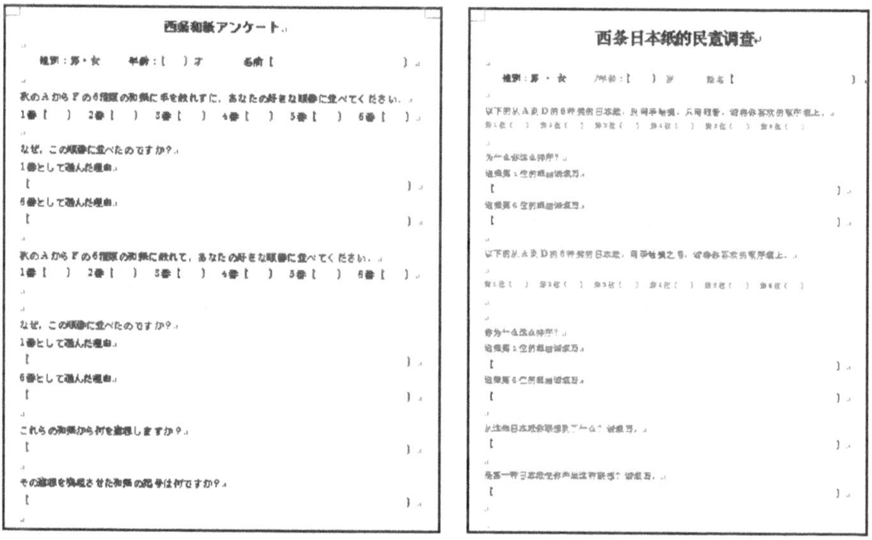

The subjects answered the matter of which it is reminded from Saijo Japanese paper from A to F. Fig. 4 shows questionnaire of Japanese and Chinese.

西条和紙アンケート

性別：男・女　年齢：[　]才　名前 [　]

次のAからFの6種類の和紙に手を触れずに、あなたの好きな順番に並べてください。
1番 [　] 2番 [　] 3番 [　] 4番 [　] 5番 [　] 6番 [　]

なぜ、この順番に並べたのですか？
1番として選んだ理由
[　]
6番として選んだ理由
[　]

次のAからFの6種類の和紙に触れて、あなたの好きな順番に並べてください。
1番 [　] 2番 [　] 3番 [　] 4番 [　] 5番 [　] 6番 [　]

なぜ、この順番に並べたのですか？
1番として選んだ理由
[　]
6番として選んだ理由
[　]

これらの和紙から何を連想しますか？
[　]

その連想を喚起させた和紙の記号は何ですか？
[　]

西条日本纸的民意调查

Fig. 4 Questionnaire of Japanese and Chinese

At first, Saijo Japanese paper was ranked according to likes and dislikes just watching. This evaluation method called visual evaluation. Then, the subjects answered the reason selected as the Japanese paper which the most favorite and 6th likes (the worst favorite) Japanese paper. Next, Saijo Japanese paper was ranked according to likes and dislikes with touching the Japanese paper. This evaluation method called contact evaluation. Then, the subjects answered the reason selected as the Japanese paper which the most favorite and 6th likes Japanese paper. Next, the words of which it is reminded from Saijo Japanese paper were asked. Average point was calculated from result of objective evaluation. Significant difference was tested by using paired t-test.

3   RESULTS AND DISCUSSIONS

3.1   Results of Structure

Fig.5 shows photos of samples. The appearance of Shibo changed with fabrication conditions. Arithmetic average roughness was shown in Fig. 6.

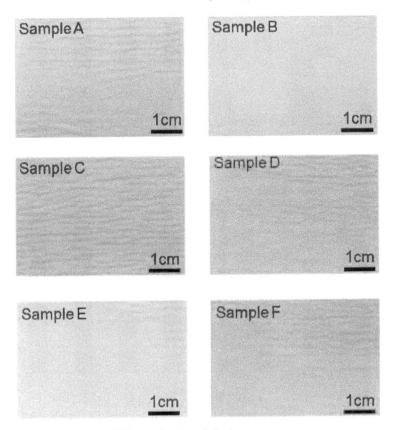

Fig. 5 Photos of various Saijo Japanese paper

Sample A was the highest value of arithmetic average roughness. Arithmetic average roughnesses of the other samples were around 10μm. Average length was shown in Fig. 7. Sample A and C show low value of average angle. These results confirmed that Sample A had the height of the highest Shibo, and the width of the narrowest Shibo. It was suggested that the appearance of Shibo of Sample A originated the height and width of Shibo. The fabrication conditions of Sample C had affected the height of Shibo. The fabrication conditions of Sample D-F made the height of Shibo low, and made width of Shibo large.

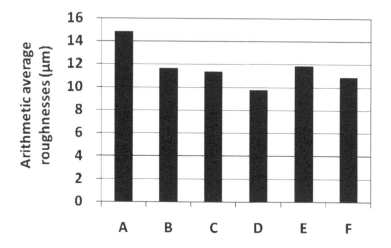

Fig. 6 Results of Arithmetic average roughness

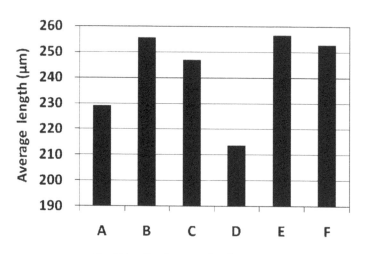

Fig. 7 Results of average length

3.2   Results of Objective Evaluation

Fig. 8 shows average evaluation point of each sample. The higher evaluation point was described, the more subject feel good. In the both of visual evaluation and contact evaluation. Sample A-E showed no significant differences between Japanese and Chinese. Visual evaluation of Sample F showed a significant difference (P<0.01) between Japanese and Chinese. In the both of Japanese and Chinese. all samples showed no significant differences between visual evaluation and contact evaluation.

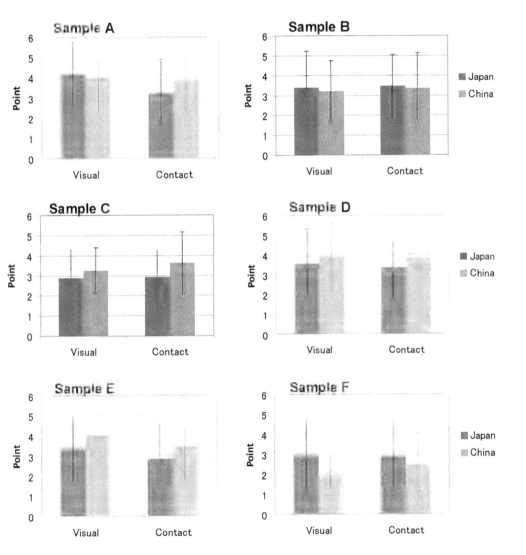

Fig. 8 average evaluation point of each sample

Average evaluation point of visual evaluation and contact evaluation shown in Fig. 9. In the visual evaluation, Japanese evaluated the sample A highly. Sample A showed no significant differences in Sample D as second highest. Chinese evaluated the sample E highly. Sample E showed no significant differences in Sample A as second highest. In the contact evaluation, Japanese evaluated the sample B highly. Sample B showed no significant differences in Sample D as second highest. Chinese evaluated the sample A highly. Sample A showed no significant differences in Sample D as second highest.

These results provided that tendencies of the evaluation to the Saijo Japanese paper of Japanese and Chinese people do not differ greatly. Furthermore, evaluation tendency changed by visual evaluation and contact evaluation has suggested that Shibo structure should be selected with product.

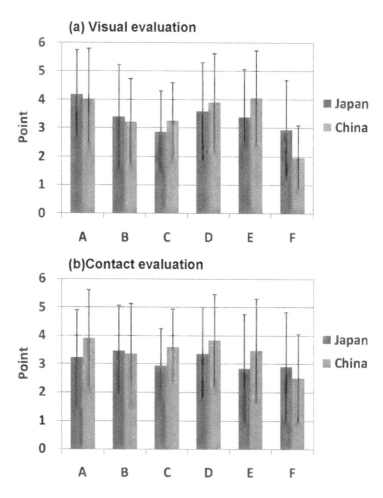

Fig. 9 Average evaluation point of (a) visual evaluation and (b) contact evaluation

4 CONCLUSIONS

The most important finding of this research can be summarized as follows;

1. Structure measurement results confirmed that sample of usual fabrication by expert had the height of the highest Shibo, and the width of the narrowest Shibo.
2. Subjective evaluation results provided that tendencies of the evaluation to the Saijo Japanese paper of Japanese and Chinese people do not differ greatly.

REFERENCES

M. Kato, 2008. SAIJO-HAN TE SUKI WASHI NO REKISI TO BUNKA, *Self-publication: M. Kato*

CHAPTER 53

One New Anthropometrical Method for Body Measurement in Motion

Y. J. WANG[12]

[1]School of Fashion Art and Engineering, Beijing Institute of Fashion Technology, China
[2]361°-BIFT High Performance Sportswear Design and Research Center
fzywyj@bift.edu.cn

ABSTRACT

Correctly acquiring body measurement in motion is very important step in high-performance sportswear design for wear ease design. Traditional manual method with tape and 3D body scanning are effective way to collect body measurements in static state. When body in motion, refereeing points of human body in motion cannot be determined. So, these two anthropometrical methods are not suitable for body measuring in motion state. In order to correctly measuring human body in motion state, one new anthropometrical method body motion analysis system is introduced and corresponding body measuring experiments in running state with this new technique are conducted in this study. Based on acquired results, the accuracy of this technique are tested by comparing with manual method and 3D body scanning. The research results indicated that body motion analysis system is one effective and accuracy method not only used in body measuring in static state, but also in body measuring in running state. Moreover, the body measurement changing in running state is analyzed based on collected body measurements in experiment and finally, correct wearing ease values designed for high-performance sportswear are determined.

Keywords: body measurement, anthropometrical method, motion state, wear ease design

1   INTRODUCTION

Running is one of the most important sports. When people run, bones and muscles work together based on the connections of joints. The positions of different joints change significantly during body movement. Meanwhile, the body skin surfaces around the joints are extended and contracted, and causes the corresponding body measurements to change. In order to provide necessary rooms for unrestricted body movements, suitable garment eases should be determined and incorporated into the clothing products. Therefore, measuring the body in running state and analyzing the differences between body measurements in static and running states are necessary for effective product development of clothing.

Many anthropometric methods have been developed over the years. Among these methods, manual tape measurement method is the traditional approach for static state measurement taking. It is also an effective method to collect body measurements when people make different body postures (Liu & Kennon, 2006). Linear body measurements, such as the circumferences of the human body, can be obtained by this method based on identified anatomical landmarks on the body skin surface. In recent years, 3D body scanning technique has been used widely in clothing product development, because of its advantages of efficiency, contactless and availability of body and shape information (Mickinnon & Istook, C., 2001, Loker et al, 2005). In 3D body scanning, normal or laser light reflection is utilized to capture data points from the body surface (Istook & Hwang, 2001). A 3D point cloud with points recorded in x, y and z coordinates is obtained from the scanning process. Many important information can be generated by processing, filtering and compressing raw data from the scan. Basically, critical anthropometric measurements can be extracted in the 3D body scanning system, which include linear circumferences of the human body and non-linear data, such as volume and cross section areas of the human body (Liu & Kennon, 2006).

A number of body measurement surveys were carried out in the last century and different sizing standards were established accordingly. As early as in 1951, a large-scale survey on British women was carried out. The USA government carried out a survey with 10,000 female subjects, in which 49 body measurements of each subject were taken between 1948 and 1959. In 1986, the Chinese government organized a national wide anthropometry survey. From the survey result, the Size Designation of clothes, GB1335-1991, was published in 1991 as a national standard in China. In the 1990s, the standard was updated, which resulted in the GB1335-1997 standard. These body measurement surveys were mainly carried out by the traditional manual measurement method, while 3D body scanning was used to collect static state body measurements in recent years.

By defining landmarks on the body skin surface, body measurements of a subject with a static posture can be easily acquired either by the manual method or the 3D body scanning. However, if a subject is in motion, for example when he is running, his body skin surface extends or contracts, therefore the landmarks on his skin surface are no longer steady. Consequently, the traditional manual method and

3D body scanning technique are not suitable to collect body measurements when the body is in dynamic movement.

In order to correctly acquire body measurements in running state, a new anthropometric method is introduced in this paper. Experiments are carried out to collect measurement data. The effectiveness of the proposed technique is compared to that of the manual method and 3D body scanning. The body measurements collected in the experiment are analyzed to calculate the variation of measurement in running, which are used for the determination of wearing ease values in running wear design.

2 A NEW ANTHROPOMETRIC METHOD FOR RUNNING STATE MEASUREMENTS

2.1 Body Motion Analysis System

A number of body motion analysis systems have been developed to record three dimensional movements and body characteristics in motion state. The analysis results are used in many applications, such as medicinal treatment, movie animation and sports design (Dobrian & Bevilacqua, 2003). In general, body motion analysis systems consist of both hardware and software, examples can be found in Figures 1 and 2. The hardware system includes passive or active markers, cameras, controller, and host computer. The passive (or active) markers are used for land-marking body surfaces. Cameras are used to capture the landmark locations during body movement. The controller provides power and links up the cameras and other devices in the system. Software is installed on the host computer. The body motion analysis system can complete the following tasks: (a) tracking landmarks on the human body in motion; (b) processing the data and displaying the motion process; (c) analyzing the body structure and the motion characteristics (Chang & Huang, 2000).

a) markers   b) camera   c) controller   d) host computer

Figure 1 Hardware system of Vicon body motion analysis system

(Source of image: http://www.vicon.com/products/)

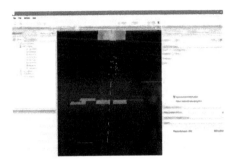

Figure 2 Software system of Vicon body motion analysis system

The process of body motion analysis is shown in Figure 3, which involved a few steps: Firstly, markers of diameter 8-12 mm are covered with reflective fabric and placed on different parts of the body. For example, in order to analyze the movement of joints in sports, markers are fixed on body joints. The body motion analysis system is calibrated and set up, including the determination of the sampling rate, the adjustment of the camera locations, and the identification of the x, y and z coordinate system and so on. Next, when subjects perform the designed body movement, the movements of the markers are tracked by cameras and location data are stored in a data-station. The data are extracted and processed, so as to define the paths of marker movement. 3D models of body motion are also created (Kapur et. al, 2005). Therefore, body characteristics in motion state can be analyzed, including segment positions, angles of joints movement and others.

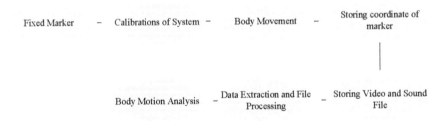

Figure 3 Procedures for body motion analysis

2.2   Anthropometry Study by Body Motion Analysis System

The body motion analysis system provides a platform to measure the body in running state, because it can accurately capture the marker locations. As given in the ISO 8559-1989 standard, land-marking is the foundation for all body measurements. The distances between landmarks are defined as various body measurements. For example, the chest girth is the maximum horizontal girth measurement passing over the shoulder blades (scapulae), under the armpits (axillae), and across the nipples. Shoulder width is the horizontal distance between the acromion extremities. If markers are placed at the same positions as the body measurement landmarks, body measurements in running state can be obtained by the body motion analysis system.

2.3   Extraction of Body Measurements

When a subject runs, the marker locations are captured by the body motion analysis system as sequential data in x, y and z coordinates. By calculating the linear distances between two markers or curve passing several markers, body measurements can be obtained in every recoding time. Finally, the variation of body measurements during the running process can be presented.

Bézier curve（B-spline curve）is often used to model a smooth curve by two or several points in 3-dimensional space (Böhm, 1977, Song & Zhang, 2001). Body measurements are defined as the straight lines or smooth curves between two marker points or passing through a few marker points. Therefore, B-spline curves are modeled to obtain the body measurements in this study.

3   RESEARCH METHODOLOGY

3.1   Experiment Design

Experiments were designed to collect the body measurements in running state. As shown in Table 1, the experiments consist of the three exercises: 1) Exercise One: static-state body measurements are collected by 3D body scanning. 2) Exercise Two: dynamic posture body measurements are acquired by manual method. 3) Exercise Three: body measurements are collected by body motion analysis system when the body is in static state and running state.

Table 1 Experiment contents and measurement methods

Number	Experimental contents	Measurement methods
1	Body measurements in static state	3D body scanning
2	Body measurements in static state and dynamic postures	Manual method
3	Body measurements in static and running state	Motion capture by body motion analysis system

3.2   Subjects

In these experiments, 5 Chinese male subjects aged 20-24 in the previous experiment were recruited as volunteers in this study. The heights of all subjects are between 170-175 cm and values of BMI (Body Mass Index) are 21-23. Before the experiment, researchers explained to all subjects about the experiment and obtained their approvals.

3.3 Body Measurements and Landmark (Marker) Definitions

According to the experimental objectives. 30 body measurements were measured in this study. These body measurements include girths. lengths and widths: (a) girth: neck base. chest. waist. hip. armscye. upper arm. elbow. forearm. wrist. thigh. mid-thigh. knee. calf as well as ankle: (b) length: front and back waist. the 7th cervical to waist. total crotch. crotch depth. central arm. inside arm. outside arm. inside leg. outside leg. front leg and back leg: (c) width: across shoulder. across chest and across back.

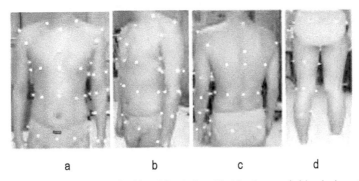

a   b   c   d

Figure 1   Marker locations on body skin: a) front view, b) side view, and c) back view of upper body, and d) back view of lower body

In order to collect these body measurements in running state. markers covered with reflective tape are fixed on the subject's body surface at landmark positions defined according to ISO8559-1989 standard. A total of 74 landmarks (markers) are defined for each subject. as shown in Figure 4. on different parts of the body: neck (4 markers). chest (16 markers). waist (8 markers). hip (8 markers). shoulder (2 markers). arm (16 markers). as well as leg (20 markers).

3.4   Experiment Instruments

In Exercise One. TC² scanner is used to collect 30 body measurements in static state and in Exercise Two. manual with tape is used to collect the same body measurements in 17 dynamic postures as shown in Figure 5 to Figure 7. In Exercise Three. the VICON 8.0 body motion analysis system (Figures 1 and 2) with 8 cameras was used to capture the location and movement of markers. Each subject was required to stand upright and remain steady to collect static state marker data. Next. he was required to run on a treadmill at a speed of 10 km/h. for a few minutes. as shown in Figure 8. The data of the 74 markers are captured every second in the running process.

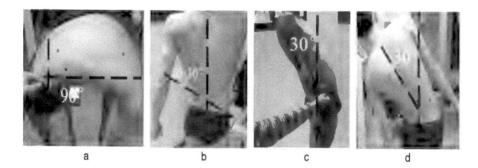

Figure 5 Dynamic postures of waist joint movements (a, b, c and d)

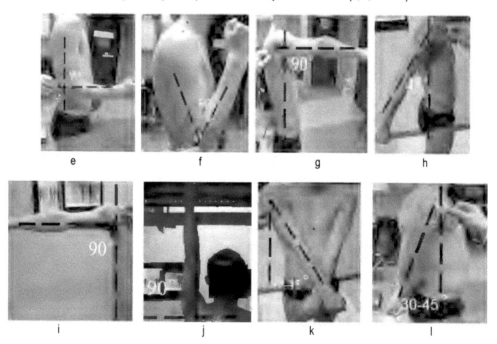

Figure 6 Dynamic postures of elbow and shoulder joint movements (e, f, g, h, i, j, k and l)

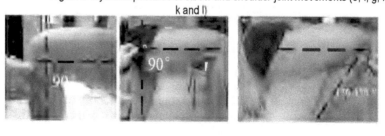

o   p   q

Figure 7 Dynamic postures of hip and knee joint movements (m, n, o, p and q)

Figure 8 Star-Trac treadmill – equipment for collecting running state measurements

3.5   Experiment Protocol

All experiments were carried out in a controlled environment with regular room temperature of 24°C, 65% relative humidity. In order to reduce the influence of air motions, wind speed is designed at less than 0.1m/s.

First, each subject is required to keep normal posture and 30 body measurements in static state are collected with 3D body scanning. After body scanning, each subject keeps required 17 dynamic postures and the same body measurements are collected with manual method. After manual measuring, 74 markers with reflective tape are fixed on the 74 landmark positions for each subject. After fixing the markers, the body motion analysis system was calibrated and set up. Each subject is required to keep the natural standing posture on the treadmill for 1 minute, and the marker positions of the standing pose (static state) are stored. Next, each subject is required to run on the treadmill at 10 km/h for 3 minutes. The marker position is recorded every second throughout the entire running exercise. In addition, video of the subject in the running exercise are recorded.

3.6   Data Collection

In the Experiment One, 30 body measurements can be collected in the 3D scanning system and In the Experiment Two, the same body measurements can be recorded directly. In Experiment Three, the data recorded by the body motion analysis system are 3-dimensional coordinates of marker positions. By processing these marker coordinates, the body measurements at the corresponding recording time can be calculated.

4 RESULTS AND DISCUSSIONS

4.1 Accuracy of New Anthropometric Method in Static State

30 Body measurements in static state acquired with the 3D body scanner and the body motion analysis system are compared in Table 2. S indicates the body measurements acquired by the 3D body scanning method, and B indicates the measurements acquired by the body motion analysis system. In order to examine the effectiveness of the body motion analysis system for the proposed application, body measurements acquired by the two methods are compared: Body measurements acquired by the 3D body scanning method are different from that by body motion analysis system -0.11 to -1.92 cm. The 7th cervical to waist length has the largest difference, -1.92 cm. The crotch depth has the smallest difference, -0.11cm. It is clear that all body measurements acquired by the body motion analysis system are slightly bigger than the values acquired by the 3D body scanning method.

T-test was carried out to examine if there is any significant difference between body measurements acquired by the two methods. The result of the t-test is also shown in Table 6-2. The p-values of all body measurements in the t-test are more than 0.05, which reveal that there are no significant difference in the body measurements acquired by the 3D body scanning method and the body motion analysis system. It means that data acquired by the body motion analysis system is accurate and the method itself is effective to collect body measurements in static state.

4.2 Accuracy of New Anthropometric Method in Running State

By analyzing the data collected with the body motion analysis system, the 30 body measurements of each subject can be obtained every second during the 3-minute running. The variation of body measurements in dynamic postures (M) and running state (B) are compared in Table 3.

T-test was carried out to examine if there is significant difference between changes of the body measurements in dynamic postures and the changes in running state. The neck base, wrist and ankle girth measurements acquired by both the manual method and the body motion analysis systems do not show any changes (0.00) in dynamic postures and running state. As shown in Table 3, 11 body measurements, including chest girth, waist girth, calf girth, across shoulder width, across back width, across chest width, front waist length, back waist length, the 7th cervical to waist length, underarm to waist length and inside arm length, have p values less than 0.05. It means that these body measurements of dynamic postures are significantly different from that of running state. The rest of the 16 body measurements have p values more than 0.05, which means that the differences between dynamic postures measurements and running state measurements are not significant.

Although 11 body measurements acquired in dynamic postures and running state are significantly different, they have similar characteristics: the measurements acquired in running state are smaller than that in dynamic postures. By comparing the running postures and the 17 designed dynamic postures (Chapter 5), it is not difficult to find that people usually keep the torso in upright posture during running to avoid falling down or being injured. The corresponding body measurements should have smaller changes compared to that of the dynamic postures, because the body torso would not have extensive movement during running motion.

The result of the t-test can prove that the body motion analysis system is an effective method to collect body measurements in running state.

Table 2 Comparison of measurement results in static state

Body measurements		S		B		S-B	t-value	P(Sig.)
		Mean (cm)	Std. Dev	Mean (cm)	Std. Dev			
Girth	Neck base	39.93	0.40	40.79	1.30	-0.86	-1.95	0.12
	Chest	96.48	1.62	97.18	1.80	-0.70	-1.33	0.25
	Waist	79.19	1.54	79.39	2.38	-0.20	-0.44	0.68
	Hip	95.95	2.58	96.98	2.49	-1.03	-1.78	0.15
	Armscye	41.50	2.17	42.26	2.93	-0.75	-1.49	0.21
	Upper arm	27.07	1.36	27.49	1.81	-0.42	-1.38	0.24
	Elbow	24.66	0.14	25.85	0.56	-0.20	-0.85	0.45
	Forearm	24.92	0.50	25.14	0.70	-0.22	-0.85	0.44
	Wrist	16.14	0.35	16.38	0.69	-0.24	-0.90	0.42
	Thigh	55.12	2.27	55.66	2.46	-0.54	-1.31	0.26
	Mid thigh	46.18	2.17	46.68	2.00	-0.50	-1.50	0.21
	Knee	37.12	1.30	37.50	1.69	-0.38	-1.20	0.30
	Calf	37.10	1.47	37.44	1.53	-0.34	-1.14	0.32
	Ankle	25.22	1.04	25.97	1.05	-0.75	-1.88	0.13
Width	Across shoulder	40.25	0.71	40.49	1.13	-0.24	-0.79	0.47
	Across chest	38.31	0.68	38.91	0.73	-0.60	-1.32	0.26
	Across back	39.11	0.60	39.69	1.51	-0.59	-1.20	0.30

Length	Front waist	45.50	0.67	46.22	1.45	-0.72 -1.60	0.19
	Back waist	49.97	0.69	50.70	1.33	-0.73 -1.42	0.23
	7th-cervical to waist	49.64	1.31	50.41	1.61	-0.77 -1.92	0.13
	Underarm to waist	45.76	1.55	46.22	1.01	-0.46 -1.14	0.32
	Total crotch	65.10	5.73	65.57	5.38	-0.48 -1.12	0.33
	Crotch Depth	19.82	0.25	19.93	0.38	-0.11 -0.42	0.70
	Central arm	54.78	1.08	55.39	1.74	-0.62 -1.27	0.27
	Inside arm	51.62	0.96	52.14	1.35	-0.52 -1.26	0.28
	Outside arm	53.51	1.28	54.10	0.62	-0.59 -1.39	0.24
	Inside leg	72.73	1.62	73.68	2.31	-0.96 -1.72	0.16
	Outside leg	102.65	1.28	103.58	2.12	-0.93 -1.39	0.24
	Front leg	93.27	1.62	93.95	2.47	-0.68 -1.16	0.31
	Back leg	99.19	0.76	100.10	0.72	-0.92 -1.88	0.13

* S=3Dbody scanning method, B= Body motion analysis system

* p≤0.05: significance

Table 3 Comparison of body measurement variations in dynamic postures and running state

Body measurements		M (in dynamic postures)		B (running state)				
		size difference (cm)	change rate (%)	Size difference (cm)	Change rate (%)	M-B	t-value	P(Sig.)
Girth	Neck base	0.00	0.00	0.00	0.00	0.00	-	-
	Chest	3.00	3.11	2.65	2.73	0.35	4.72	0.01*
	Waist	2.00	2.54	1.59	2.01	0.41	3.98	0.02*
	Hip	5.60	5.86	5.82	6.00	-0.22	-0.73	0.51
	Armscye	4.5	10.92	4.76	11.26	-0.26	-0.97	0.39
	Upper arm	3.10	11.31	3.28	11.93	-0.18	-0.91	0.42
	Elbow	7.40	30.08	7.52	30.25	-0.12	-0.90	0.42
	Forearm	1.10	4.33	0.80	3.18	-0.30	0.69	0.53
	Wrist	0.00	0.00	0.00	0.00	0.00	-	-

	Thigh	2.80	5.08	3.42	6.14	-0.62	-1.79	0.15
	Mid thigh	1.80	3.91	2.52	5.40	-0.72	-1.41	0.23
	Knee	5.60	15.05	6.54	17.44	-0.94	-1.36	0.25
	Calf	1.40	3.77	1.22	3.26	0.18	0.484	0.65
	Ankle	0.00	0.00	0.00	0.00	0.00	-	-
Width	Across shoulder	2.80	6.93	1.56	3.85	1.24	2.95	0.04*
	Across chest	4.00	10.36	2.22	5.71	1.78	2.73	0.05*
	Across back	5.00	12.76	2.24	5.64	2.76	8.23	0.00*
Length	Front waist	4.40	9.65	1.10	2.38	3.30	15.06	0.00*
	Back waist	4.60	9.20	1.41	2.78	3.19	7. 64	0.00*
	7th-cervical to waist	4.20	9.11	1.06	2.10	3.14	5.40	0.01*
	Underarm to waist	4.40	14.24	0.82	1.77	3.58	13.59	0.00*
	Total crotch	1.70	2.58	1.83	2.79	-0.13	-0.97	0.39
	Crotch Depth	1.10	5.50	1.14	5.72	-0.04	-0.28	0.79
	Central arm	5.40	9.84	5.58	10.07	-0.18	-0.81	0.46
	Inside arm	3.50	6.80	3.46	6.64	0.04	0.25	0.82
	Outside arm	3.50	6.53	3.54	6.54	-0.04	-0.22	0.84
	Inside leg	1.00	1.38	1.06	1.44	-0.06	-0.24	0.82
	Outside leg	2.80	2.72	3.42	3.30	-0.62	-1.53	0.20
	Front leg	2.20	2.36	2.57	2.74	-0.37	-1.19	0.30
	Back leg	5.50	5.55	5.72	5.71	-0.22	-0.98	0.38

M=by manual method, B= by body motion analysis system, * p≤0.05: significance

5 CONCLUSIONS

In this paper, a new anthropometric method of human body in running state has been proposed and systematically verified. The research results indicated that the body motion analysis system is an effective and accurate method for measuring the body in static state and in running state. In static state, body measurements obtained by 3D body scanning and body motion analysis systems are similar, and all p-values in t-test are more than 0.05, which means that there is no significant change between the two methods. For running state measurements, p-values in t-test indicate that there are significant changes in 9 measurements of the upper body including chest girth, waist girth, across shoulder width, across back width, across chest width, front waist length, back waist length, the 7th cervical to waist length and underarm to waist length. But there is no significant change found in other body measurements. By analyzing people's running posture and the measurement results, it is found that the body motion analysis system can correctly reflect the changing characteristics of the human body in running state. Furthermore, measurements of the upper body have small changes, because of the upright body posture in running. Therefore, the result has indicated that the body measurements in running state acquired by the body motion analysis system are accurate.

ACKNOWLEDGMENTS

The author would like to acknowledge Beijing Institute of Fashion Technology and 361°(China) Limited Company for financial support in this research project (project code:HXKY05110324).

REFERENCES

Böhm W., 1977. Cubic B-spline curves and surfaces in computer aided geometric design, *Computing,* 19, Pp. 29-34.

Chang I.C. & Huang C. L., 2000. The Model-based Human Body Motion Analysis System. *Image and Vision Computing* 18, Pp.1067-1083.

Dobrian C. & Bevilacqua F. A., 2003. Gestural Control of Music Using the VICON 8 Motion Capture System, available onlin at :http://music.arts.uci.edu/dobrian/motioncapture/

Istook C. & Huang S. J., 2001, 3D Body Scanning System with Application of Garment Design and Fit, *International Journal of Clothing Science and Technology,* Vol. 5 (2), Pp.120-132.

Kapur A., Tzanetakis G., Babul N.V., Wang G. & Cook P.R., 2005, *The 8th International Conference on Digital Audio Effects (DAFX-05),* Madrid, Spain.

Liu C. & Kennon, R., 2006. Body scanning of dynamic posture. *International Journal of Clothing Science and Technology* 18 (3), Pp.166-178.

Loker S., Ashdown S., & Schoenfelder K., 2005. Size-specific Analysis of Body Scan Data to Improve Apparel Fit. *Journal of Textile and Apparel, Technology and Management,* Vol. 4 (3), Pp.1-15.

Mickinnon L. & Istook C., 2001, Comparative Analysis of the Image Twin System and the 3T6 Body Scanner, *Journal of Textile and Apparel, Technology and Management,* 1 (2), p.1-7.

Song K. & Zhang W. Y., 2001. Approximation of the Apparel Construction Curves with Beizer Curves. *Journal of International Textile,* Vol.4, Pp.76-78.

CHAPTER 54

Experimental Measurement of Body Size and Practice of Teaching and Research

Zhou Xiang   Hu Shou-zhong

Shanghai University of Engineering Science
Shanghai, China
zx1898@126.com

ABSTRACT

Ergonomics of human based on human experiments to measure the size of the teaching and practice, based on the basic conditions for Science and Technology national science and technology platform as an opportunity to focus on the work of the project, according to the state juvenile body size measurement of the technical requirements, relying on key subjects of Shanghai (apparel design and engineering) building, well-designed experimental teaching projects and practices. Experimental measurement of the human dimension of teaching is leading on the research frontier disciplines of the human body and hot issue in a more accurate grasp of the human body can carry out a number of scientific researches in ergonomics.

Keywords: ergonomics, the body size measurement, experimental teaching, research and practice

PREFACE

According to requirements of the key work item of the Ministry of Science and Technology of China National Science and technology platform, Clothing research center of Shanghai University of Engineering Science, used 3D body scanner, successfully completed the sampling body size measurement experiment of 4000 Shanghai juveniles from March 2007 to June 2008.

1 THE EXPERIMENTAL SIGNIFICANCE OF BODY MEASUREMENT

1.1 Current situation of the research at home and abroad

The underlying technology support of various measurement means is based on the 3D body measurement system. In foreign, there are Loughborough body scanner, [TC] ² (American Textile and Apparel Technology Center) stratified contour measurement method, the automatic body measurement instrument of the agency of British Defense Clothing and Textile, and etc. The 3D body scanner takes only a few seconds to produce countless linear and non linear measurement data of a body. Secondly, the measurement data is more accurate than traditional physical measurement result. The measurement can be repeated and the data can be modified by using software. Thirdly, it can produce results of a digital format which can be automatically integrated into clothing CAD system [1].

Human body measurement research was started relatively late in China. Some studied the classification criteria of current woman shape in different countries, some studied the classification method of woman shape in China which was still in the stage of study[2]; some extracted samples from females university students at the age of 19 ~ 25 in Xi'an College of Engineering Science and Technology, receiving 212 data samples, and choose the objects who grow up in the northeast, north and west China, to get 175 data samples to research the classification of the somatotype of female college students in the three regions[3]; some studies the older body shape which the body measurement data is derived from the State General Administration of Sport (group company) which investigate Chinese human body physique, that is the "2000 national physical fitness monitoring report". The age is big span, from 15 to 59. And the sampled population is much more [4]. However, the human body data acquisition hardware and processing software is unknown. Some studied 400 samples from North to northeast China in 10 provinces and autonomous regions, age ranging from 19 to 65. The measuring tools had tape, Martin in vivo measurement instrument, angle meter, and so on [5].

1.2 The significance of body measurement

Human body measurement was an experimental tool to obtain anthropometric data. In accordance with the uniform sampling method and technical specifications, in the measurement site, it collected 3D (or 2D) scan data of minor systemic and each body segment, formatted the human data measurement platform and standardization process, centralize to process unified data, built the nation foundation database of Chinese juvenile body size that is represent the characteristics of China population. It can be positive to service for the formulation of national standards and management, and provide first hand information to produce school uniforms, public facilities, sports facilities, public construction; it was the technical basis of the whole society the basic. Which can be set up and join in "China human size measurement network", to complete the research of "human body database" of development of Shanghai municipal disciplines.

2   THE BASIS OF EXPERIMENTS

2.1   Hardware configuration

The software and hardware equipment had been centralized purchased in line with the" human dimension measurement specification requirement to measuring equipment" (see Table 1). Such as 3D body scanner, scanner manual software V2, portable non-contact three-dimensional imaging measurement system 3DSS-STD-11, seat lifting device, Martin measuring instrument, and so on. The 3D body scanner was based on laser optical triangulation measurement principle, fast, precise and non-contact completed automatic measurement that is more than 100 human key size , according to the measuring scheme, it can automatic export the anthropometric data and adjust the nonstandard data. It would give strong technical support to the research units that need human data acquisition and processing for the standard of human body dimensions, anatomy and ergonomics, professional selection, fashion design.

Table 1 Body Measurement Experiment Measuring Equipment

	Equipment	Precision Requirement
1	Height gauge	1mm
2	Weight gauge	0.1kg lever scale
3	Touch gauge	1cm
4	Two-dimensional color scanner	150dpi color
5	Three-dimensional body scanner	2mm
6	The chair lift using measurement	without back; chair surface size>50cm×50cm; stepless lifting, run length >15cm; maximum height>50cm;
7	Angle gauge/ Calipers	1mm
8	soft tape	1mm

2.2   The training of laboratory personnel

The testers of the human body measurement experiment who control the rhythm of experiment, master the management and organization of the platform and image technology of the whole measurement experiment, shall be trained and assessed by the body measurement experts from China National Institute of standardization in accordance with the "technical specification of human dimension measurement" (scanning image request, the dress requirements of measured person, the requirements of measuring postural and 21 marks paste, data naming conventions and other operation).

2.3 Experimental guide books

The experiment of human body measurement required students to understand the importance, science significance and experimental feasibility of the experiment. Experimental guide books clearly required students to master the testing technology, in accordance with the "technical specification of human dimension measurement", to understand and grasp the whole measurement experimental platform, the final image, data requirements, write the experiment report according to the specification requirements.

3 TEST METHODS AND STEPS

3.1 Experimental schemes

According to the requirement of "human body measurement technology standard", before the experiment, it must be in accordance with guidance of teacher and experiment guide book, to write experiment scheme and arrange different time for boys and girls, fill the meter in order to gather statistics and processing of data.

Table 2 Measurement Record

Name		Sex		Nationality	
Birth	__/__/____ (dd/mm/yy)		Birth place		
Education Grade	1 Kindergarten□ 2 Primary school (□one □two □three □four □five □six) 3 Junior middle school(□one □two □three) 4 Senior high school (secondary school, vocational school) (□one □two □three) 5 others □				
Measurement Location	_____(city)_____(state/province)_____(county/district)				
Height: ____cm	Toes high: ____cm		Middle finger point for high:____m		
Weight: ____kg	Boll of foot girth: ____cm		Vertical foot girth: ____cm		
Tester Signature			Test date		

3.2 Experimental steps and requirements (see Table 3)

Table 3 Minor body measurement chart

Test Procedure	Test Requirement and Content
Before the test	Test introduction, explanation and demonstration
Hand test	Net right hand, palm flat instrument, fingers extended (1. four fingers together 2. fingers separated)
Posture demonstration	During scanning there are two standing postures and a sitting posture. Keep hands, waist and legs in 90 degrees.
Weight test	Light in weight measuring instrument, hold your head up high
Height test	Hold your head up high, eyes look straight ahead
Touch test	Eyes look straight ahead, right middle finger touches instrument
Foot test	Barefoot, right foot wai, vertical foot wai, Toe high
Wear test	No light lace underpants , no liner bra, no jewelry, belt measuring cap
Paste identification point	21 identification points must be precisely regulated paste in place
Three dimensional scanning	During scanning there are two standing postures and a sitting posture. Keep hands, waist and legs in 90 degrees. Sequence accurately, norms in place
Storage test	Check stand, sit, complete specifications, no shadow

3.3   Main measurement parameters (see Table 4)

Table 4 Minor Body Dimensions of Main Measurement Parameters

Height	Waist - small hip
Head height	Hip waistband
Cervical height	Waist - hip
Hip height	Pelvis - waistband length
Maximum Hip Height	Lower trunk length
Belly height	Waist girth
Maximum Belly Height	Hip girth
Breast height	Buttock girth
Neck front height	Maximum Hip Girth
Distance neck - vertical	Belly circumference
Neck front to vertical	Maximum Belly Circumference
Distance Scapula- vertical	Arm length left
Mid neck girth	Arm length right

3.4   The processing results of experiments data

(1) Classifying the shape as the basic parts of height, chest, each body was required to calculate the two element linear regression equation of the entire control site to two physical parameters. The site parameters had immanent connection. Using the method of regression analysis, it described the relationships among the data of different parts of human body in statistically, which was very important to computer aided design (CAD) and inference of new cutting formulas [6] .

(2) According to the different body shape, the garment industry should use the prototype that was very closely conforming to the body in the production of minor clothes. 8 prototypes were obtained by using the physical research, fuzzy clustering, and the relation analysis of the relevant parts. They included standard type and chunky type of children under the age of 14, high thin, standard, and chunky type of above 14 years old boys, high lean development standard, and weight standard development standard, chunky mature type of above 14 years old girls.

(3) Submitting the collected data and technical information to the China National Institute of standardization, can build the latest human dimensions database suitable for the social development our country, fill the blank and formulate relevant national standards, as basis of a whole society to share technical, promote the fruition.

4   CONCLUSION

After the conclusion and summary of the experimental data of human body measurement by the project researchers, the corresponding academic viewpoints and research conclusion, had been published 3 academic papers in the national Chinese core journals. The head of experimental project invited to the second ergonomics International Symposium in 2008 in United States, and read out the "The analysis of the minor's bodily form based on fuzzy cluster in Shanghai area" and "Measurement and Analysis of the Minor's Bodily Form Based on Fussy Recognition" published by ISPT.

The experiment teaching of Human body measurement is the leading domestic, it had a more accurate grasp about the frontier science and hot issues in the field of research of human body, can carry out a number of human ergonomics science research. Secondly, it comprehensively enhanced the experimental management levels and the utilization rate of large equipments, built experimental platform of human size measurement. And made full use of the laboratory, the hardware and software of the authority of research institutions and the base of cooperation among enterprise, college and research institution, enhanced the strength of undertaking the high level experiment project. Finally, the experimental data and the technical information bad been submitted to the China National Institute of standardization and been processed, to establish " The human body measurement database" which can represent the characteristic of Chinese shape, formulate relevant national standards, and can be the basic technology for the community sharing, to service for the economy of East China and the whole country, and promote the fruition.

REFERENCES

[1]Translated by Wang Qiming. The comparison of application potential of 3D body scanner, Foreign textile technology, 2002 (7): 35-38

[2]He Li, Zhang Haiquan. The comparison of classification of woman shape. Progress in Textile Science & technology, 2007 (5): 95-98

[3]Zheng Yan, Zhang Xin. The somatotype research of female college students in three regions. Journal of Xi'an College of Engineering Science and Technology, 2004 (9): 210-214

[4]Liu Yu, Zhang Zufang. The research of garment size based on senile somatotype characteristics. Journal of Donghua University (SOCIAL SCIENCES EDITION), 2004 (3) 2

[5]Wang Aihua, Chen Mingyan, Yang Zitian. The body analysis of adult male in north area. Journal of Donghua University (NATURAL SCIENCE EDITION), 2004 (4): 49-55

A Study of Shoe Sizing Systems: Foot Anthropometry of Filipino Children Aged 7-12

Aura C. Matias, R.M. Macaranday, J. Mangubat, R.B. Reyes and J.A. Tan

University of the Philippines
Quezon City, PHILIPPINES
aura.matias@coe.upd.edu.ph

ABSTRACT

The Philippines lacks an anthropometric measurement database for the Filipino foot. The currently used shoe sizing systems in the Philippines are (1) based on the foot anthropometry of Westerners, (2) has no "intrinsic" meaning to the shoe buyer, and (3) still based on English units, and not on the standard metric system. This study compiled measurements of 11 foot dimensions of Filipino children aged 7 to 12 years old. Males have generally bigger foot measurements than females. The measurements start to significantly differ between males and females from ages 10 to 12 – the period where physiological differences between genders start to appear. Foot length, foot width and foot height are the components that contribute most to foot size variability. A Filipino shoe sizing system based on a child's age is proposed. The proposed system can be used by manufacturers of school shoes made of black leather such as those currently used by Filipino grade school students.

Keywords: anthropometry, shoe sizing, foot measurements

1   INTRODUCTION

Shoes are an essential need to protect people's feet. For suitable design of shoes, foot dimensions of consumers are required. Length, widths and heights of feet should match with shoes in order for footwear to be comfortable. Wearing unfitted shoes leads to foot disorders, mental impacts, dexterity reduction, comfort reduction, increasing energy consumption and decreasing efficiency for doing tasks.

There are several different shoe-size systems that are used worldwide (International Shoe Size Conversion Charts, 2010). These systems differ in what they measure, what unit of measurement they use, and where the size 0 (or 1) is positioned. Only a few systems also take the width of the feet into account. Some regions use different shoe-size systems for different types of shoes (e.g., men's, women's, children's, sport, or safety shoes). Sizing systems differ in what units of measurement they use. This also results in different increments between shoe sizes because usually, only "full" or "half" sizes are made. Due to the different units of measurements, converting between different sizing systems results in round-off errors and unusual sizes (Everman, 2009).

Currently, the Philippines lack an anthropometric measurement database for the Filipino foot. This oftentimes leads to difficulty in finding the right shoe-size. Shoemakers adopt different international sizing systems such as those of the US, UK, and Asian size. To address this problem, the study will gather foot measurement data of Filipino children, aged 7 to 12, and use this data to propose a Filipino shoe sizing system.

2   METHODOLOGY

A total of 240 students from Grade 1 and Grade 2 students were the subjects of this study. The name, age, grade level, and current shoe size of every student were recorded. None of the subjects had any foot illness or foot abnormalities.

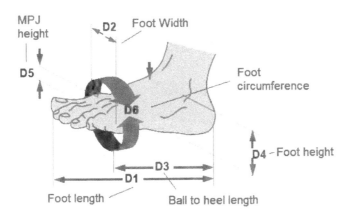

Figure 1 Six foot dimensions measured and recorded.

A total of 11 foot dimensions was recorded for each individual: foot length (measured as the distance between the two longest points of the foot), arch length or ball-to-heel length (measured as the distance between the heel to the foot's ball), food width (measured as the distance between the two widest points of the foot), foot height or dorsal arch height (measured as the straight line distance from the sole of the foot to the point where the top of the foot meets the front of the leg), MPJ height (metatarsophalangeal joint) at the 1st toe, the length of the five toes, and foot circumference. An anthropometer was used to measure the first five foot dimensions while a measuring tape will be used to measure the last six. MiniTab 14.0 and SPSS 13.0 were the computer applications used for statistical analysis.

3   RESULTS AND DISCUSSION

A total of 240 samples were collected. Normality tests confirmed that the data set of foot anthropometric data follows a normal distribution. The data is divided into six groups – one group for each age: 7, 8, 9, 10, 11, and 12. These groups are further subdivided into two subgroups based on gender, female or male to see if foot measurements between male and female children are significantly different.

Results show that the measurements for each dimension gradually increase as the age increases. Filipino male grade school children are also observed to have generally bigger foot measurements than their female counterparts. Based on a two-tailed t-test, foot length measurements significantly differ between males and females from ages 10-12. This is intuitive since this is the period of puberty where physiological differences between males and females start to manifest.

The following graphs in Figures 2 to 5 summarizes the comparison between the foot measurements of females (in red) and males (in green) group per age level. The range of ages where the measurements between the two genders are highlighted by the shaded circle.

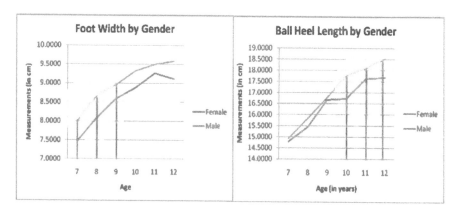

Figure 2 Significant differences (highlighted portions of the graphs) between the mean values of female and male measurements per age in terms of foot length and ball-heel length

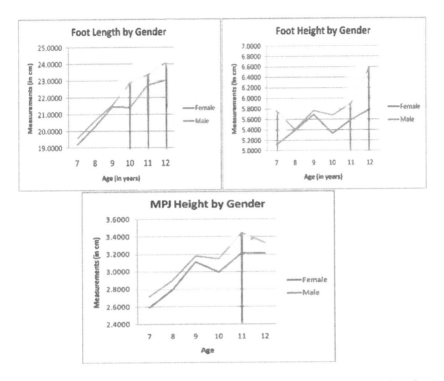

Figure 3 Significant differences (highlighted portions of the graphs) between the mean values of female and male measurements per age in terms of foot length, foot height and MPJ height.

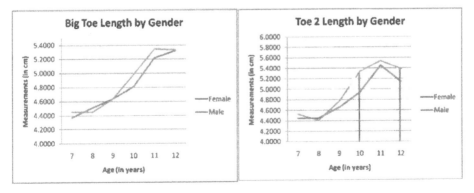

Figures 4. Significant differences (highlighted portions of the graphs) between the mean values of female and male measurements per age in terms of big toe length and toe 2 length

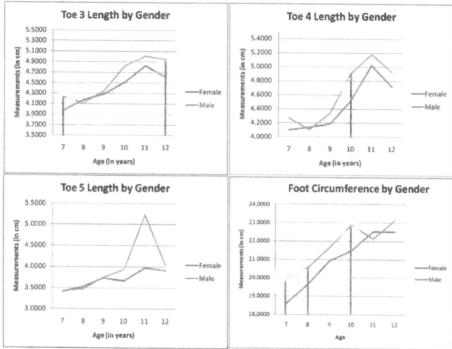

Figures 5. Significant differences (highlighted portions of the graphs) between the mean values of female and male measurements per age in terms toe 3 length, toe 4 length, toe 5 length, and foot circumference.

For economic reasons, most systems use foot length as the only variable in shoe-sizing. This study, however, measured eleven dimension to describe foot size. Factor Analysis was used to identify the factors that contribute the most to the variability of foot size. The top five contributors to the variance are the foot length, ball-heel length, foot width, foot height and MPJ height. This means that both foot length and foot width must be accounted for in shoe sizes to ensure good fit. The other components — ball-heel length, foot height and MPJ height — should also be considered in shoe designs. Table 1 presents the results of the Factor Analysis in detail.

Table 1 Results of Factor Analysis

Variables	Component	
	1	2
Foot Length	**0.147**	0.122
Ball-Heel Length	**0.146**	0.158
Foot Width	0.130	**0.261**
Foot Height	0.080	**0.291**
MPJ Height	0.109	**0.330**
Big Toe	0.133	-0.229
Toe 2	0.141	-0.272
Toe 3	0.131	-0.321
Toe 4	0.128	-0.330
Small Toe	0.037	-0.207
Foot Circ	0.114	0.230

4   PROPOSED FILIPINO SHOE SIZING SYSTEM

A Filipino shoe-sizing system is proposed because of observed problems in the use of foreign shoe-sizing systems. First, common shoe size labels such as the EU and US are based on the foot anthropometry of Westerners. Also, the current numbering system has no intrinsic meaning to the shoe buyer. Lastly, the EU/US systems are still based on English units, and not on the standard metric system. As an answer to these problems, the Filipino shoe-sizing system will be based on foot anthropometric data of Filipino children as Asians generally have smaller and narrower feet compared to westerners. The system will also consider the inherent buying behavior of Filipino parents where children's sizes are intuitively based on a child's age and buying shoes of larger sizes is acceptable because the child's feet are still growing.

4.1   The Proposed Filipino Shoe-Sizing System

The proposed Filipino shoe-sizing system will use the child's age as the basis for shoe size. For example, a size 7 will be appropriate for children aged 7 years old, while a size 8 will be appropriate for children aged 8 years old. The shoe lengths are based on actual foot length measurements collected per age level. The increment to be used is at 0.5 cm (such as in Mondopoint), which is between the step size of the Parisian and the English system. Shoe length measurements will differ between boys and girls from age 10 to 12, where physiological differences between genders start to appear. Standard shoe width for each length is based on proportion of the given shoe length.

This system is recommended for use for closed school shoes such as those made of black leather often used by children in school. It already incorporates an allowance of at least 0.5 centimeters between shoe length and foot length for socks, comfort and future growth.

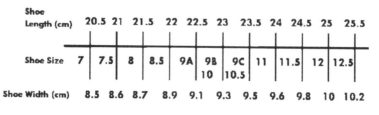

Figure 6 Filipino shoe-sizing system for girls aged 7-12.

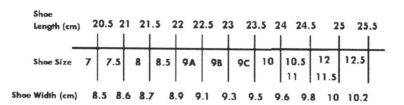

Figure 7 Filipino shoe-sizing system for boys aged 7-12.

Suppose that an eight year-old Filipino girl with foot length of 20 cm and foot width of 8.5 cm is to buy a new pair of school shoes. Using the proposed shoe-sizing system, she would be a size 8. Her new shoes have an allowance of 1.5 cm (21.5 cm - 20 cm) for length.

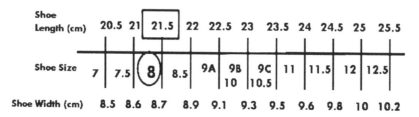

Figure 8 Using the Filipino shoe-sizing system for an 8 year-old girl.

4.2   Validation of the Proposed System

Foot length measurements from the actual samples grouped by age are compared with the proposed shoe size length measurements from the system to determine the system's accuracy. All 240 samples were evaluated. A "fit" is counted when the foot length measurement of the sample (given the child's age) is different by only 0.5 centimeters to 3 centimeters from the shoe lengths proposed by the system (for that age). This allowance is for socks, comfort, and future growth. Table 2 presents the percentages of fit per age level.

Table 2 Data fit of the samples to the proposed sizing system

Age	7	8	9	10-F	10-M	11-F	11-M	12-F	12-M
Fit (%)	83%	81%	87%	89%	68%	81%	83%	91%	66%

The prescribed foot length measurements from the proposed system accommodate 68% to 91% of the actual children sampled. Resulting low percentage of fit from 10-M and 12-M can be attributed to small sample size for these groups.

5   CONCLUSIONS

The following conclusions can be drawn from this foot anthropometric study:
- Foot length, foot width and foot height components contribute most to foot size variability. As such, these variables were incorporated in the proposed Filipino shoe-sizing system.
- The use of a Filipino shoe-sizing system based on Filipino children's anthropometric data per age level is recommended. The Filipino shoe-sizing system will use a numbering system based on Filipinos' inherent buying behavior of basing an item size on the child's age.
- Female and male foot lengths significantly differ for children ages 10 to 12. Therefore, a separate sizing system for each gender should be created for these age groups.

REFERENCES

Everman, Victoria. (2009). *Determining Your Shoe Size Measuring Your Feet to Know Your Shoe Size.* Accessed January 6, 2010 from How to Do Things: http://www.howtodothings.com/fashion-and-personal-care/a2774-how-to-determine-your-shoe-size.html

"International Shoe Size Conversion Charts - Mens & Ladies." International Shoe Size Conversion Charts - Mens & Ladies. N.p., n.d. Web. 8 Mar. 2010.

CHAPTER 56

The Customer Requirements for Sandals-Inspirations for Design

Steve N.H. Tsang[1], John K.L. Ho[2], Alan H.S. Chan[1]

[1]Department of Systems Engineering and Engineering Management
[2]Department of Mechanical and Biomedical Engineering
City University of Hong Kong
Hong Kong, China
nhtsang2@gapps.cityu.edu.hk

ABSTRACT

Kansei Engineering is a consumer-oriented technology for ergonomics product development and has been successfully implemented by many companies around the world. Sandals are one of the most popular footwear for teenagers. In the market, there are many different kinds of sandals designed for various purposes such as beach sandals, health sandals, fashion sandals, etc. With the increasing demands of fashion sandals from teenagers, in this study, the fashion sandals were targeted and Kansei Engineering was applied to study the relationship between the requirements of fashion sandals and the desirable product features.

Keywords: Kansei, Kansei Engineering, Design, Sandals

1   INTRODUCTION

In the past decades, new product development was mainly conducted internally by manufacturers according to their own perception and concept towards the market needs. However, with the advancement of living, consumers concern not only the price and durability of products, but also the matching between product design and their personal feelings and requirements for the specific products. To cope with the increasing demands of more 'personalized' product design from consumers, Kansei Engineering was proposed by Nagamachi about 40 years ago (Nagamachi, 1986,

1989, 1997). Kansei is a Japanese word meaning the consumer's psychological feeling and image to a product, and Kansei engineering is an approach to translate the consumer's Kansei into the product design elements (Nagamachi, 1995, 2002). Therefore, Kansei Engineering is a consumer-oriented technology for ergonomics product development. Since its first invention, Kansei Engineering has been widely applied to different industries such as automotive, construction machine, electric home appliance, costume, cosmetic, etc. (Jindo and Hirasago, 1997; Schütte and Eklund, 2005; Hunag, Tsai, and Hunag, 2011; Nagamachi, 2011). Till now, there are six types of Kansei Engineering methods. Type I is Category Classification which translates Kansei from zero to nth category. Type II is Kansei Engineering Computer System in which Kansei Engineering is aided by computer systems. Type III is Kansei Engineering Modeling which uses mathematical framework for selection of the most appropriate ergonomics design. Type IV is Hybrid Kansei Engineering constructed by forward and backward reasoning. Type V is virtual Kansei Engineering which combines Kansei Engineering with virtual reality. Type VI utilizes a Collaborative Kansei Designing System to provide a platform for designers to make a new design via internet using Kansei databases (Nagamachi, 1999). The typical procedures of conducting Kansei Engineering starts from collecting consumers' Kansei in specific product domain, and then analyzes the collected Kansei data using some statistical or engineering methods. Afterwards, the process continues with interpretation of the analyzed data and converting those data to a new product domain for designing a new Kansei product (Nagamachi, 2011).

Sandals are a kind of footwear consisting of a sole fastened to the foot by either thongs or straps. There are many types of sandals ranging from open toe to non open toe, and from high heel to flat base. Regarding serving purpose, sandals can be further classified into beach sandals, health sandals, fashion sandals, home sandals, etc. In this study, we tried to implement the Kansei Engineering method to design a pair of sandals which can fulfill most teenagers' demands and requirements. With the Kansei Engineering method as a basis, this sandal design study was separated into four phases; they were 1) collection of Kansei words for the requirements of sandals, 2) identification of major Kansei words as key factors for sandal design, 3) investigation of the relationship between the selected Kansei words and product features, and finally, 4) establishment of the desirable product features for new sandal design.

2   METHOD

2.1   Collection of the Kansei words for the requirements of sandal

The Kansei words were mainly collected from experienced sandal wearers aged 18 to 25 with phone interviews in this study. Apart from phone interviews, some Kansei words were taken from advertisements, brochures, and websites of sandals. In total, 50 Kansei words for sandal requirements were collected and they are summarized in Table 1.

Table 1 Kansei words collected for sandal design

No.	Kansei word	No.	Kansei word
1	Comfortable	26	Help displaying frank and energetic personality
2	Nice appearance	27	Give a neutral look and gait
3	Cheap price	28	Match with casual wear
4	Material	29	Possible to wear in rain
5	Durable	30	Moisture absorption
6	Safe	31	Easily dry material
7	Be healthy to foot	32	Abrasion resistant outsole
8	Color	33	Outsole tough enough to stand with sharp objects
9	Light in weight	34	Tough straps
10	Balance in hardness and softness	35	Surface area of straps
11	Allowing ventilation	36	Not easily thrown off during walking or running
12	Gas permeable	37	Skin abrasion prevented
13	Soft texture	38	Open toe
14	Give a soft feeling when stepping on it	39	No abrasion of skins in between toes by flip-flop kind of bindings
15	Easy to put on	40	Not easily lead to twisting of ankles
16	Carefree style	41	Skidproof outsole
17	Fashionable	42	Skidproof footbed
18	Special in style	43	With heel cup
19	Youngish in style	44	Contoured footbed
20	Ever popular in style	45	With medial arch and toe bar
21	Simple in style	46	Prevent foot fatigue
22	Easy to mix and match with dressing	47	Reduce the pressure of heel
23	Finely decorated	48	Shock absorbing
24	Fancy design	49	Height of the shoe heel
25	Elegant	50	Thickness of the outsole

2.2 Identification of major Kansei words as key factors for sandal design

Knowing the relevant Kansei words for sandal design, a questionnaire including a 5-point Semantic Differential Scale (SD-Scale) (1 for extremely not important and 5 for extremely important) with 50 Kansei words was prepared. The questionnaire was then distributed to 140 students of City University of Hong Kong for responses. All the participants, 72 males and 68 females, of ages between 18 and 30 years, had previous experience in wearing sandal.

In order to reduce the number of Kansei words, correlated Kansei words were grouped together as a factor (component) using principal component analysis (PCA). PCA is a variable reduction procedure for rationalizing the observed variables into a smaller number of principal components that will account for most of the variance in the observed variables.

2.2.1 PCA for male participants

PCA was performed, and it was stopped until no single variable was loaded to a component. After a series of PCA iterations, the number of variables was reduced from 50 to 42. The 42 variables of the Kansei words were subjected to PCA again. The results showed the presence of 10 components with eigenvalues greater than 1, and a scree test (Figure 1) also suggested that only the first 10 components were meaningful; thus, only the first 10 components were retained for rotation. The 10 components explained a total of 73.7% of the variance. The components and the corresponding variables are presented in Table 2. A variable was said to load on a given component if the factor loading was greater than 0.3 for that component and it was statistically significant when the loading value was greater than 0.65. Using these criteria, 7 variables with 6 showing significant values were loaded on the first component, which accounted for 25% of the variance and was labeled as 'Fashionable' component to represent the characteristic of the loading variables. For the second component, there were also 7 variables loaded, this component explained 13.6% of the variance and it was named as 'Durable material of the outsole to protect the foot' component.

Table 2 The principal components of the Kansei words for male participants (only the first component is shown)

Principal component	Variables	Factor loadings	% of variance explained		The interpretation of the component
			By each component	Cumulative	
1	*17 Fashionable	0.852	24.985	24.985	Fashionable
	*2 Nice appearance	0.796			
	*19 Youngish in style	0.792			
	*8 Color	0.785			
	*18 Special in style	0.716			
	*28 Match with casual wear	0.673			
	20 Ever popular in style	0.606			

* Statistically significant with factor loading greater than 0.65

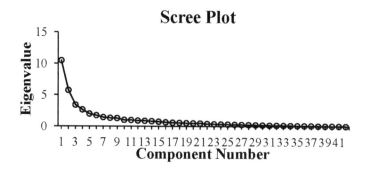

Figure 1 The scree plot for male sandal wearers

2.2.2 PCA for female participants

For female participants, the same method as that for male participants was used for analysis. Figure 2 shows a scree plot for the results for female participants. There were 48 components after eliminating 2 components from the first PCA. The 48 variables of the Kansei words were then subjected to PCA again. 12 components were found with eigenvalues greater than 1 and the scree test also showed that the eigenvalue begin to level off after component 12. Therefore, the first 12 components were extracted for further investigation. The components and their corresponding factors are shown in Table 3. The 12 components in total explained 77.4% of variance. The first component, loaded by 6 variables, explained 26.3% of variance and was named as 'Being ventilated, soft and light in material' component to characterize those loaded variables.

Table 3 The principal components of the Kansei words for female participants (only the first component is shown)

| Principal component | Variables | Factor loadings | % of variance explained | | The interpretation of the component |
			By each component	Cumulative	
1	*12 Gas permeable	0.845	26.261	26.261	Being ventilated, soft and light in material
	*11 Allowing ventilation	0.826			
	*9 Light in weight	0.780			
	*13 Soft texture	0.691			
	10 Balance in hardness and softness	0.597			
	14 Give a soft feeling when stepping on it	0.515			

* Statistically significant with factor loading greater than 0.65

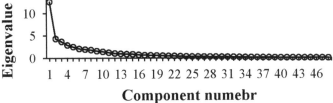

Component numebr

Figure 2 The scree plot for female sandal wearers

2.3   Investigation of the relationship between the selected Kansei words and product features

After the identification of the principal components for sandal requirements, the relationships between those components and product features were analyzed using multiple regression analysis.

In this part, only the results obtained from female participants were used for analysis because the limitation of time and resources. Since the first component explained the largest proportion of the variance, we only considered the significant variables inside the first component of their relationships with sandal features. The first component, named as 'being ventilated, soft and light in material', contained 4 significant variables (factor loading > 0.65). Amongst these four variables, three of them, 'gas permeable', 'allowing ventilation' and 'soft texture', were related to the material for straps and footbed of sandals, while the other variable 'light in weight' was influenced not only by straps and footbed, but also the outsole. To narrow the scope of this study, only the design of straps and footbed were considered; therefore, the variable 'light in weight' was excluded for analysis.

20 sandal samples, with the combinations of two different strap designs and ten different materials, were given to participants for evaluations. The questionnaire consisted of the pictures of sandal designs and the real samples of ten different materials, and participants had to evaluate different sandal samples in terms of 'allowing ventilation', 'gas permeable', and 'soft texture' using a 5-point scale. 97 female students from City University of Hong Kong, aged 18 to 23 years, took part in the evaluations.

The multiple regression equation takes the form $y = c + b_1x_1 + b_2x_2 + b_3x_3 + b_4x_4$, where $b_1, b_2, ..., b_n$ are the regression coefficients measuring how strongly each independent variable (predictor) influences the dependent variable (criterion), and c is a constant (intercept) of the equation. In both cases, design of straps (flip-flop or double) showed no significant contribution to the dependent variables ('allowing ventilation', 'gas permeable', and 'soft texture'); therefore they were eliminated from the equations. The regression equations describing the relationships between the independent variables and the dependent variables are established and shown below ((1), (2), and (3)). In a bid to compare the contribution of each independent variable, we converted the unstandardized coefficients to same scale (standardized coefficients, Beta) for comparison. Table 4 lists the beta values for each independent variable to the dependent variables, and the adjusted R square for each model is also shown in the table. Of those significant variables for 'allowing ventilation', all variables had positive contribution to the dependent variable. The variables of sm(f) (0.781) and si(s) (0.611) made comparatively larger contribution, while pEVA(f) (0.156) and l(s) (0.122) contributed less to the dependent variable. The results imply that using straw mat for footbed and silk for straps can increase the perceived feeling of allowing ventilation more than other materials do. For 'gas permeable', l(f) (-0.104) and pPVC(s) (-0.245) showed negative beta values suggesting that they had negative contribution to the dependent variable. Amongst other variables with positive beta values, sm(s) (0.715) and si(f) (0.425) showed the highest values for straps and footbed respectively indicting they were the favorable materials leading to 'gas permeable'. Regarding the sentiment on 'soft texture', pPE (-0.232), co (-0.360), and pPVC (-0.438) showed negative beta values, while the other variables showed positive values. Amongst the 8 variables to 'soft texture', m(s) (0.487) and si(f) (0.381) were of the highest contribution; thus, mesh and silk should be utilized for the straps and footbed respectively if softness of the sandal texture is of great concern. All the models exhibited a high adjusted R square (R^2 = 0.968) showing that each model could explain 96.8% variance of the dependent variable.

$$AV = 2.115 + 1.725sm(f) + 1.640m(f) + 0.850n(f) + 0.345pEVA(f) + 1.350si(s) + 1.095c(s) + 0.207l(s) \dots\dots(1)$$

$$GP = 2.487 + 0.888si(f) + 0.258n(f) - 0.217l(f) + 1.493sm(f) + 1.183m(s) + 0.708c(s) - 0.512pPVC(s) \dots\dots(2)$$

$$ST = 2.998 + 0.808si(f) + 0.428c(f) + 0.298pEVA(f) - 0.928pPVC(s) - 0.762co(s) - 0.492pPE(s) + 1.033m(s) + 0.633l(s) \dots\dots(3)$$

Where AV=allowing ventilation, GP=gas permeable, ST=soft texture, (f)=footbed, (s)=strap, m=mesh, n=nylon, sm=straw mat, c=canvas, co=cork, l=leather, pPE=plastic PE, pEVA=plastic EVA, pPVC=plastic PVC, and si=silk.

Table 4 Results of multiple regression analysis for 'allowing ventilation', 'gas permeable', and 'soft texture'

	Product Material	Standardized Coefficient (Beta)	Sig.	Adjusted R²
Allowing Ventilation	sm(f)	0.781	0.000	0.968
	m(f)	0.743	0.000	
	si(s)	0.611	0.000	
	c(s)	0.496	0.000	
	n(f)	0.365	0.000	
	pEVA(f)	0.156	0.005	
	l(s)	0.122	0.018	
Gas Permeable	sm(s)	0.715	0.000	0.968
	m(s)	0.566	0.000	
	si(f)	0.425	0.000	
	c(s)	0.339	0.000	
	n(f)	0.124	0.017	
	l(f)	-0.104	0.039	
	pPVC(s)	-0.245	0.000	
Soft Texture	m(s)	0.487	0.000	0.968
	si(f)	0.381	0.000	
	l(s)	0.298	0.000	
	c(f)	0.202	0.001	
	pEVA(f)	0.140	0.013	
	pPE(s)	-0.232	0.000	
	co(s)	-0.360	0.000	
	pPVC(s)	-0.438	0.000	

(f) = footbed, (s) = strap

2.4 Establishment of the desirable product features of new design of sandal

After conducting the multiple regression analysis, the contributing variables to the sentiments on 'allowing ventilation', 'gas permeable', and 'soft texture' were identified. According to the findings, strap type was not influential to those sentiments; therefore, the design of the sandal focused only on the material use for the footbed and straps. For the footbed, the materials attaining the highest contribution to the three sentiments were applied. Straw mat was good at 'allowing ventilation', and silk was 'gas permeable' and with 'soft texture'. To make use of these advantages, both straw mat and silk were used for the footbed and they were arranged alternatively. For the strap design, both flip-flop and double strap styles were selected for the new sandal designs. Regarding the use of materials, the upper part of the strap was made of straw mat to enhance the feeling of 'gas permeable', while the inner part was made of mesh to provide 'soft texture' for wearers. To achieve the perception of 'allowing ventilation', silk was used for the edges and the thong of the strap. The final designs of the sandals are shown in Figure 3.

516

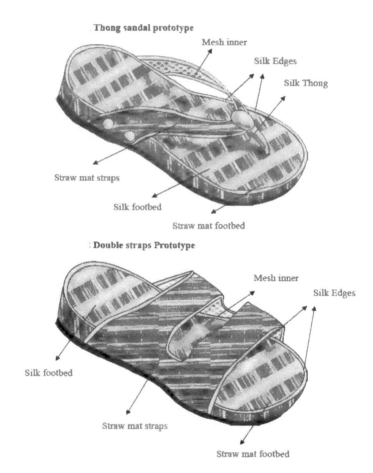

Figure 3 The prototypes for the sandal design

3. CONCLUSION

Kansei Engineering is a customer-oriented method for new product development. To design a pair of sandals which can fulfill wearers' expectations and desires, Kansei Engineering was applied in our study. The whole study adopted the procedures suggested by Nagamachi for Kansei Engineering implementation. It started by collecting relevant Kansei words from interviews and relevant advertising materials. Afterwards, Principal Component Analysis (PCA) was utilized to load the related Kansei words to a corresponding component. Knowing which Kansei words (component) were (was) most concerned by teenagers, multiple regression analysis was used to identify the relationship between the Kansei words and the product features. Finally, a pair of Kansei sandals was successfully designed.

ACKNOWLEDGEMENT

We would like to thank the student, Ms. T C Fung, conducting the survey and analyzing the data for the paper.

REFERENCES

Huang, M.S., H.C. Tsai, and T.H. Huang. 2011. Applying Kansei engineering to industrial machinery trade show booth design. *International Journal of Industrial Ergonomics* 41: 72-78.

Jindo, T. and K., Hirasago. 1997. Application studies to car interior of Kansei engineering. *International Journal of Industrial Ergonomics* 19: 105-114.

Nagamachi, M. 1986. Image technology and its application. *The Japanese Journal of Ergonomics* 29: 196-197.

Nagamachi, M. 1989. Kansei Engineering. Tokyo: Kaibundo Publishing.

Nagamachi, M. 1995. Kansei Engineering: A new ergonomic consumer-oriented technology for product development. *International Journal of Industrial Ergonomics* 15: 3-11.

Nagamachi, M. 1997. Kansei Engineering: the framework and methods. In. Kansei Engineering, ed. M. Nagamachi. Kaibundo Publishing, Kure, pp1-9.

Nagamachi, M. 1999. Kansei engineering: the implications and applications to product development. *Proceedings of IEEE International Conference on Systems, Man and Cybernetics* 6: 273-278.

Nagamachi, M. 2002. Kansei engineering as a powerful consumer-oriented technology for product development. *Applied Ergonomics* 33: 289-294.

Nagamachi, M. 2011. Kansei/affective engineering., Boca Raton, FL, USA: CRC Press.

Schütte, S. and J. Eklund. 2005. Design of rocker switches for work-vehicles-an application of Kansei Engineering. *Applied Ergonomics* 36: 557-567.

Author Index

Milton Keynes UK
Ingram Content Group UK Ltd.
UKHW031124141024
449569UK00006B/454